Birkhäuser

Trends in Mathematics

Research Perspectives CRM Barcelona

Volume 11

Since 1984 the Centre de Recerca Matemàtica (CRM) has been organizing scientific events such as conferences or workshops which span a wide range of cutting-edge topics in mathematics and present outstanding new results. In the fall of 2012, the CRM decided to publish extended conference abstracts originating from scientific events hosted at the center. The aim of this initiative is to quickly communicate new achievements, contribute to a fluent update of the state of the art, and enhance the scientific benefit of the CRM meetings. The extended abstracts are published in the subseries Research Perspectives CRM Barcelona within the Trends in Mathematics series. Volumes in the subseries will include a collection of revised written versions of the communications, grouped by events. Contributing authors to this extended abstracts series remain free to use their own material as in these publications for other purposes (for example a revised and enlarged paper) without prior consent from the publisher, provided it is not identical in form and content with the original publication and provided the original source is appropriately credited.

More information about this subseries at http://www.springer.com/series/13332

Andrei Korobeinikov · Magdalena Caubergh ·
Tomás Lázaro · Josep Sardanyés
Editors

Extended Abstracts Spring 2018

Singularly Perturbed Systems, Multiscale Phenomena and Hysteresis: Theory and Applications

Editors
Andrei Korobeinikov
Centre de Recerca Matemàtica
Universitat Autònoma de Barcelona
Barcelona, Spain

Magdalena Caubergh
Facultat de Ciències
Universitat Autònoma de Barcelona
Barcelona, Spain

Tomás Lázaro
Departament de Matemàtiques
Universitat Politècnica de Catalunya
Barcelona, Spain

Josep Sardanyés
Centre de Recerca Matemàtica
Universitat Autònoma de Barcelona
Barcelona, Spain

ISSN 2297-0215 ISSN 2297-024X (electronic)
Trends in Mathematics
ISSN 2509-7407 ISSN 2509-7415 (electronic)
Research Perspectives CRM Barcelona
ISBN 978-3-030-25260-1 ISBN 978-3-030-25261-8 (eBook)
https://doi.org/10.1007/978-3-030-25261-8

Mathematics Subject Classification (2010): 34-06, 35-06, 37-06, 39-06, 74-06, 76-06, 78-06, 92-06

This book is published under the imprint Birkhäuser, www.birkhauser-science.com by the registered company Springer Nature Switzerland AG
The registered company address is: Gewerbestrasse 11, 6330 Cham, Switzerland

Preface

This volume of the Trends in Mathematics: Research Perspectives CRM-Barcelona, offers to your attention a selection of short papers based on the presentations that were made at the joint *9th International Workshop on MUlti-Rate Processes and HYSteresis* (MURPHYS) and *4th International Workshop on Hysteresis and Slow-Fast Systems* (HSFS). The workshop was jointly organized by the Centre de Recerca Matemàtica, Barcelona, and the Collaborative Research Center 910, Berlin, and hosted by the Centre de Recerca Matemàtica, Barcelona, from May 28 to June 1, 2018. This meeting, MURPHYS-HSFS-2018, continued a successful series of biennial multidisciplinary conferences on Multi-Rate Processes and Hysteresis, that previously took place in Cork (Ireland, 2002–2008), Pécs (Hungary, 2010), Suceava (Romania, 2012), Berlin (Germany, 2014), and Barcelona (2016), as well as the series of workshops on Hysteresis and Slow-Fast Systems held in Lutherstadt, Wittenberg, Berlin, and Barcelona.

MURPHYS-HSFS-2018 workshop, dedicated to mathematical theory and applications of the singularly perturbed systems, systems with hysteresis and recent general trends in dynamical systems, brought together over 60 researchers working on hysteresis and multi-scale phenomena from Europe, USA, Russia, and other countries. Participants shared and discussed recent developments of analytical techniques in several areas of common interest. Topics in this volume include analysis of hysteresis phenomena, multiple scale systems, self-organizing nonlinear systems, singular perturbations, and critical phenomena, as well as applications of the hysteresis and the theory of singularly perturbed systems to fluid dynamics, chemical kinetics, cancer modeling, population modeling, mathematical economics, and control. This volume is intended to give to the contributors an opportunity to quickly report their latest research findings: the most of the short articles in this volume are brief preliminary summaries presenting new results that were not yet published in regular research journals.

We are happy to acknowledge support to the Workshop by the *Agència de Gestió d'Ajuts Universitaris i de Recerca (AGAUR) of the Generalitat de Catalunya, Collaborative Research Center 910: Control of self-organizing non-linear systems (Germany), Ministerio de Economía, Industria y Competitividad of the Spanish government*, and the *Centre de Recerca Matemàtica*. We also would like to express our gratitude to the CRM leadership and members of administrative staff whose enthusiastic work contributed a lot to the workshop success.

Bellaterra, Barcelona, Spain
October 2018

Andrei Korobeinikov
Magdalena Caubergh
Tomás Lázaro
Josep Sardanyés

Contents

Constructive Method of Decomposition in Singularly Perturbed Problems of Non-holonomic Mechanics

Alexander Kobrin and Vladimir Sobolev

Abstract The method of decomposition of singularly perturbed differential systems is developed and application of the method to the problems of non-holonomic mechanics is considered.

1 Introduction

The main object of our consideration is the following system of differential equations:

$$\dot{x} = f(x, y, t, \varepsilon), \qquad \varepsilon\dot{y} = g(x, y, t, \varepsilon), \tag{1}$$

where x and f are vectors in Euclidean spaces $\mathbb{R}^m$, y and g are vectors in $\mathbb{R}^n$, $t \in \mathbb{R}$, and ε is a small positive parameter. The goals of the paper are to construct a transformation reducing (1) to the system

$$\dot{v} = F(v, t, \varepsilon), \qquad \varepsilon\dot{z} = G(v, z, t, \varepsilon),$$

and to discuss some applications to the problems of non-holonomic mechanics. In this paper, the model of Chaplygin [1] is considered as an example of the investigation

A. Kobrin was funded by the Russian Foundation for Basic Research (grant 16-01-00429). V. Sobolev was supported by the Russian Foundation for Basic Research and the Government of the Samara Region (grant 16-41-630524-p) and the Ministry of Education and Science of the Russian Federation under the Competitiveness Enhancement Programme of Samara University (2013–2020).

A. Kobrin (✉)
Moscow Power Engineering Institute, Moscow, Russia
e-mail: kobrinai@yandex.ru

V. Sobolev
Department of Differential Equations and Control Theory,
Samara National Research University, Moskovskoye shosse, 34,
Samara 443086, Russian Federation
e-mail: v.sobolev@ssau.ru

A. Korobeinikov et al. (eds.), *Extended Abstracts Spring 2018*,
Trends in Mathematics 11, https://doi.org/10.1007/978-3-030-25261-8_1

of the non-holonomic problem by the methods integral manifolds. Some theoretical and applied results along these lines were obtained in [2, 4].

2 Splitting Transformation

Suppose that system (1) confirms with the following hypothesis:

I. the equation $g(x, t, y, 0) = 0$ has an isolated solution $y = h_0(x, t)$ for $t \in \mathbb{R}$, $x \in \mathbb{R}^m$;
II. the functions f, g, and h_0 are $(k + 2)$ times continuously differentiable $(k \geq 0)$ in $\Omega_0 = \{(x, y, t, \varepsilon) \mid x \in \mathbb{R}^m,\ \|y - h_0(x, t)\| < \rho,\ t \in \mathbb{R},\ 0 \leq \varepsilon \leq \varepsilon_0\}$;
III. the eigenvalues $\lambda_i(x, t)$ $(i = 1, \ldots, n)$ of the matrix $B(x, t) = \frac{\partial g}{\partial y}(x, h_0(x, t), t, 0)$ satisfy the inequality $Re\lambda_i(x, t) \leq -2\gamma < 0$.

Under such assumptions, the system (1) has the slow integral manifold $y = h(x, t, \varepsilon)$. The flow on this manifold is described by the m-dimensional system $\dot{x} = f(x, h(x, t, \varepsilon), t, \varepsilon)$. An exact calculations of $h(x, t, \varepsilon)$ is generally impossible, and various approximations are necessary. One possibility is the asymptotic expansions of $h(x, t, \varepsilon)$ in the integer power of small parameter ε:

$$h(x, t, \varepsilon) = h_0(x, t) + \varepsilon h_1(x, t) + \varepsilon^2 h_2(x, t) + \cdots$$

To consider the behaviour in some neighbourhood of the slow integral manifold $y = h(x, t, \varepsilon)$, let us introduce a new variable z by the formula $y = z + h(x, t, \varepsilon)$. Furthermore, we introduce an additional variable v, which satisfies the equation $\dot{v} = f(v, h(v, t, \varepsilon), t, \varepsilon) = F(v, t, \varepsilon)$, which describes the flow on the slow integral manifold. As the next step, we introduce a variable w by the formula $x = v + w$. For z, v and w, we have the differential equations

$$\dot{v} = F(v, t, \varepsilon), \qquad \dot{w} = W(v, w, z, t, \varepsilon), \qquad \varepsilon\dot{z} = B(v, t)z + Z(v, w, z, t, \varepsilon),$$

where

$$Z(v, w, z, t, \varepsilon) = g\,(v + w, z + h(v + w, t, \varepsilon), t, \varepsilon) - B(v, t)z - \varepsilon\frac{\partial h}{\partial t}(v + w, t, \varepsilon)$$

$$-\varepsilon\frac{\partial h}{\partial x}(v + w, t, \varepsilon) f(v + w, z + h(v + w, t, \varepsilon), t, \varepsilon),$$

$$W(v, w, z, t, \varepsilon) = f(v + w, z + h(v, w, t, \varepsilon), t, \varepsilon) - F(v, t, \varepsilon).$$

If all assumptions I–III hold for $k \geq 1$, then there exists $\varepsilon_2, 0 < \varepsilon_2 \leq \varepsilon_1$, such that for all $\varepsilon \in (0, \varepsilon_2]$, the last system possesses the integral manifold $w = \varepsilon H(v, z, t, \varepsilon)$, the flow on which is described by the differential system

$$\dot{v} = F(v, t, \varepsilon), \qquad \varepsilon\dot{z} = G(v, z, t, \varepsilon), \tag{2}$$

$$G(v, z, t, \varepsilon) = B(v, t)z + Z(v, \varepsilon H(v, z, t, \varepsilon), z, t, \varepsilon).$$

In many cases, H can be found as an asymptotic expansion

$$H(v, z, t, \varepsilon) = H_0(t, v, z) + \varepsilon H_1(t, v, z) + \varepsilon^2 H_2(t, v, x) + \cdots$$

from the corresponding *invariance equation*

$$\varepsilon\frac{\partial H}{\partial t} + \varepsilon\frac{\partial H}{\partial v}F(v, t, \varepsilon) + \frac{\partial H}{\partial z}[B(v, t)z + Z(v, \varepsilon H, z, t, \varepsilon)] = W(v, \varepsilon H, z, t, \varepsilon). \tag{3}$$

Our main goal is the constructing of the transformation

$$x = v + \varepsilon H(v, z, t, \varepsilon), \qquad y = z + h(x, t, \varepsilon), \tag{4}$$

which reduces the original system (1) to the form

$$\dot{v} = F(v, t, \varepsilon), \qquad \varepsilon\dot{z} = G(v, z, t, \varepsilon). \tag{5}$$

Let $(x(t),\ y(t))$ be a solution to (1) with an initial condition $x(t_0) = x_0,\ y(t_0) = y_0$. There exists a solution $(v(t),\ z(t))$ of (5) with the initial condition $v(t_0) = v_0$, $z(t_0) = z_0$, such that

$$x(t) = v(t) + \varepsilon H(v(t), z(t), t, \varepsilon), \quad y(t) = z(t) + h(x(t), t, \varepsilon). \tag{6}$$

It is sufficient to show that (6) takes place under $t = t_0$. Substituting $t = t_0$ in (6) we obtain $x_0 = v_0 + \varepsilon H(v_0, z_0, t_0, \varepsilon),\ \ y_0 = z_0 + h(x_0, t_0, \varepsilon)$, and, therefore, $z_0 = y_0 - h(x_0, t_0, \varepsilon)$. For v_0, we have the equation $v_0 = x_0 - H(v_0, z_0, t_0, \varepsilon) = V(v_0)$, which has a unique solution for any $x_0 \in \mathbb{R}^m$ and fixed z_0 and t_0, where $\|z_0\| = \|y_0 - h(x_0, t_0, \varepsilon)\| \leq \rho_2$, for some ρ_2.

The following statement is true; see [5]:

Theorem 1 *Suppose the assumptions I–III hold. Then, there exist numbers* ε_2 *and* ρ_2 *such that for all* $\varepsilon \in (0, \varepsilon_2]$ *any solution* $x = x(t, \varepsilon),\ y = y(t, \varepsilon)$ *of system (1) with the initial condition* $x(t_0, \varepsilon) = x_0,\ \ y(t_0, \varepsilon) = y_0$, *where* $\|y_0 - h(x_0, t_0, \varepsilon)\| \leq \rho_2$, *can be represented in form of (6).*

This proposition means that in the ρ_2-neighbourhood of the slow integral manifold $y = h(x, t, \varepsilon)$ of system (1) can be reduced to the form (5) by the splitting transformation (4). Thus, system (1) is split into two subsystems, the first of which is independent and contains a small parameter in a regular manner. Note that the initial value v_0 can be calculated in a form of an asymptotic expansion: $v_0 = v_{00} + \varepsilon v_{01} + \varepsilon^2 v_{02} + \cdots$. For example, $v_{00} = z_0$, $v_{01} = -H(x_0, z_{00}, t_0, 0)$, where $z_{00} = y_0 - h(x_0, t_0)$.

3 Systems that Are Linear with Respect to Fast Variables

Consider the differential system

$$\dot{x} = f_0(x,t,\varepsilon) + F_1(x,t,\varepsilon)y, \tag{7}$$

$$\varepsilon\dot{y} = g_0(x,t,\varepsilon) + G_1(x,t,\varepsilon)y, \tag{8}$$

where $x \in \mathbb{R}^m,\ y \in \mathbb{R}^n,\ t \in \mathbb{R}$.

Suppose that the following representations take place

$$F_1(x,t,\varepsilon) = \sum_{j\geq 0}\varepsilon^j F_{1,j}(x,t), \qquad G_1(x,t,\varepsilon) = \sum_{j\geq 0}\varepsilon^j G_{1,j}(x,t),$$

$$f_0(x,t,\varepsilon) = \sum_{j\geq 0}\varepsilon^j f_{0,j}(x,t), \qquad g_0(x,t,\varepsilon) = \sum_{j\geq 0}\varepsilon^j g_{0,j}(x,t).$$

Here, $G_{1,0} = G_{1,0}(x,t)$ plays the role of matrix $B(x,t)$. The formulae for the coefficients of asymptotic expansions of slow integral manifold $h = h(x,t,\varepsilon)$ take the form

$$h_0 = G_{1,0}^{-1} g_{0,0},$$

$$h_k = G_{1,0}^{-1}[\frac{\partial h_{k-1}}{\partial t} + \sum_{p=0}^{k-1}\frac{\partial h_p}{\partial x}(f_{0,k-1-p} + \sum_{j=0}^{k-1-p} F_{1,j}h_{k-p-1-j}) - g_{0,k} - \sum_{j=1}^{k} G_{1,j}h_{k-j}],$$

$k \geq 1$. The invariance Eq. (3) for the fast integral manifold $H = H(v,z,t,\varepsilon)$ in this case takes the form

$$\varepsilon\frac{\partial H}{\partial t} + \varepsilon\frac{\partial H}{\partial v}[f_0(v,t,\varepsilon) + F_1(v,t,\varepsilon)h(v,t,\varepsilon)] + \frac{\partial H}{\partial z}[G_1(v+\varepsilon H,t,\varepsilon)$$

$$-\varepsilon\frac{\partial h}{\partial x}(v+\varepsilon H,t,\varepsilon)F_1(v+\varepsilon H,t,\varepsilon)]z = f_0(v+\varepsilon H,t,\varepsilon) - f_0(v,t,\varepsilon)$$

$$+F_1(v+\varepsilon H,t,\varepsilon)(z + h(v+\varepsilon H,t,\varepsilon)) - F_1(v,t,\varepsilon)h(v,t,\varepsilon).$$

Setting $\varepsilon = 0$, we obtain

$$\frac{\partial H_0}{\partial z}G_{1,0}(v,t)z = F_{1,0}(v,t)z.$$

It is possible to represent $H_0(v,t,z)$ in the form $H_0(v,z,t) = D_0(v,t)z$, where matrix $D_0(v,t)$ satisfies the equation

$$D_0(v,t)G_{1,0}(v,t) = F_{1,0}(v,t),$$

and, therefore,

$$H_0(v,z,t) = F_{1,0}(v,t)G_{1,0}^{-1}(v,t)z.$$

Neglecting terms of order $o(\varepsilon)$, we use the transformation

$$x = v + \varepsilon H_0(v,z,t), \quad y = z + h_0(x,t) + \varepsilon h_1(x,t) \tag{9}$$

to reduce system (7) to a nonlinear block triangular form

$$\begin{aligned}
\dot{v} &= f_{0,0}(v,t) + F_{1,0}(v,t)h_0(v,t) + \varepsilon[f_{0,1}(v,t) \\
&\quad + F_{1,0}(v,t)h_1(x,t) + F_{1,1}(v,t)h_0(v,t)] + O(\varepsilon^2), \\
\varepsilon\dot{z} &= [G_{1,0}(v,t) + \varepsilon(G_{1,1}(v,t) + \frac{\partial G_{1,0}}{\partial x}(v,t)H_0(v,z,t) \\
&\quad - \frac{\partial h_0}{\partial x}(v,t)F_{1,0}(v,t))]z + O(\varepsilon^2).
\end{aligned} \tag{10}$$

4 Chaplygin Sleigh

As an example of an investigation of a non-holonomic problem by the integral manifolds methods, we consider the problem of Chaplygin [1]. The following differential system

$$\dot{u} = \omega\left(\varepsilon V + h\omega\right), \qquad I_c\,\dot{\omega} = -Fh, \qquad \varepsilon\dot{F} = -\frac{I_c + mh^2}{mI_c}F + m\omega u.$$

can be considered as the model of the Chaplygin [1, 3]. This $3D$ system is linear with respect to the fast variable F, and, hence, we can reduce this $3D$ system by the transformation

$$F = z + h_0(u,\omega) + \varepsilon h_1(u,\omega), \qquad u = v_1, \qquad \omega = v_2 + \varepsilon\frac{h}{I_c + mh^2}z,$$

to the following two subsystems: the slow subsystem

$$\dot{v}_1 = mhv_1^2 - \varepsilon v_2\frac{I_c m v_1 v_2}{I_c + mh^2} + O(\varepsilon^2), \qquad \dot{v}_2 = -\frac{h}{I_c}\left[\frac{I_c m v_1 v_2}{I_c + mh^2} + \varepsilon h_1(v_1,v_2)\right] + O(\varepsilon^2),$$

and the fast subsystem

$$\varepsilon \dot{z} = -\left[\frac{I_c + mh^2}{I_c} - \varepsilon \frac{mh}{I_c + mh^2} v_1 + O(\varepsilon^2)\right] z.$$

References

1. S.A. Chaplygin, On the theory of motion of nonholonomic systems. The reducing multiplier theorem. Math. Collection. **28**(1911), 303–314
2. J. Eldering, Realizing nonholonomic dynamics as limit of friction forces. Regul. Chaotic Dyn. **21**, 390–409 (2016)
3. A. Kobrin, V. Sobolev, Integral manifolds of fast-slow systems in nonholonomic mechanics. J. Proc. Eng. **201**, 556–560 (2017)
4. S. Koshkin, V. Jovanovic, Realization of non-holonomic constrains and singular perturbation theory for plane dumbbels. J. Eng. Math. **1**, 1–19 (2017)
5. V.A. Sobolev, Integral manifolds and decomposition of singularly perturbed systems. Syst. Control Lett. **5**, 169–179 (1984)

Analysis of Temporal Dissipative Solitons in a Delayed Model of a Ring Semiconductor Laser

Alexander Pimenov, Andrei G. Vladimirov and Shalva Amiranashvili

Abstract Temporal dissipative solitons are short pulses observed in periodic time traces of the electric field envelope in active and passive optical cavities. They sit on a stable background, so that their trajectory comes close to a stable steady state solution between the pulses. A common approach to predict and study these solitons theoretically is based on the use of Ginzburg–Landau-type partial differential equations, which, however, cannot adequately describe the dynamics of many realistic laser systems. Here, for the first time, we demonstrate the formation of temporal dissipative soliton solutions in a time-delay model of a ring semiconductor cavity with coherent optical injection, operating in anomalous dispersion regime, and perform bifurcation analysis of these solutions.

1 Introduction

Temporal localized structures (TLS) of light propagating along the axial direction in nonlinear cavities attracted significant theoretical and experimental attention in the past decade due to their potential applications for optical data storage and transmission [3, 5, 8, 9]. Similarly to the solitons of nonlinear Schrödinger equation [15], dissipative optical TLS known also as temporal cavity solitons are localized in time and can be studied with the help of complex Ginzburg–Landau-type equations in the co-moving reference frame as stationary solutions of a properly constructed ordinary differential equations [6]. Although this approach allows a detailed bifurcation

A. Pimenov and A.G. Vladimirov acknowledge the support of SFB 787 of the DFG. Sh. Amiranashvili acknowledges the support of the DFG under Project 389251150.

A. Pimenov (✉) · A. G. Vladimirov · S. Amiranashvili
Weierstrass Institute, Mohrenstr. 39, 10117 Berlin, Germany
e-mail: pimenov@wias-berlin.de

A. G. Vladimirov
e-mail: andrei.vladimirov@wias-berlin.de

S. Amiranashvili
e-mail: shalva.amiranashvili@wias-berlin.de

A. Korobeinikov et al. (eds.), *Extended Abstracts Spring 2018*,
Trends in Mathematics 11, https://doi.org/10.1007/978-3-030-25261-8_2

analysis of TLS solutions, complex Ginzburg–Landau models are hardly applicable to account accurately for certain important physical effects in realistic laser devices, such as those containing intracavity semiconductor medium [10]. On the other hand, traveling-wave-type PDEs can be used to model any laser device [12], though TLS are periodic solutions of these equations, and analysis of such solutions is usually limited to direct numerical simulations. In the last few years, delay differential equations (DDEs), that can be derived from traveling wave equations under certain non-restrictive simplifying physical assumptions [13], proved to be a viable alternative to PDEs. In DDE models, periodic regimes such as mode-locked pulses and TLS (or regimes that are both [11]) can be studied using well-developed Floquet theory and software packages such as DDE-BIFTOOL, [2].

In this paper, using a time-delay model, we investigate dissipative solitons (DSs) first predicted theoretically in the anomalous dispersion regime in the Lugiato–Lefever equation (LLE) [7], which is equivalent to the driven damped nonlinear Schrödinger equation. This equation describes qualitatively the dynamics of the electric field envelope in a passive optical cavity subject to weak coherent optical injection, when the injection frequency is close to a resonant frequency of the cavity. However, far enough from the resonance, one can observe a bistability between two branches of dissipative solitons corresponding to different longitudinal cavity modes. This phenomenon is missing in the LLE and can be studied using a traveling-wave-type equation, [4]. Similarly to the traveling wave equation, DDE models account fully for the multimode nature of the optical cavities, and, in addition, the anomalous dispersion of the fiber waveguide can be described by including a distributed delay term into model equations [10]. Here, we develop an DDE model to study dissipative soliton in an optically injected ring cavity laser containing semiconductor optical amplifier (SOA), long dispersive fiber delay line, and a narrow bandpass spectral filter. We perform stability analysis of the injection-locked steady states in the limit of large delay [14] and demonstrate analytically the appearance of modulational instability and cavity solitons. Finally, we reduce fully distributed DDE model to a simplified DDE model that preserves the effect of the chromatic dispersion on the dynamics of the ring laser, and perform numerical continuation and stability analysis of the periodic cavity soliton solutions in this model using DDE-BIFTOOL, [2].

2 Delayed Model of a Dispersive Semiconductor Ring Laser

Let us consider a laser consisting of an SOA as an amplifying medium, a spectral filter, and a dispersive fiber delay line in a ring cavity (see Fig. 1) subject to a single-mode optical injection [10]. We assume that the chromatic dispersion of the delay line material is caused by a Lorentzian absorption line with the full-width at half-maximum Γ the central frequency Ω detuned with respect to the reference frequency associated with the central wavelength of the amplification line of the SOA. The normal dispersion regime corresponds to $\Omega > 0$ and the anomalous dispersion to $\Omega < 0$. We consider the following set of DDEs for the complex envelope of the

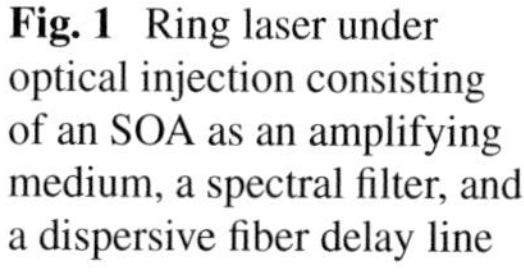

Fig. 1 Ring laser under optical injection consisting of an SOA as an amplifying medium, a spectral filter, and a dispersive fiber delay line

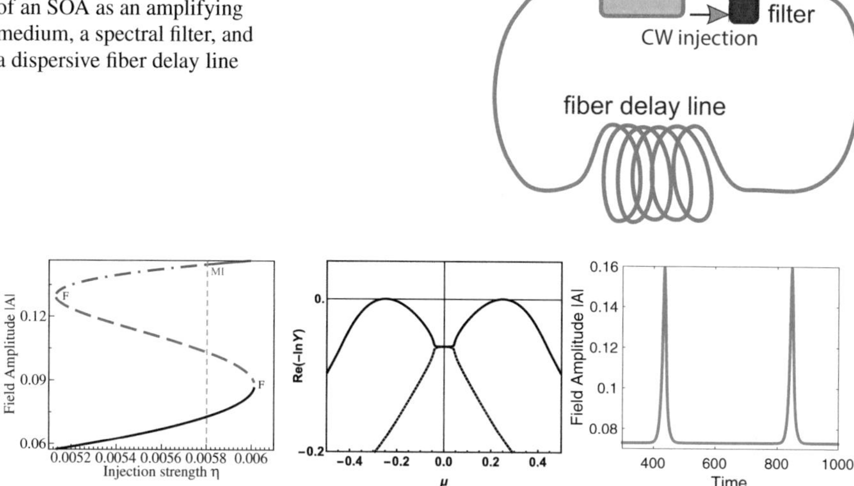

Fig. 2 Hysteresis loop formed by two bistable branches of injection-locked steady states of (1)–(3) obtained by varying optical injection strength η for $\sigma = 0$ (left) and destabilization of the top branch via modulational instability (MI) for $\sigma = 2000$ (center) leading to the formation of periodic dissipative solitons with the period close to $T = 400$ (right). Other parameters are $\Omega = -13$, $\Gamma = 0.001$, $\alpha = 5$, $\kappa = 0.3$, $\gamma = \gamma_g = 1$, $g_0 = 1.19$, $\eta = 0.0058$, $w = w_0 = 0$, and $\phi = -0.2 + \frac{\sigma(\alpha\Gamma-\Omega)}{\Gamma^2+\Omega^2} - \frac{\alpha \log \kappa}{2}$

electric field $A(t)$ at the entrance of the SOA, polarization $P(t)$, and the saturable gain of the SOA $G(t)$

$$\frac{dA}{dt} + (\gamma - iw)A = \gamma\sqrt{\kappa}e^{(1-i\alpha)G/2+i\varphi}\left[A_T + P_T\right] + \eta e^{iw_0 t}, \tag{1}$$

$$\frac{dG}{dt} = \gamma_g\left[g_0 - G - (e^G - 1)\left|A_T + P_T\right|^2\right], \tag{2}$$

$$P(t) = -\sigma L\int_{-\infty}^{t} e^{-(\Gamma+i\Omega)(t-s)}\frac{J_1\left[\sqrt{4\sigma(t-s)}\right]}{\sqrt{\sigma(t-s)}}A(s)ds, \tag{3}$$

where $A_T = A(t-T)$, $P_T = P(t-T)$, T is the cavity round trip time, γ and w describe the width and the central frequency of the filter, η is the strength of the optical injection at the frequency w_0, and σ is the total dispersion strength. The parameters κ, φ, α, γ_g, and g_0 describe, respectively, linear attenuation and phase shift per cavity round trip, linewidth enhancement factor, carrier relaxation rate, and the pump current.

3 Stability Analysis in the Limit of Large Delay, Modulational Instability, and Dissipative Solitons

We take w_0 as the reference frequency ($w_0 = 0$). Therefore, DDEs (1)–(3) have solutions in form of the injection-locked steady state $A(t) = A_0 e^{i\varphi_0}$, $G(t) = G_0$, where $P(t)$ can be expressed as $P(t) = P_0 = \left(e^{-\sigma/[\Gamma+i\Omega]} - 1\right) A_0$. Similarly to the case of LLE [6], we are interested in the situation, where the branch of steady states shown in the left panel of Fig. 2 exhibits a bistable behavior due to the presence of strong nonlinear phase–amplitude coupling introduced by the linewidth enchancement factor α.

We look for DSs in the vicinity of the bistability curve when the upper steady state is destabilized at large enough dispersion strength σ via a modulational instability in the anomalous dispersion regime ($\Omega < 0$). Linearizing equations (1)–(3) near the injection-locked steady state solution, assuming that linear perturbations evolve exponentially in time $\delta A, \delta P, \delta G \propto e^{\lambda t}$, where λ is the eigenvalue, evaluating the integral (3) as $\delta P \propto e^{\lambda t}\left(e^{-\sigma/[\Gamma+\lambda+i\Omega]} - 1\right)$, and taking determinant of the Jacobian of the resulting system we obtain a transcendental characteristic equation in the following form:

$$c_1(\lambda)Y^2 + c_2(\lambda)Y + c_3(\lambda) = 0, \tag{4}$$

where $Y = e^{-\lambda T}$ is the exponential term that appears from the delayed variables A_T, P_T. We look for instability of the steady state in the limit of large delay $T \gg 1$, when the eigenvalues with vanishing real parts ($\mathrm{Re}\,\lambda \sim 1/T \ll 1$) that belong to the so-called pseudo-continuous spectrum [14] cross imaginary axis. To this end, we neglect $\mathrm{Re}\,\lambda \ll 1$ in coefficients c_1, c_2, c_3 and obtain $\mathrm{Re}\,\lambda$ as the function of $\mathrm{Im}\,\lambda$

$$T\,\mathrm{Re}\,\lambda = -\,\mathrm{Re}\log Y_k(\mu), \qquad \mu = \mathrm{Im}\,\lambda,$$

where Y_1 and Y_2 represents two roots of quadratic equation (4) and two curves of pseudo-continuous spectrum (see Fig. 2, center). We note that the algorithm of stability analysis in the limit of large delay was developed for conventional DDEs [14], and it is valid in the presence of distributed delay term (3) as well. In the absence of optical injection $\eta = 0$, the necessary analytical condition for the appearance of the modulational instability of the relative steady state of the passive system with the rotation frequency ν is

$$\alpha D_2 < -\frac{1}{\gamma^2}, \tag{5}$$

where the second-order dispersion coefficient is given by $D_2 = \mathrm{Im}\,\frac{\mathrm{d}^2}{\mathrm{d}\nu^2}\left(\frac{-\sigma}{\Gamma+i(\Omega+\nu)}\right)$. We have used this condition to locate modulational instability of the injection-locked steady state at $\sigma = 2000$ for anomalous dispersion $\Omega = -13$ (see Fig. 2, center) and found a stable dissipative soliton using direct numerical simulation of (1)–(3) (see Fig. 2, right).

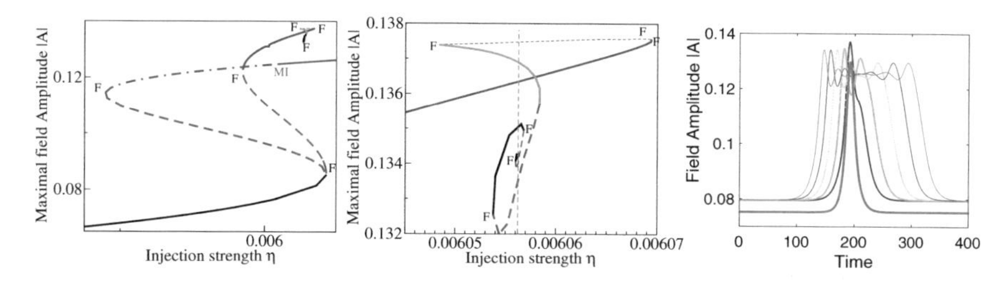

Fig. 3 Bifurcation diagram for steady states and periodic DS solutions of (6)–(8) obtained by varying η (left) and magnified region of multistability (center) between various branches of DSs, accompanied by the profiles (right) of stable periodic DSs from different branches for $\eta = 0.006$ (thick solid) and $\eta \approx 0.00606$ (thin lines). Here, $\sigma = 9$, $\Omega = -2$, $\Gamma = 0$, $\kappa = 0.25$, $g_0 = 1.33$, $\phi = -0.252 - \frac{\alpha \log \kappa}{2}$, and other parameters are as shown in Fig. 2

4 Reduced DDEs and DDE-BIFTOOL

Using Padé approximant of the kernel (3) in the frequency domain, we can simplify system (1)–(3) and obtain an approximate DDE model with a single fixed delay that is still capable of describing the formation of DSs in the anomalous dispersion regime

$$\frac{dA}{dt} + (\gamma - iw)A = \frac{\gamma\sqrt{\kappa}e^{(1-i\alpha)G/2+i\varphi}}{1+\frac{i\sigma}{2(\Omega-i\Gamma)}}\left[\left(1-\frac{i\sigma}{2(\Omega-i\Gamma)}\right)A_T + P_T\right] + \eta e^{iw_0 t}, \tag{6}$$

$$\frac{dG}{dt} = \gamma_g\left[g_0 - G - (e^G - 1)\frac{\left|\left(1-\frac{i\sigma}{2(\Omega-i\Gamma)}\right)A_T + P_T\right|^2}{1+\frac{\sigma(\sigma-4\Gamma)}{4(\Gamma^2+\Omega^2)}}\right], \tag{7}$$

$$\frac{dP}{dt} = -\left[i(\Omega - i\Gamma) + \frac{\sigma}{2\left(1+\frac{i\sigma}{2(\Omega-i\Gamma)}\right)}\right]P - \frac{\sigma}{1+\frac{i\sigma}{2(\Omega-i\Gamma)}}A. \tag{8}$$

Similarly to the previous subsection, we have performed stability analysis of the injection-locked states of (6)–(8) in the limit of large delay, found the parameter values where the top branch looses the stability due to anomalous dispersion. We used DDE-BIFTOOL [2] to confirm our prediction, perform continuation and stability analysis of DSs (see Fig. 3), and demonstrate very good qualitative agreement with the full model (1)–(3). Unlike to that of the LLE model [6], the DS branch shown in Fig. 3 demonstrates a spiraling behavior, which bares similarities to the spatial cavity soliton branches calculated earlier in the models of semiconductor devices [1].

References

1. M. Brambilla, L.A. Lugiato, F. Prati, L. Spinelli, W.J. Firth, Spatial soliton pixels in semiconductor devices. Phys. Rev. Lett. **79**(11), 2042 (1997)
2. K. Engelborghs, T. Luzyanina, G. Samaey, DDE-BIFTOOL v.2.00: A MATLAB package for bifurcation analysis of delay differential equations, Department of Computer Science, K.U.Leuven (2001)
3. B. Garbin, J. Javaloyes, G. Tissoni, S. Barland, Topological solitons as addressable phase bits in a driven laser. Nat Commun **6** (2015)
4. Y.V. Kartashov, O. Alexander, D.V. Skryabin, Multistability and coexisting soliton combs in ring resonators: the Lugiato-Lefever approach. Opt. Express **25**(10), 11550–11555 (2017)
5. F. Leo, S. Coen, P. Kockaert, S.P. Gorza, P. Emplit, M. Haelterman, Temporal cavity solitons in one-dimensional Kerr media as bits in an all-optical buffer. Nat Photonics **4**(7), 471 (2010)
6. F. Leo, L. Gelens, P. Emplit, M. Haelterman, S. Coen, Dynamics of one-dimensional Kerr cavity solitons. Opt. Express **21**(7), 9180–9191 (2013)
7. L.A. Lugiato, R. Lefever, Spatial dissipative structures in passive optical systems. Phys. Rev. Lett. **58**(21), 2209 (1987)
8. M. Marconi, J. Javaloyes, S. Balle, M. Giudici, How lasing localized structures evolve out of passive mode locking. Phys. Rev. Lett. **112**(22), 223901 (2014)
9. M. Marconi, J. Javaloyes, S. Barland, S. Balle, M. Giudici, Vectorial dissipative solitons in vertical-cavity surface-emitting lasers with delays. Nat. Photonics **9**(7), 450–455 (2015)
10. A. Pimenov, S. Slepneva, G. Huyet, A.G. Vladimirov, Dispersive time-delay dynamical systems. Phys. Rev. Lett. **118**(19), 193901 (2017)
11. C. Schelte, J. Javaloyes, S.V. Gurevich, Dynamics of temporally localized states in passively mode-locked semiconductor lasers. Phys. Rev. A **97**(5), 053820 (2018)
12. A.G. Vladimirov, A.S. Pimenov, D. Rachinskii, Numerical study of dynamical regimes in a monolithic passively mode-locked semiconductor laser. IEEE J. Quantum Electron. **45**(5), 462–468 (2009)
13. A.G. Vladimirov, D. Turaev, Model for passive mode-locking in semiconductor lasers. Phys. Rev. A **72** (2005). 13 pages
14. S. Yanchuk, M. Wolfrum, A multiple time scale approach to the stability of external cavity modes in the Lang-Kobayashi system using the limit of large delay. SIAM J. Appl. Dyn. Syst. **9**, 519–535 (2010)
15. V.E. Zakharov, A.B. Shabat, Exact theory of two-dimensional self-focusing and one-dimensional self-modulation of waves in nonlinear media. Sov. Phys. JETP **34**(1), 62–69 (1972)

Multi-scale Problem for a Model of Viral Evolution with Random Mutations

Aleksei Archibasov

Abstract The model of viral dynamics with random mutations is considered. This model describes the cells' population dynamics with significantly different life cycles. The presence of different timescales leads to a singularly perturbed system. The latter makes it possible to apply the technique of separating timescales and thereby reducing the dimensionality of the model.

1 Introduction

It is well known that for the singularly perturbed systems with several small parameters Tikhonov's theorem is applicable [3]. In this theorem, the passage to the limit of the solution to a degenerate problem in a system with several small parameters multiplying derivatives is justified. In [1], a similar theorem is formulated and proved for the system of singularly perturbed partial integrodifferential equations with one small parameter. This theorem can be generalized to the case of several small parameters.

Let us consider the singularly perturbed system of integrodifferential equations with two small parameters

$$\begin{aligned} \varepsilon x_t' &= f(x, \int_\Omega g(s, v)ds), \\ \varepsilon \nu v_t' &= h(y, v), \\ y_t' &= w(s, x, y, v, \int_\Omega q(s, r, y, v)dr), \end{aligned} \tag{1}$$

This work is supported by the Russian Foundation for Basic Research and Samara region (grant 16-41-630529-p) and the Ministry of Education and Science of the Russian Federation as part of a program to increase the competitiveness of Samara University in the period 2013–2020.

A. Archibasov (✉)
Department of Applied Mathematics and Physics, Samara National Research University, Samara, Russia
e-mail: aarchibasov@gmail.com

A. Korobeinikov et al. (eds.), *Extended Abstracts Spring 2018*,
Trends in Mathematics 11, https://doi.org/10.1007/978-3-030-25261-8_3

with the initial conditions $x(0) = x^0,\ v(0,s) = v^0(s),\ y(0,s) = y^0(s)$, where x, v, $y \in R$, $\varepsilon \ll 1$, $\nu \ll 1$ are the small positive parameters. We assume that system (1) satisfies the following conditions:

(i) The functions $f(x, z_1)$, $g(s, v)$, $h(y, v)$, $w(s, x, y, v, z_2)$, and $q(s, r, y, v)$ and their partial derivatives with respect to all variables are uniformly continuous and bounded in the respective domains $D_1 = \{|x| \le a,\ |z_1| \le b_1\}$, $D_2 = \{s \in \Omega,\ \ |v| \le c\}$, $D_3 = \{|y| \le d,\ |v| \le c\}$, $D_4 = \{s \in \Omega,\ |x| \le a,\ |y| \le d,\ |v| \le c,\ |z_2| \le b_2\}$, and $D_5 = \{s, r \in \Omega,\ \ |y| \le d,\ |v| \le c\}$.

(ii) The equation $h(y, v) = 0$ has an isolated root $v = \varphi(y)$ in the domain $\{|y| \le d\}$ and in this domain the function $v = \varphi(y)$ is continuously differentiable.

(iii) The inequality $h_v(y, \varphi(y)) \le -\alpha < 0$ holds for $|y| \le d$. This condition implies, that the stationary point $\hat{v} = \varphi(y)$ of the first-order associated equation $\hat{v}'_\tau = h(y, \hat{v})$, which contains y as a parameter, is Lyapunov asymptotically stable as $\tau \to +\infty$ uniformly with respect to y, $|y| \le d$.

(iv) There exist a solution $\hat{v} = \hat{v}(\tau, s)$ of the initial value problem $\hat{v}'_\tau = h(y^0(s), \hat{v})$, $\hat{v}(0, s) = z^0(s)$ for $\tau \ge 0$, $\forall s \in \Omega$. Furthermore, this solution tends to the stationary point $\varphi(y^0(s))$ as $\tau \to +\infty$ $\forall s \in \Omega$, i.e., $v^0(s)$ belongs to the domain of attraction of the stable stationary point $\varphi(y^0(s))$.

(v) The equation $f(x, z_1) = 0$ has an isolated root $x = \psi(z_1)$ in domain $|x| \le a$ and in this domain function $x = \psi(z_1)$ is continuously differentiable.

(vi) The inequality $f_x(\psi(z_1), z_1) \le -\beta < 0 (z_1 = \int_\Omega g(s, \varphi(y))ds)$ holds for $|y| \le d$, i.e., the stationary point $\hat{x} = \psi(z_1)$ of the second-order associated equation $\hat{x}'_\tau = f(\hat{x}, \int_\Omega g(s, \varphi(y))ds)$, which contains y as a parameter, is Lyapunov asymptotically stable as $\tau \to +\infty$ uniformly with respect to y, $|y| \le d$.

(vii) There exists a solution $\hat{x}(\tau)$ to the problem $\hat{x}'_\tau = f(\hat{x}, \int_\Omega g(s, \varphi(y^0(s)))ds)$, with initial value $\hat{x}(0) = x^0$ for $\tau \ge 0$. Further, this solution tends to the stationary point $\psi(\int_\Omega g(s, \varphi(y^0(s)))ds)$ as $\tau \to +\infty$, i.e., x^0 belongs to the domain of attraction of the stable stationary point.

(viii) The truncated system

$$\begin{aligned} y'_t &= w(s, \psi(z_1), y, \varphi(y), \int_\Omega q(s, r, y, \varphi(y))dr), \\ x &= \psi(z_1), \\ v &= \varphi(y), \\ z_1 &= \int_\Omega g(s, \varphi(y))ds \end{aligned} \tag{2}$$

with initial condition $y(0, s) = y^0(s)$ has a unique solution $\bar{y}(t, s)$, $\bar{x}(t) = \psi(\int_\Omega g(s, \varphi(\bar{y}(t, s)))ds)$, $\bar{v}(t, s) = \varphi(\bar{y}(t, s))$.

Theorem 1 *If conditions (i)–(viii) are satisfied, then, for sufficiently small* ε *and* ν, *for some* $T > 0$, *the problem (1) has a unique solution* $x(t,\varepsilon,\nu)$, $v(t,s,\varepsilon,\nu)$, $y(t,s,\varepsilon,\nu)$, *which is related to the solution* $\bar{x}(t)$, $\bar{v}(t,s)$, $\bar{y}(t,s)$ *of the truncated problem (2) by the limit formulas*

$$\lim_{\varepsilon\to+0,\nu\to+0} x(t,\varepsilon,\nu) = \bar{x}(t) = \psi\Big(\int_\Omega g(s,\varphi(\bar{y}(t,s)))ds\Big),\ 0 < t \le T,$$
$$\lim_{\varepsilon\to+0,\nu\to+0} v(t,s,\varepsilon,\nu) = \bar{v}(t,s) = \varphi(\bar{y}(t,s)),\ 0 < t \le T,\ s \in \Omega,$$
$$\lim_{\varepsilon\to+0,\nu\to+0} y(t,s,\varepsilon,\nu) = \bar{y}(t,s),\ 0 \le t \le T,\ s \in \Omega.$$

Note that the limiting equalities for the variables x and v are not uniform for $t \ge 0$. The boundary layer phenomenon occurs [4].

2 Model

Let us consider the next model of viral dynamics with random mutations.

$$x'_t = b - \sigma x(t) - \int_\Omega \alpha(s)x(t)v(t,s)ds,$$
$$y'_t = \int_\Omega p_1(s,r)\alpha(r)x(t)v(t,r)\,dr - m(s)y(t,s),$$
$$v'_t = k(s)y(t,s) - c(s)v(t,s).$$

In this model, $x(t)$ is the concentration of uninfected (susceptible) cells at the time t, $y(t,s)$, $v(t,s)$ are the density distributions of infected target cells (CD4+ cells, or T helper cells, or Th cells) and free virus particles, respectively, in a one-dimensional phenotype space $s \in \Omega$ at the time t. The uninfected cells susceptible to the virus are produced at a constant rate b and die of natural reasons unrelated to the virus infection at a rate $\sigma x(t)$, $\sigma > 0$. The factors α, m, k and c are characteristics of the virus phenotype, and hence, they are functions of the variable s or r. It is assumed that mutations occur in the process of cell infection. Function $p_1(s,r)$ describes the probability that the infected by virus of phenotype r cell produces exclusively virus of phenotype s.

It should be noted that there are three very different timescales: life cycles of uninfected and infected cells and free virus particles. The presence of considerably different timescales indicates that the model can be significantly simplified. Following, for example [2], let us introduce the dimensionless variables and parameters

$$t = T\bar{t},\ s = S\bar{s},\ x(t) = X\bar{x}(\bar{t}),\ y(t,s) = Y(\bar{s})\bar{y}(\bar{t},\bar{s}),\ v(t,s) = V(\bar{s})\bar{v}(\bar{t},\bar{s}), \tag{3}$$

$$T = 1/(\mu m_0),\ S = 1,\ X = b/\sigma,\ V = (k_0/c_0)Y,\ Y = b/m_0, \tag{4}$$

where m_0, k_0, c_0 are $m(s)$, $k(s)$, $c(s)$ of the wild (initial or any fixed) strain. T is measured in the units of time, while X, Y, and V are in the units of concentrations of target cells and free virus.

Substituting (3), (4) into the model and denoting $R_0(\bar{s}) = b\alpha(s)k(s)/(\sigma m(s)c(s))$ (the basic reproduction ratio), $\bar{m}(\bar{s}) = m(s)/m_0$, $\varepsilon = \mu m_0/\sigma$ è $\nu = \sigma/c_0$, we get singularly perturbed ("slow-fast") system with two small parameters:

$$\begin{aligned}
\varepsilon\bar{x}'_{\bar{t}} &= 1 - \bar{x}(\bar{t}) - \int_\Omega R_0(\bar{s})\bar{x}(\bar{t})\bar{v}(\bar{t},\bar{s})\,d\bar{s},\\
\bar{y}'_{\bar{t}} &= \bar{m}(\bar{s})/\mu\left(\int_\Omega p_1(\bar{s},\bar{r})R_0(\bar{r})\bar{x}(\bar{t})\bar{v}(\bar{t},\bar{r})\,d\bar{r} - \bar{y}(\bar{t},\bar{s})\right),\\
\varepsilon\nu\bar{v}'_{\bar{t}} &= c(\bar{s})/c_0\left(\bar{y}(\bar{t},\bar{s}) - \bar{v}(\bar{t},\bar{s})\right).
\end{aligned}$$

Setting $\nu = 0$, we obtain the first-order degenerate system

$$\begin{aligned}
\varepsilon\bar{x}'_{\bar{t}} &= 1 - \bar{x}(\bar{t}) - \int_\Omega R_0(\bar{s})\bar{x}(\bar{t})\bar{v}(\bar{t},\bar{s})\,d\bar{s},\\
\bar{y}'_{\bar{t}} &= \bar{m}(\bar{s})/\mu\left(\int_\Omega p_1(\bar{s},\bar{r})R_0(\bar{r})\bar{x}(\bar{t})\bar{v}(\bar{t},\bar{r})\,d\bar{r} - \bar{y}(\bar{t},\bar{s})\right),\\
0 &= c(\bar{s})/c_0\left(\bar{y}(\bar{t},\bar{s}) - \bar{v}(\bar{t},\bar{s})\right).
\end{aligned}$$

The third equation is algebraic and has root $\bar{v} = \bar{y}$. For the first-order associated equation $\hat{v}'_\tau = c(\bar{s})/c_0\left(\hat{v}(\tau,\bar{s}) - \bar{y}\right)$, where $\bar{y}$ enters as a parameter, the root $\hat{v} = \varphi(\bar{y}) = \bar{y}$ is the asymptotically stable (in the sense of Lyapunov) stationary point. Let us add the initial conditions $\bar{x}(0) = x^0$, $\bar{y}(0,\bar{s}) = y^0(\bar{s})$ and $\bar{v}(0,\bar{s}) = v^0(\bar{s})$. At the initial value of the parameter $\bar{y}$, i.e., at $\bar{y} = y^0(\bar{s})$, the first-order associated equation with the initial condition $\bar{v}(0,\bar{s}) = v^0(\bar{s})$ has a unique solution $\hat{v} = y^0(\bar{s}) + \left(v^0(\bar{s}) - y^0(\bar{s})\right)\exp\left(-c(\bar{s})/c_0\tau\right)$, and $\hat{v}(\tau,\bar{s}) \to \varphi(y^0(\bar{s})) = y^0(\bar{s})$ as $\tau \to +\infty\ \forall\bar{s} \in \Omega$. Thereby the initial point $v^0(\bar{s})$ of the first-order associated equation belongs to the domain of attraction of the stable stationary point $\varphi(y^0(\bar{s}))$.

Then let us $\varepsilon = 0$. We obtain the second-order degenerate system

$$\begin{aligned}
0 &= 1 - \bar{x}(\bar{t}) - \int_\Omega R_0(\bar{s})\bar{x}(\bar{t})\bar{v}(\bar{t},\bar{s})\,d\bar{s},\\
\bar{y}'_{\bar{t}} &= \bar{m}(\bar{s})/\mu\left(\int_\Omega p_1(\bar{s},\bar{r})R_0(\bar{r})\bar{x}(\bar{t})\bar{v}(\bar{t},\bar{r})\,d\bar{r} - \bar{y}(\bar{t},\bar{s})\right),\\
0 &= c(\bar{s})/c_0\left(\bar{y}(\bar{t},\bar{s}) - \bar{v}(\bar{t},\bar{s})\right).
\end{aligned}$$

first equation in which is algebraic with respect to $\bar{x}$ and has a root $\bar{x} = \Psi(\bar{v}) = \left(1 + \int_{\Omega} R_0(\bar{s})\bar{v}(\bar{t}, \bar{s})\, d\bar{s}\right)^{-1}$. This root is the asymptotically stable stationary point (in the sense of Lyapunov) of second-order associated to the equation

$$\hat{x}'_{\tau} = -\big(1 + \int_{\Omega} R_0(\bar{s})\bar{v}(\bar{t}, \bar{s})\, d\bar{s}\big)\hat{x}(\tau) + 1.$$

The latter equation with the initial condition $\bar{x}(0) = x^0$ at the initial value of the parameter $\bar{v}$ $\bar{v} = v^0(\bar{s})$ has a unique solution $\hat{x}(\tau) = \left(x^0 - 1/f\right)\exp(-f\tau) + 1/f$, where $1/f = \Psi(v^0(\bar{s})) = 1 + \int_{\Omega} R_0(\bar{s})v^0(\bar{s})\, d\bar{s}$, for all $\tau \geq 0$, and $\hat{x}(\tau) \to \Psi(v^0(\bar{s}))$ as $\tau \to +\infty$. Thus, the initial point x^0 of the second-order associated equation belongs to the domain of attraction of the stable stationary point. Thus, all the conditions of the theorem are satisfied and, consequently, the limiting equalities hold (under the assumption of the existence and uniqueness of the solution of truncated problem).

Thus, the original system can be reduced to a single integrodifferential equation

$$\bar{y}'_{\bar{t}} = \bar{m}(\bar{s})/\mu\left(\int_{\Omega} p_1(\bar{s}, \bar{r})R_0(\bar{r})\bar{y}(\bar{t}, \bar{r})\, d\bar{r} \Big/ \left(1 + \int_{\Omega} R_0(\bar{r})\bar{y}(\bar{t}, \bar{r})\, d\bar{r}\right) - \bar{y}(\bar{t}, \bar{s})\right).$$

3 Conclusion

In this paper, we considered the model of viral dynamics with random mutations that contain the population dynamics of uninfected cells, infected cells, and free virus particles. Using the analog of Tikhonov's theorem, timescale separation procedure is carried out. As a result, the original systems of three integrodifferential equations are reduced to a single one. This fact can be used to simplify the numerical simulation of such complex systems. As a rule in evolutionary biology, mathematical models are usually formulated as integrodifferential equations and the same technique can be employed to the ones as well.

References

1. A.A. Archibasov, A. Korobeinikov, V.A. Sobolev, Pasage to the limit in a singularly perturbed partial integro-differential system. Differ. Equ. **52**(9), 1115–1122 (2016)
2. A. Korobeinikov, A.A. Archibasov, V.A. Sobolev, Multi-scale problem in the model of RNA virus evolution. J. Phys. Conf. Series **727** (2016)
3. A.N. Tikhonov, Systems of differential equations with small parameters multiplying the derivatives. Mat. Sb. **31**, 575–586 (1952)
4. A.B. Vasilieva, V.F. Butuzov, L.V. Kalachev, *The boundary function method for singular perturbation problems* (SIAM, Philadelphia, 1995), p. 221

A Discrete Variant Space Model of Cancer Evolution

Andrei Korobeinikov and Stefano Pedarra

Abstract In this paper, we suggest a discrete variant space model of cancer evolution. The model is reasonably simple, deterministic, and is formulated as a system of ordinary differential equations. The model is based on the concept of "multi-strain modeling" (or quasi-species), which is successfully applied in modeling of the infectious disease dynamics and viral dynamics. The model constructed in this paper is mechanistic; that is, it is based upon a set of explicitly stated assumptions and hypothesis ("the first principles"). This implies that model's parameters, as well as results obtained, can be immediately interpreted, and that a further model development, e.g., incorporation into the model factors such as anticancer therapies, immune response, etc., is a reasonably straightforward procedure. To illustrate this model applicability, results of numerical simulations, as well as their biological interpretations, are provided.

1 Introduction

The term "cancer" refers to a group of diseases, which can affect almost any tissue and organ and are characterized by the uncontrolled growth of abnormal cells, which cell cycle is much faster than that of the normal cells. Cancer appears as a result of a series of mutations of normal cells, which occur during the DNA replication process or as a result of a somatic mutation. Cancer cells are usually characterized by their genome instability, and as a consequence of this, by extremely high levels

A. Korobeinikov is supported by Ministerio de Economía y Competitividad of Spain via grant MTM2015-71509-C2-1-R.

A. Korobeinikov (✉)
Centre de Recerca Matemàtica, Campus de Bellaterra, Edifici C, 08193 Barcelona, Spain
e-mail: akorobeinikov@crm.cat

Departament de Matemàtiques, Universitat Autònoma de Barcelona, Barcelona, Spain

S. Pedarra
Dipartimento di Matematica "Tullio Levi-Civita", Università degli Studi di Padova, Padova, Italy
e-mail: stefano.pedarra@studenti.unipd.it

A. Korobeinikov et al. (eds.), *Extended Abstracts Spring 2018*,
Trends in Mathematics 11, https://doi.org/10.1007/978-3-030-25261-8_4

of mutability and evolvability. The genome instability, as well as the mutability and evolvability of cancer, is one of the cancer hallmarks [3–5]. As a result of this very high mutability, a typical tumor is composed of a very large number of cancer genotypes. Moreover, the mutability, evolvability, and the resulting genetic diversity of cancer make its treatment very difficult.

The mentioned genome instability, high mutability, and evolvability of cancer makes its study from the point of view of evolutionary biology essential. Accordingly, there is a growing interest and a certain progress in mathematical modeling of cancer evolution. Usually, a mathematical model of cancer evolution utilizes the idea of quasi-species (or multi-strain modeling) and is formulated in the form of a system of ordinary differential equations, partial differential equations, or integrodifferential equations (see, e.g., [9–11, 14, 15], and bibliography therein; the same conceptual ideas were also developed and applied for mathematical modeling of viral evolution [1, 2, 6–8, 13]). In this paper, we use these ideas to construct a reasonably simple mechanistic model of cancer evolution on the basis of the model of cancer and normal cells competition.

2 Model

In order to model cancer evolution, let us assume that there is a system composed of the normal cells and cancer cells of n different genotypes, where $n \to \infty$ or is a very large number. Let us denote the size of cell population of the ith genotype at time t by $C_i(t)$ and the size of the normal cells population by $C_0(t)$. We assume that (i) all the cells reproduce and die; (ii) there are limited resources, which limit populations growth through inhibiting the reproduction and accelerating the death; (iii) cells of the different genotypes have to compete for these limited resources; (iv) in the process of mitosis, with some probability p_{ij}, a cell of the ith genotype can produce a mutant daughter cell of the jth genotype, which, subsequently, goes to the jth population; and (v) as a result of somatic mutation, with probability $q_{i,j}$ a cell of the ith genotype can move to the jth genotype.

We start at the Lotka–Volterra model of competing populations which is a usual basis for cancer modeling [12]:

$$\dot{C}_i(t) = r_i C_i \left(1 - \frac{1}{K} \sum_{j=0}^{n} b_{ij} C_j \right), \tag{1}$$

where $i = 0, 1, 2, \ldots, n$. In these equations, $r = (r_i)_{i=0}^{n}$ represents the vector of the growth rates, the elements of matrix $B = (b_{ij})_{i,j=0}^{n}$ represent the relative competitive capabilities of the genotypes and K is the carrying capacity of the system. To introduce into this model, a possibility of mutations that occur during DNA replication, it is necessary to separate the birth and the death rates, as these mutations occur in the process cell reproduction. For these two processes, we have

$$\text{birth rate of the } i\text{-th population} = a_i C_i \left(1 - \frac{h_i}{K} \sum_{k=0}^{n} b_{ik} C_k \right), \tag{2}$$

$$\text{death rate of the } i\text{-th population} = -d_i C_i \left(1 + \frac{g_i}{K} \sum_{j=0}^{n} b_{ik} C_k \right). \tag{3}$$

Here, a_i and d_i are the per capita birth and death rates of the ith genotype cells, and h_i and g_i are weights that fine-tune the relative impacts of the lack of resources and competition on the proliferation and death rates, respectively. These parameters are related to r_i by the equalities $r_i = a_i - d_i$ and $r_i = a_i h_i + d_i g_i$.

Then, using an approach suggested in [9], we can introduce a possibility of mutation into the model modifying the birth term (2). Then the growth of the ith genotype population is represented by the following equation:

$$\dot{C}_i = \sum_{j=0}^{n} \left(p_{ji} a_j C_j \left(1 - \frac{h_j}{K} \sum_{k=0}^{n} b_{jk} C_k \right) \right) - d_i C_i \left(1 + \frac{g_i}{K} \sum_{k=0}^{n} b_{ik} C_k \right) + \sum_{j=0}^{n} q_{ji} C_j - \sum_{j=0}^{n} q_{ij} C_i . \tag{4}$$

Here, the diffusion-like term $\sum_{j=0}^{n} q_{ji} C_j - \sum_{j=0}^{n} q_{ij} C_i$ represents the somatic mutations. All the parameters of this model are positive real numbers, except for the weights h_i and g_i and the elements of probability matrices $P = (p_{ij})$ and $Q = (q_{ij})$, which can be zero.

This discrete variant space model is formulated as a system of ODEs. An equivalent continuous variant space model is formulated in [11].

To non-dimensionalize the system (4), we introduce nondimensional variables $x_i(\tau)$ and τ as the following:

$$x_i = b_{ii} C_i / K , \qquad \tau = T t , \qquad T = p_{00} a_0 h_0 + d_0 g_0 . \tag{5}$$

Please note that d_0 and p_{00} are always positive, whereas h_0 and g_0 cannot be equal to 0 simultaneously and, hence, T is always positive. Substituting these variables into the system (4) and separating the linear and the nonlinear parts of the equations, we rewrite the system in the following form:

$$\frac{dx_i}{d\tau} = \sum_{j=0}^{n} u_{ij} x_j - \sum_{j=0}^{n} \sum_{k=0}^{n} v_{ij} f_{jk} x_j x_k , \quad i = 0, \ldots, n . \tag{6}$$

Here,

$$u_{ij} = \begin{cases} \frac{(p_{jj} a_j + q_{jj} - d_j - \sum_{k=0}^{n} q_{jk}) e_{ij}}{T} & \text{if } j = i , \\ \frac{(p_{ji} a_j + q_{ji}) e_{ij}}{T} & \text{if } j \neq i , \end{cases} \qquad v_{ij} = \begin{cases} \frac{(p_{jj} a_j h_j + d_j g_j) e_{ij}}{T} & \text{if } j = i , \\ \frac{p_{ji} a_j h_j e_{ij}}{T} & \text{if } j \neq i , \end{cases}$$

and $f_{ij} = b_{ij} / b_{jj}$ and $e_{ij} = b_{ii} / b_{jj}$.

3 Simulations

To illustrate model behavior, we run numerical simulations. In these simulations, we assume that $h_i = 0$ for all $i = 0, \ldots, n$. (That is, a shortage of the resources does not affect the reproduction; it is equivalent to an assumption that a decrease of new births is attributed to an increment of deaths.) Furthermore, we assume that $Q = (q_{ij}) = 0$ for all $i, j = 0, \ldots, n$. (That is, we disregards somatic mutations.) We assume that a cell of the ith genotype can produce a daughter cell only of ith, $(i-1)$th, or $(i+1)$th genotypes with probabilities given by matrix P:

$$P = \begin{bmatrix} 0.9 & 0.1 & 0 & \cdots & \cdots & 0 \\ 0.1 & 0.8 & 0.1 & 0 & \cdots & 0 \\ \vdots & \vdots & \vdots & \vdots & \vdots & \vdots \\ 0 & \cdots & 0 & 0.1 & 0.8 & 0.1 \\ 0 & \cdots & \cdots & 0 & 0.1 & 0.9 \end{bmatrix}.$$

In the simulations, the environment carrying capacity $K = 10^5$ cells, and time t is measured in days. Values of the other parameters, as well as, the initial conditions used in the three simulations are summarized the following table:

Parameter	Simulation # 1	Simulation # 2	Simulation # 3
n	50	50	200
a_i	2	10	$4 - 2e^{-i}$
d_i	0.2	0.2	$0.2 - 0.1e^{-i}$
g_i	9	51	19
b_{ij}	$2 - i/n$	$1 + i/n$	$2 - i/n$
$C_i(0)$	$(K-1, 1, 0, \ldots, 0)$	$(K-1, 1, 0, \ldots, 0)$	$(K-1, 1, 0, \ldots, 0)$

Results of the simulations are depicted in Figs. 1, 2, and 3, respectively. In simulation #1, we used $b_{ij} = 2 - i/n$ for all j. This implies that the Darwinian fitness of the genotypes grows as i increases. The initial conditions in the simulation #1 implies that initially, there was present only one mutant cell of the first mutant genotype. The formation of a traveling wave moving in the direction of increasing i is clearly seen in Fig. 1. This implies that in this simulation an average fitness of the tumor population steadily increases.

Please note that in Fig. 1, the populations of genotypes $i = 47$ to 50 remain approximately constant after $t \approx 120$ days: for this simulation, we consider a system of 50 genotypes and, hence, in this simulation, the Darwinian fitness of the 50th genotype is maximal. This enables this genotype, as well as close genotypes 47th to 49th, to eventually prevail in the system. Of course, in a real-life system, the number of possible mutant genotypes n is significantly higher, and, hence, no such steady prevalence of a particular mutant genotypes can be observed.

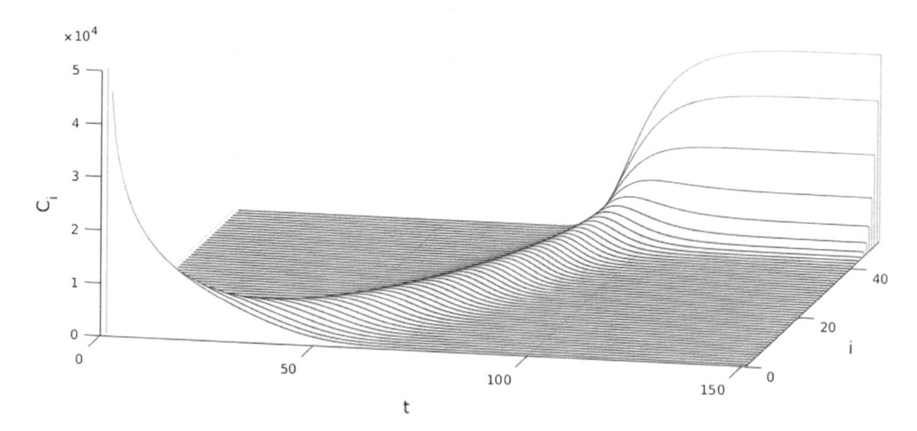

Fig. 1 Simulation #1: variation of the genotype abundance in time

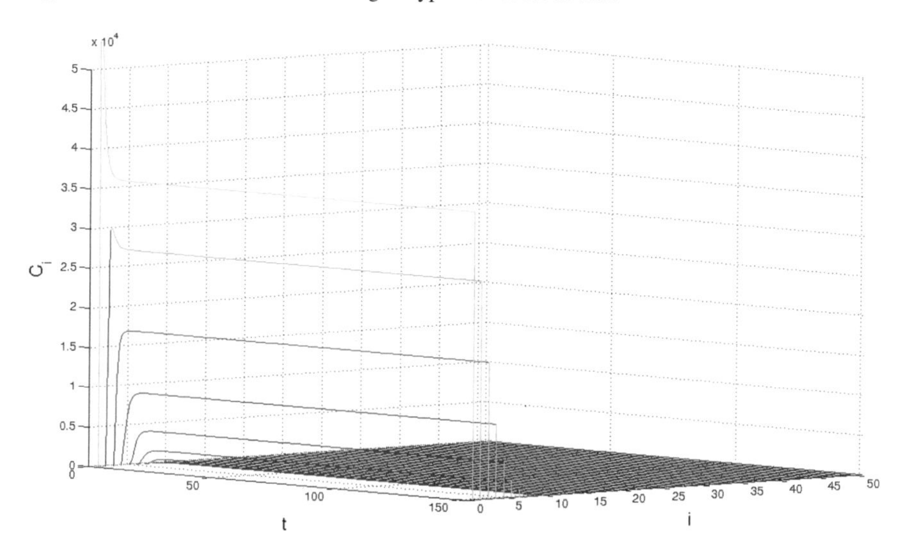

Fig. 2 Simulation #2: variation of the genotype abundance in time

Figure 2 shows the results of simulation #2. For this simulation, in contrast to simulation #1, we take $b_{ij} = 1 + i/n$. This implies that the Darwinian fitness decreases as i grows. The simulation confirms an intuitive expectation that in such a case, for whatever large levels of mutation probabilities, the mutations are unable to fix in the system and the mutant cells will be eventually removed from the tissue.

In the simulation #3, we consider the impact of the proliferation rates on the evolution. It is a well-known fact that cancer cells proliferate faster than the healthy cells and that the proliferation rates depend, above all, on the degree of differentiation.

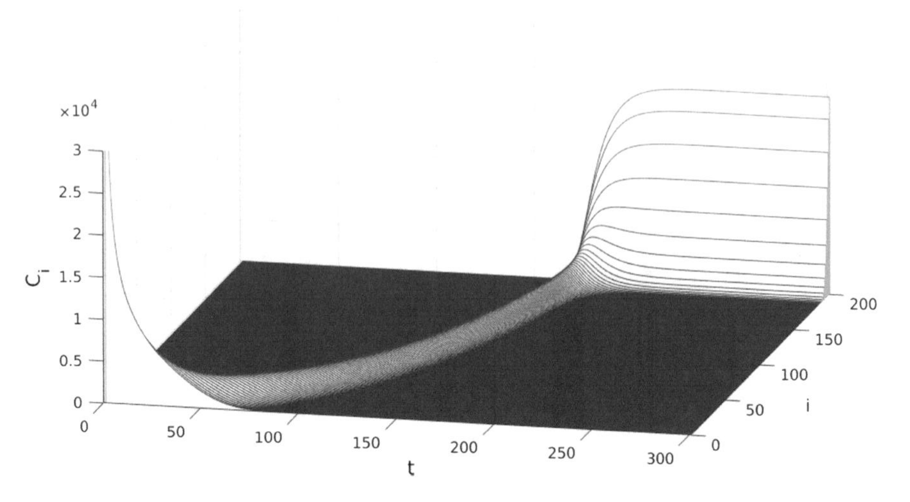

Fig. 3 Simulation #3: variation of the genotype abundance in time

Accordingly, in the simulation #3, we used $a_i = 4 - 2e^{-i}$ and $d_i = 0.2 - 0.1e^{-i}$, in combination with the same b_{ij} as in simulation #1. Figure 3 shows the results of simulation #3. As one can expect, in this case, a traveling wave of evolution is forming as well. However, it moves faster than in the simulation #1, where the proliferation and death rates were constant. Moreover, it is easy to see that in this case the speed of the traveling wave notably decreases as i grows (and as a_i and d_i grow).

It is hardly surprising that the system behavior in the simulations #1 and #3 is very similar: the only difference is the higher speed of the traveling wave in simulation #3. In both these simulations, the mean genotype number grows converging to the last genotype that has the highest level of the Darwinian fitness. The variance of the genotype distribution in the population initially grows until it reaches, at a certain time t^*, its maximum value, and then it slowly decreases. Such a behavior is intriguing and counterintuitive.

The analysis of the simulation results suggests that the comparative values of the competition factors b_{ij} mostly determine the system behavior and that changes of these values can change the system qualitative behavior.

References

1. A.A. Archibasov, A. Korobeinikov, V.A. Sobolev, Asymptotic expansions of solutions for a singularly perturbed model of viral evolution. Comput. Math. Math. Phys. **55**(2), 240–250 (2015)
2. A.A. Archibasov, A. Korobeinikov, V.A. Sobolev, Passage to the limits in a singularly perturbed partial integro-differential system. Differ. Equ. **52**(9), 1115–1122 (2016)

3. Y.A. Fouad, C. Aanei, Revisiting the hallmarks of cancer. Am. J. Cancer Res. **7**(5), 1016–1036 (2017)
4. D. Hanahan, R.A. Weinberg, The hallmarks of cancer. Cell **100**(1), 57–70 (2000)
5. D. Hanahan, R.A. Weinberg, Hallmarks of cancer: the next generation. Cell **144**(5), 646–674 (2011)
6. A. Korobeinikov, Immune response and within-host viral evolution: immune response can accelerate evolution. J. Theor. Biol. **456**, 74–83 (2018)
7. A. Korobeinikov, A. Archibasov, V. Sobolev, Order reduction for an RNA virus evolution model. Math. Biosci. Eng. **12**(5), 1007–1016 (2015)
8. A. Korobeinikov, A. Archibasov, V. Sobolev, Multi-scale problem in the model of RNA virus evolution. J. Phys. Conf. Ser. **727**, 012007 (2016)
9. A. Korobeinikov, K.E. Starkov, P.A. Valle, Modeling cancer evolution. J. Phys. Conf. Ser. **811**, 012004 (2017)
10. D. Masip, A. Korobeinikov, A continuous phenotype space model of cancer evolution. J. Phys. Conf. Ser. **811**, 012005 (2017)
11. D. Moreno-Martos, A. Korobeinikov, A mathematical model of cancer evolution. Trends Math. Res. Perspect. CRM Barcelona **11**, Summer 2018, Birkhäuser, Basel (2018)
12. J.D. Murray, *Mathematical Biology I, An Introduction*, 3rd edn. (Springer, New York, 2002)
13. S. Pagliarini, A. Korobeinikov, A mathematical model of marine bacteriophage evolution. R. Soc. Open Sci. **5**, 171661 (2018)
14. J. Sardanyés, R. Martínez, C. Simó, R.V. Solé, Abrupt transitions to tumor extinction: a phenotypic quasispecies model. J. Math. Biol. **74**, 1589–1609 (2017)
15. R.V. Solé, S. Valverde, C. Rodríguez-Caso, J. Sardanyés, Can a minimal replicating construct be identified as the embodiment of cancer? Bioessays **36**, 503–512 (2014)

Mathematical Modelling of HIV Within-Host Evolution

Anna Maria Riera-Escandell and Andrei Korobeinikov

Abstract The majority of hypotheses suggested to explain the human immunodeficiency virus (HIV) progression explains the particularities of the disease by a very high mutability and evolvability of the virus. For HIV, several mechanisms of mutation are possible, and it is reasonable to assume that a mechanistic mathematical model of HIV evolution reflect these mechanisms. In this contribution, we formulate three different mathematical models of within-host HIV evolution that corresponds to three different virus mutation mechanisms. Simulations demonstrate that, for realistic rates of evolution, either of the three models leads to a very similar (and for some models to identical) outcome and, hence, only one of the mechanisms (presumably, the simplest one) can be used in simulations.

1 Models of HIV Evolution

Developing a mathematical model of HIV evolution requires a mechanistic approach that considers a combination of factors responsible for natural selection of the virus and a mechanism that describes random. Factors responsible for natural selection usually act on the same timescale as the lifespan of the organism involved. In contrast,

A. Korobeinikov is supported by Ministerio de Economía y Competitividad of Spain via grant MTM2015-71509-C2-1-R.

A. M. Riera-Escandell (✉) · A. Korobeinikov
Centre de Recerca Matemàtica, Campus de Bellaterra, Edifici C, 08193 Barcelona, Spain
e-mail: annamaria.riera.escandell@gmail.com

A. M. Riera-Escandell
Mathematical Institute, Oxford University, Oxford, UK

A. Korobeinikov
Departament de Matemàtiques, Universitat Autònoma de Barcelona, 08193 Barcelona, Spain
e-mail: akorobeinikov@crm.cat

A. Korobeinikov et al. (eds.), *Extended Abstracts Spring 2018*,
Trends in Mathematics 11, https://doi.org/10.1007/978-3-030-25261-8_5

evolution progresses at a much slower rate. This implies the inclusion in the model of multiple timescales, which can differ one from one another by several orders of magnitude. Furthermore, the viral and host generation timescales also differ by orders of magnitude, thus adding an additional layer of complexity to the mathematical analysis of HIV evolution [3, 5, 6, 8].

In order to describe HIV evolution, we formulate mathematical models of HIV dynamics, which are based on the Nowak and May model of within-host HIV dynamics [7]. The Nowak–May model describes the interaction of three populations; namely, the susceptible target cells, the infected cells and free virus particles. Let us assume that $X(t)$ and $Y(t)$ denote the current concentrations of the susceptible and the infected target cells, respectively, and $V(t)$ is the current concentration of the free virus particles. The Nowak–May model postulates that the susceptible cells enter into the system with a constant rate λ and die at a per capita rate m; that the susceptible cells are infected by the free virus particles at rate $\alpha X V$; that the infected cells produce, at a per capita rate k, the free virus particles and die at a per capita rate a; the free virions are removed from the system at a per capita rate u.

In order to describe mutations in this model, we, first, have to assume the existence of a multitude of viral genotypes; each genotype corresponds to a viral subpopulation. Let us assume that there are n $(n \to \infty)$ different viral types (strains) of concentrations $V_i(t)$ and, hence, different infected subclasses $Y_i(t)$ (where $i = 1, 2, \ldots, n$). Moreover, properties of each type are described by the variant-specific parameters α_i, a_i, k_i and u_i. Please note that parameters λ and m characterise the target cells and, hence, are variant-independent.

In the framework of the Nowak–May model, three mechanisms of random mutations are possible. By the first mechanism, an error can occur in the process of virus production. That is, a cell infected by the jth viral type can, with probability $\rho_{i,j}^{(1)}$, produce the ith-type virion. The corresponding *Model 1* is

$$\begin{aligned}
\dot{X} &= \lambda - mX - \sum_{i=1}^{n} \alpha_i X V_i\,, \\
\dot{Y}_i &= \alpha_i X V_i - a_i Y_i\,, \qquad\qquad (1)\\
\dot{V}_i &= \sum_{j=1}^{n} k_j \rho_{i,j}^{(1)} Y_j - u_i V_i\,.
\end{aligned}$$

By the second mechanism, an error occurs in the process of transcript of the viral genomic material to the cell nuclear. That is, due to such an error, a cell, infected by the jth-type virion, produces exclusively the ith-type virus. Denoting $\rho_{i,j}^{(2)}$, the probability that an infection by the jth- type results in production of the ith type, we obtain equations of *Model 2*

$$\dot{X} = \lambda - mX - \sum_{i=1}^{n} \alpha_i X V_i \, ,$$
$$\dot{Y}_i = \sum_{j=1}^{n} \alpha_j \rho_{i,j}^{(2)} X V_j - a_i Y_i \, , \tag{2}$$
$$\dot{V}_i = k_i Y_i - u_i V_i \, .$$

Finally, the third mechanism assume that a cell infected with the ith viral type produces virus of this type for some time and then, as a result of an error, switches to the production of the j type virus exclusively. Assuming that γ is the rate of such mutations and that $\rho_{i,j}^{(3)}$ is a probability that a mutation changes the virus production from the jth type to the ith type, we obtain equations of *Model 3* as follows:

$$\dot{X} = \lambda - mX - \sum_{i=1}^{n} \alpha_i X V_i \, ,$$
$$\dot{Y}_i = \alpha_i X V_i - a_i Y_i + \gamma \Big(\sum_{j=1}^{n} \rho_{i,j}^{(3)} Y_j - Y_i \Big) , \tag{3}$$
$$\dot{V}_i = k_i Y_i - u_i V_i \, .$$

We assume that for all models, the mutation matrix $P = ||\rho_{i,j}||$ is strongly diagonal-prevalent and can be written as $P^{(l)} = I - \mu^{(l)} Q^{(l)}$ $(l = 1, 2, 3)$, where I is the identity matrix, $\mu^{(l)}$ a small parameter and $Q^{(l)} = ||q_{i,j}^{(l)}||$ a matrix whose diagonal elements are of order 1.

For all three models, $R_i^0 = \lambda \alpha_i k_i / m b_i u_i$ is the basic reproduction number of the ith viral variant.

1.1 Models Non-dimensionalisation and Order Reduction

Following [1, 2, 4, 5], we conduct the models non-dimensionalisation and order reduction.

For *Model 1*, we define $\varepsilon = \mu^{(1)} b_1 / m$, $\epsilon_i = m/b_i$ and $\bar{c}_i = u_i / b_1$. Please note that $\varepsilon \ll 1$, and, also, $\epsilon_i < 1$ for all i. Then, assuming that $\varepsilon = 0$ and $\epsilon_i = 0$ and denoting the non-dimensional variables $\bar{v}_i(\bar{t})$ and $\bar{t}$, we obtain equations

$$\frac{d\bar{v}_i}{d\bar{t}} = \frac{\bar{c}_i}{\mu^{(1)}} \cdot \frac{R_i^0 - 1}{1 + \sum_{j=1}^{n} R_j^0 \bar{v}_j} \bar{v}_i \left(1 - \frac{\sum_{j=1}^{n} R_j^0 \bar{v}_j}{R_i^0 - 1} \right) - \frac{\bar{c}_i}{1 + \sum_{j=1}^{n} R_j^0 \bar{v}_j} \sum_{j=1}^{n} q_{i,j}^{(1)} \frac{b_i}{k_i} \frac{k_j}{b_j} R_j^0 \bar{v}_j . \tag{4}$$

Table 1 Values of the parameters and initial conditions used in simulations

Parameter	Value	Parameter	Value	Parameter	Value
λ	20	α_a	$8 \cdot 10^{-6}$	$Y_i(0)$, $i = 1, \ldots, n$	0
m	0.02	α_b	$1.8 \cdot 10^{-8}$	$V_1(0)$	250
$b_i \equiv b$	0.8	γ	0.48	$V_i(0)$, $i = 2, \ldots, n$	0
$k_i \equiv k$	10^3	$\mu^{(i)}$, $i = 1, 2, 3$	$2.5 \cdot 10^{-4}$	n	1000
$u_i \equiv u$	8	$X(0)$	1000		

For *Model 2*, we define $\bar{b}_i = b_i/b_1$, $\varepsilon = \mu^{(2)} b_1/m$ and $\nu_i = m/u_i$. Assuming that $\nu_i = 0$ and $\varepsilon = 0$, we obtain

$$\frac{d\bar{v}_i}{d\bar{t}} = \frac{\bar{b}_i}{\mu^{(2)}} \cdot \frac{R_i^0 - 1}{1 + \sum_{j=1}^n R_j^0 \bar{v}_j} \bar{v}_i \left(1 - \frac{\sum_{j=1}^n R_j^0 \bar{v}_j}{R_i^0 - 1}\right) - \frac{\bar{b}_i}{1 + \sum_{j=1}^n R_j^0 \bar{v}_j} \sum_{j=1}^n q_{i,j}^{(2)} R_j^0 \bar{v}_j. \tag{5}$$

For *Model 3*, we define $\varepsilon = \mu^{(3)} b_1/m$, $\nu_i = m/u_i$, $\bar{b}_i = b_i/b_1$ and $\bar{\gamma} = \gamma/b_1$, and, as above, assume that $\nu_i = 0$ and $\varepsilon = 0$. We get

$$\frac{d\bar{v}_i}{d\bar{t}} = \frac{\bar{b}_i}{\mu^{(3)}} \cdot \frac{R_i^0 - 1}{1 + \sum_{j-1}^n R_j^0 \bar{v}_j} \bar{v}_i \left(1 - \frac{\sum_{j=1}^n R_j^0 \bar{v}_j}{R_i^0 - 1}\right) - \bar{\gamma} \sum_{j=1}^n q_{i,j}^{(3)} \frac{b_i}{b_j} \bar{v}_j. \tag{6}$$

2 Results

Our goal is to compare the outcomes for these three models. In order to do this, we run numerical simulations for these models with the following assumptions made about the models parameters. We set $\rho_{1,1} = \rho_{n,n} = 1 - \mu$, $\rho_{i,i} = 1 - 2\mu$ and $\rho_{i,i+1} = \rho_{i,i-1} = \mu$, $\forall i \neq 1, n$. For simplicity, we assume that the only variant-depending parameter is α_i, postulating that $\alpha_i = \alpha_a + \alpha_b(i - 1)$, where α_a and α_b are constants. All other parameters are assumed to be the same for all viral variants. Their values used in the simulations are shown in Table 1; we used these values for all Models except for the reduced Model 1, where we scaled parameters u and k to have $\bar{c}_i = 1$.

Figure 1a shows how concentrations of viral variants change in time. The process displayed in this figure is equivalent to a travelling wave in a continuous space. Figure 1b shows the typical distribution of the viral variants in a quasi-species at specific moments for Model 1. (For all other models both pictures are very similar).

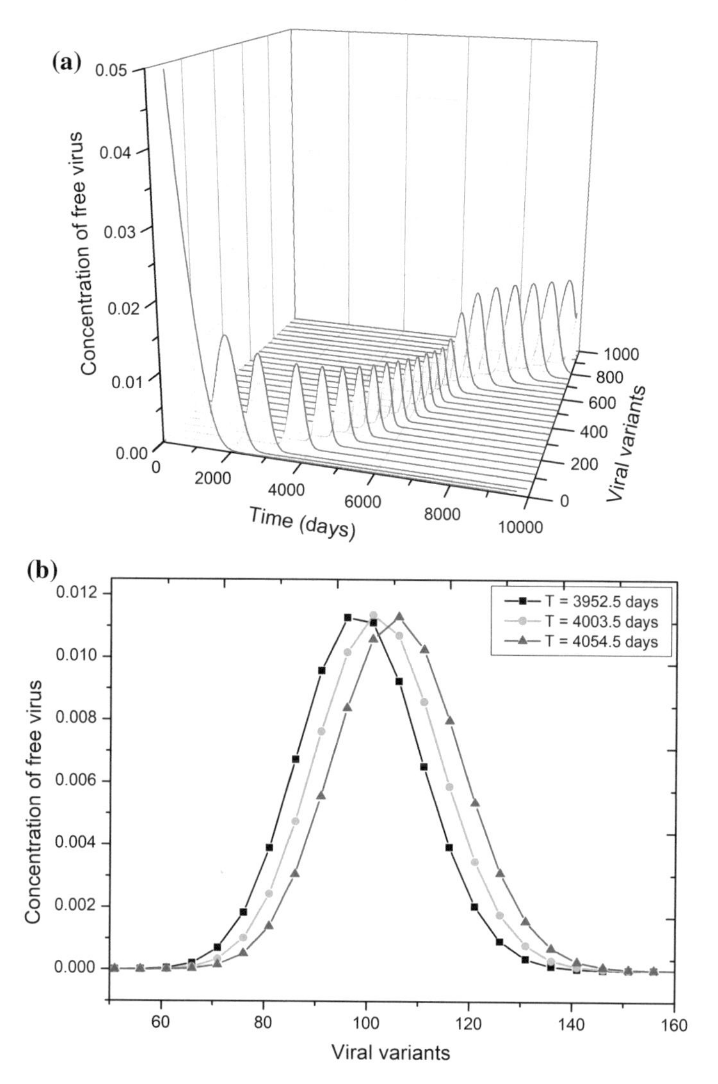

Fig. 1 Viral concentrations in time (**a**) and by variants for $t = 3952.5$, 4003.5 and 4054.5 days (**b**) for the reduced Model 1. The parameters in the simulations are from Table 1, apart from $\mu = 0.025$, $b = 0.266$ and $u = 24$

Figure 2 shows the mean value $i_{mean}(t) = (\sum_{i=1}^{n} i \cdot v_i)/(\sum_{i=1}^{n} v_i)$ in time for the three original (Fig. 2a) and three reduced (Fig. 2b) models. It is easy to see that all three models produce very similar outcomes. Moreover, the results for Models 1 and 2 are virtually indistinguishable for both the full and the reduced systems.

Figure 3 shows the changes of the concentration of healthy cells and the total viral load $v(t) = \sum_{i=1}^{n} v_i(t)$ in time for Models 1, 2 and 3. Results for Models 1 and 2

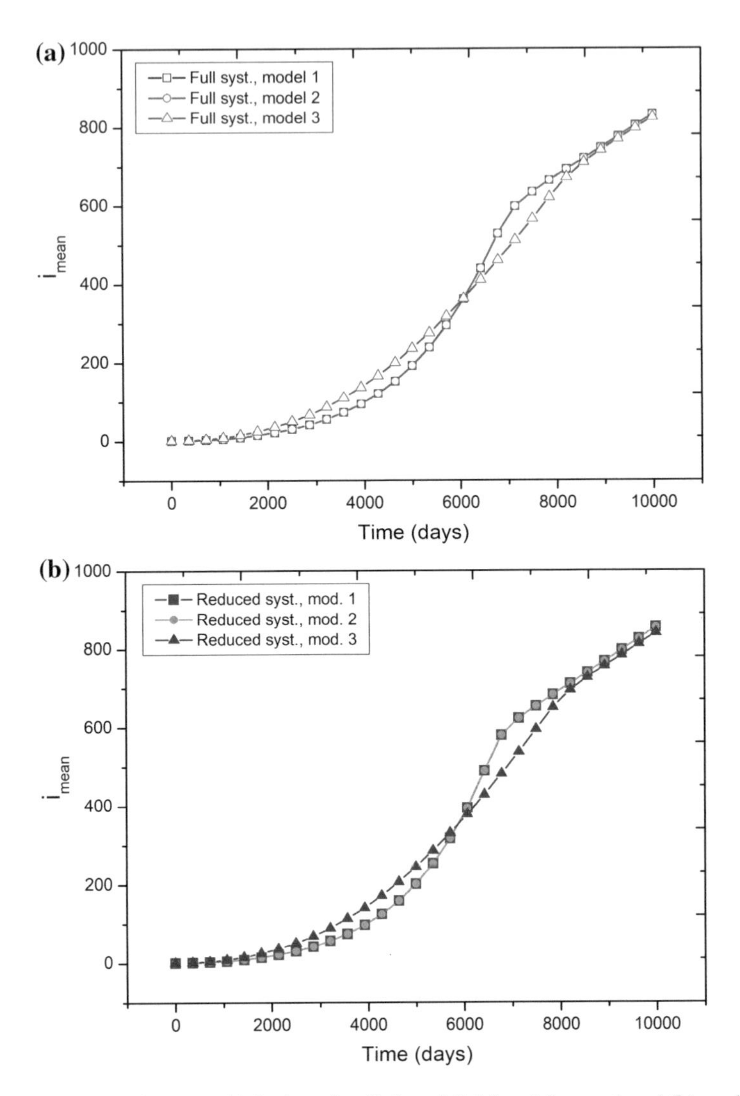

Fig. 2 Mean viral variant $i_{mean}(t)$ in time, for all three full (**a**) and three reduced (**b**) models

are identical. Results for Model 3 are very close to Models 1 and 2, and after some time nearly coincide with these.

The simulations that we run for viral evolution models with the three different mutation mechanisms show that if the speed of evolution is sufficiently slow (as it is in real life), all these models produce equivalent or very similar results. Moreover, simulations demonstrate that each of these models can be reduced to a simpler system, which also produces the very similar results.

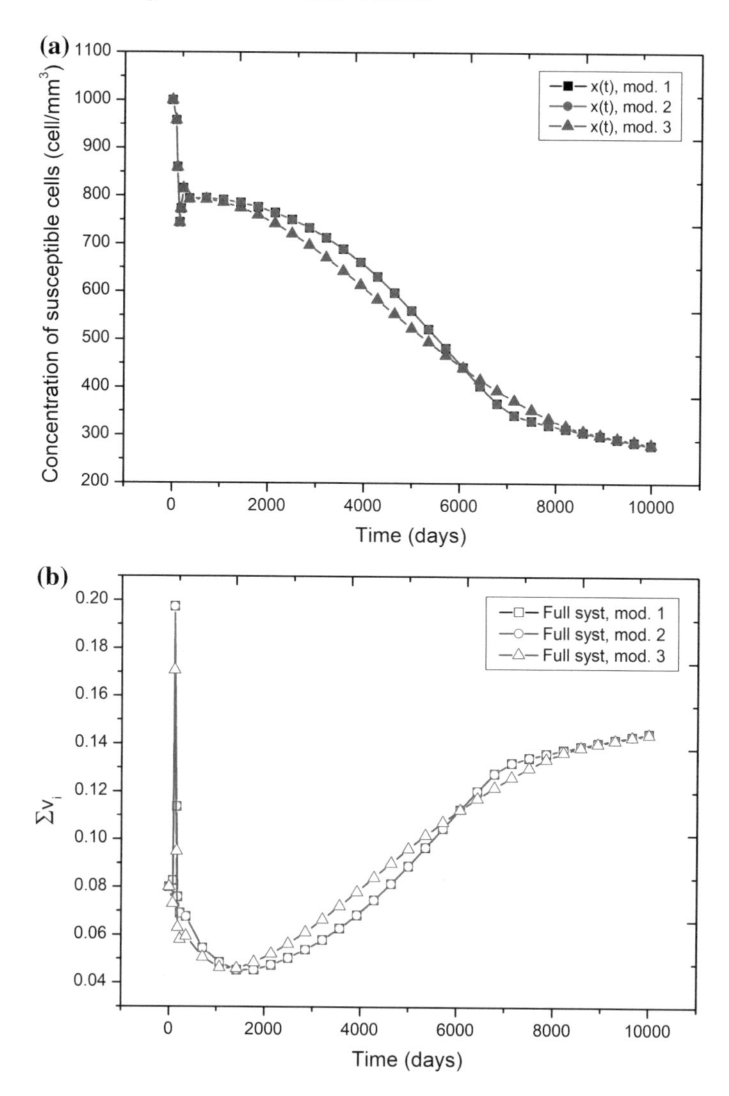

Fig. 3 Total concentrations of the susceptible cells $X(t)$ (**a**) and the total viral load (**b**) for Models 1, 2 and 3

References

1. A. Archibasov, A. Korobeinikov, V. Sobolev, Asymptotic expansions of solutions in a singularly perturbed model of virus evolution. Comput. Math. Math. Phys. **55**(2), 240–250 (2015)
2. A. Archibasov, A. Korobeinikov, V. Sobolev, Passage to the limit in a singularly perturbed partial integro-differential system. Differ. Equ. **52**(9), 1115–1122 (2016)
3. A. Korobeinikov, Immune response and within-host viral evolution: immune response can accelerate evolution. J. Theor. Biol. **456**, 74–83 (2018)

4. A. Korobeinikov, A. Archibasov, V. Sobolev, Order reduction for an RNA virus evolution model. Math. Biosci. Eng. **12**(5), 1007–1016 (2015)
5. A. Korobeinikov, A. Archibasov, V. Sobolev, Multi-scale problem in the model of RNA virus evolution. J. Phys. Conf. Ser. **727**(1), 012007 (2016)
6. A. Korobeinikov, C. Dempsey, A continuous phenotype space model of RNA virus evolution within a host. Math. Biosci. Eng. **11**(4), 919–927 (2014)
7. M. Nowak, M.R. May, *Virus dynamics* (Oxford University Press, 2000)
8. S. Pagliarini, A. Korobeinikov, A mathematical model of marine bacteriophage evolution. R. Soc. Open Sci. **5**, 171661 (2018)

Optimal Control for Anticancer Therapy

Evgenii N. Khailov, Anna D. Klimenkova and Andrei Korobeinikov

Abstract In this report, the controlled Lotka–Volterra competition model is used to describe the interaction of the concentrations of healthy and cancer cells. For this controlled model, the minimization problem of the terminal functional is considered, which is a weighted difference of the concentrations of cancerous and healthy cells at the final moment of the treatment period. To analyze the optimal solution of this problem, which consists of the optimal control and the corresponding optimal solutions of the differential equations that determine the model, the Pontryagin maximum principle is applied. It allows to highlight the values of the model parameters under which the optimal control corresponding to them is a piecewise-constant function with at most one switching. Also, the values of the model parameters are found, under which the corresponding optimal control is either a bang–bang function with a finite number of switchings, or in addition to the bang–bang-type portions (nonsingular portions), it also contains a singular arc. Further, only numerical investigations of the optimal control are possible. Therefore, the report presents the results of numerical calculations performed using the software BOCOP-2.1.0 that lead us to the conclusions about the possible type of the optimal control and the corresponding optimal solutions.

Research Perspectives CRM Barcelona, Summer 2018, vol. 11, in Trends in Mathematics Springer-Birkhäuser, Basel.

E. N. Khailov (✉) · A. D. Klimenkova
Faculty of Computational Mathematics and Cybernetics, Lomonosov Moscow State University, 119991 Moscow, Russia
e-mail: khailov@cs.msu.su

A. D. Klimenkova
e-mail: klimenkovaad@mail.ru

A. Korobeinikov
Centre de Recerca Matemàtica, Campus de Bellaterra, 08193 Bellaterra, Barcelona, Spain
e-mail: akorobeinikov@crm.cat

A. Korobeinikov et al. (eds.), *Extended Abstracts Spring 2018*,
Trends in Mathematics 11, https://doi.org/10.1007/978-3-030-25261-8_6

1 Introduction

In recent decades, a significant progress has been made in identifying and explaining the processes that arise in the development of cancer, as well as in developing methods and tools for its earlier diagnosis and treatment. A significant contribution to this progress was made by the use of mathematical modeling, which allowed to simulate a likely behavior of cells and organs before the actual disease develops. The most common are the mathematical models, describing the development of a cancerous tumor. The description of the tumor volume dynamics is possible in terms of the dynamics of competing populations of healthy and cancer cells. For this purpose, the classical Lotka–Volterra competing population model can be used [5, 6]. To find effective in some sense (that should be determined) treatment strategies, the optimal control theory can be applied.

2 Model

We consider the following nonlinear control system of differential equations:

$$\begin{cases} \dot{x}(t) = r(1-\kappa_1 w(t))(1-x(t)-a_{12}y(t))x(t) - m_1 u(t)x(t), \\ \dot{y}(t) = \quad (1-\kappa_2 w(t))(1-y(t)-a_{21}x(t))y(t) - m_2 u(t)y(t), \quad t \in [0,T], \\ x(0) = x_0, \;\; y(0) = y_0; \;\; x_0, y_0 > 0. \end{cases} \tag{1}$$

This model describes the interaction between the tumor cells, of population size or concentration $y(t)$, and normal cells, of population size or concentration $x(t)$. Functions $u(t)$ and $w(t)$ are bounded controls that represent the intensity of the therapies. These can be, for instance, drug concentration or intensity the radiotherapy. We assume that control $u(t)$ kills the cells (cytotoxic therapy), whereas control $w(t)$ inhibits their proliferation (cytostatic therapy), and that both controls are bounded:

$$0 \leqslant u(t) \leqslant u_{\max} \leqslant 1, \quad 0 \leqslant w(t) \leqslant w_{\max} < \min\{\kappa_1^{-1}, \kappa_2^{-1}\}.$$

In this model, r is the intrinsic growth rate of the normal cells; a_{12} and a_{21} represents the comparable compatibility of the tumor cells and healthy cells; m_1 and m_2 are the efficacy (killing rates) of the therapy with respect to the normal and tumor cells, respectively; κ_1 and κ_2 are the efficacies of the therapy in inhibiting the normal and tumor cells proliferation, respectively.

In the absence of the controls, model (1) is the classical Lotka–Volterra model of two competing populations. Qualitative behavior of such a system is completely determined by mutual location of lines $x + a_{12}y = 1$ and $y + a_{21}x = 1$. Figure 1 shows four possible robust scenarios of the system dynamics. (In this figure, we disregard the fifth case, where these two lines coincide, as this case occurs on a subset of the parameter space of measure zero.) It is easy to see that for these robust

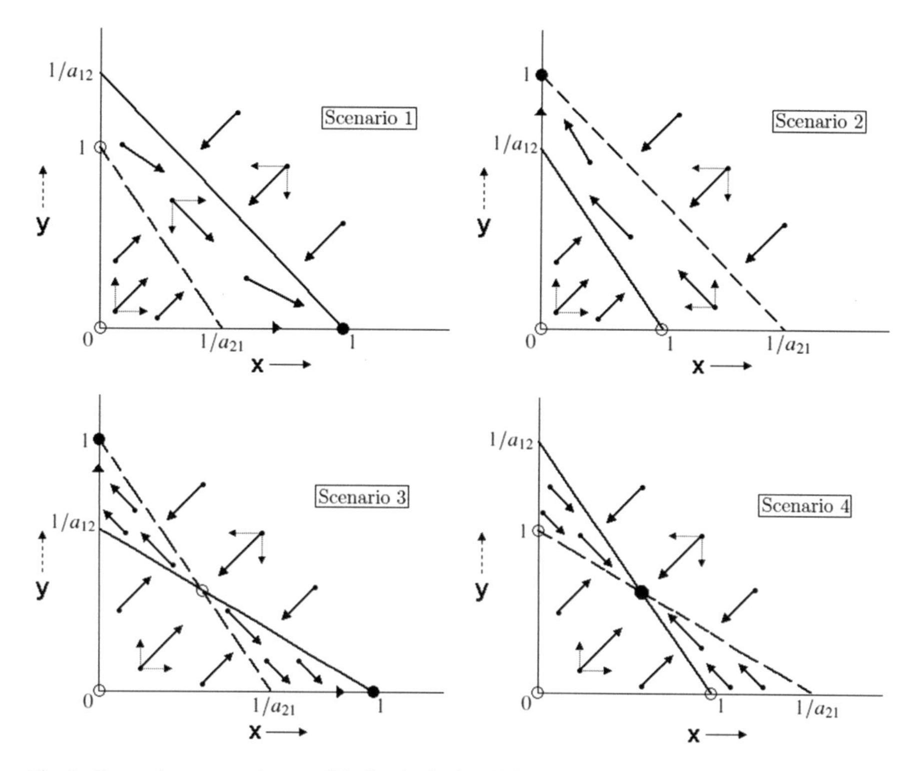

Fig. 1 Four robust scenarios possible for the Lotka–Volterra model of two competing populations. (Adopted from [7])

cases, the system has up to four nonnegative equilibrium states, namely (0, 0), (0, 1), (1, 0), and $\big((1-a_{12})/(1-a_{12}a_{21}),\ (1-a_{21})/(1-a_{12}a_{21})\big)$. The origin is always an unstable node, whereas types of the other points depend on the model parameters and can be either saddles (marked by circles in Fig. 1), or attracting nodes (marked by dots).

Figure 1 implies that cancer can appear and develop either in Scenario 2, or in Scenario 4, as in Scenarios 1 and 3, point (0, 1) is asymptotically stable, and, if a small number of malicious cells appear as a result of a mutation, these are to be eliminated by competition with the normal cells. This figure also suggests that the objective of a therapy is the transition of the system to Scenario 1 (ideally), or, at least, to Scenario 3, where cancer cells will be driven to extinction.

Let us assume that inequalities

$$a_{12} \cdot a_{21} \neq 1, \quad m_2 > m_1, \quad \kappa_2 > \kappa_1 \tag{2}$$

hold, and that control $w(t)$ is constant, $w(t) \equiv const$. Let us denote $q_1 = r(1-\kappa_1 w)$ and $q_2 = 1-\kappa_2 w$. Then we obtain the following system:

$$\begin{cases} \dot{x}(t) = q_1(1 - x(t) - a_{12}y(t))x(t) - m_1u(t)x(t), \\ \dot{y}(t) = q_2(1 - y(t) - a_{21}x(t))y(t) - m_2u(t)y(t), \quad t \in [0, T], \\ x(0) = x_0, \ y(0) = y_0; \ x_0, y_0 > 0. \end{cases} \tag{3}$$

Please note that under the above-made assumption, expressions $m_1q_2a_{21} - m_2q_1$ and $m_1q_2 - m_2q_1a_{12}$ cannot be equal zero at the same time:

$$m_1q_2a_{21} - m_2q_1 \neq 0, \ m_1q_2 - m_2q_1a_{12} \neq 0. \tag{4}$$

The set of admissible controls $\Omega(T)$ is formed by all Lebesgue measurable functions $u(t)$, which for almost $t \in [0, T]$ satisfy the constraints: $0 \leqslant u(t) \leqslant u_{\max} \leqslant 1$. The boundedness, positiveness, and continuation of solutions for system (3) are established by the following lemma.

Lemma 1 *For any admissible control $u(\cdot) \in \Omega(T)$, the corresponding solutions $x(t)$, $y(t)$ to system (3) are defined on the entire interval $[0, T]$ and satisfy inclusion*

$$(x(t), y(t)) \in \Theta = \left\{(x, y) : 0 < x < x_0\mathrm{e}^{q_1T}, \ 0 < y < y_0\mathrm{e}^{q_2T}\right\}, \quad t \in [0, T]. \tag{5}$$

For system (3), on the set of admissible controls $\Omega(T)$, we consider the problem of minimization of a terminal functional, which is a weighted difference of the concentrations of cancerous and normal cells at the final moment of the therapy:

$$J(u) = y(T) - \alpha x(T), \tag{6}$$

where $\alpha > 0$ is the given weighted coefficient. Lemma 1 guarantees the existence of the optimal solution for the minimization problem (6): for optimal control $u_*(t)$, $x_*(t)$, $y_*(t)$ are corresponding optimal solutions of system (3); see [2].

3 Pontryagin Maximum Principle

To analyze the optimal control $u_*(t)$ and the corresponding optimal solutions $x_*(t)$, $y_*(t)$, we apply the Pontryagin maximum principle [3]. We define Hamiltonian

$$H(x, y, u, \psi_1, \psi_2) = (q_1(1 - x - a_{12}y)x - m_1ux)\psi_1 + (q_2(1 - y - a_{21}x)y - m_2uy)\psi_2,$$

where ψ_1, ψ_2 are the adjoint variables. By the Pontryagin maximum principle, for optimal control $u_*(t)$ and optimal solutions $x_*(t)$, $y_*(t)$, there exists vector function $\psi_*(t) = (\psi_1^*(t), \psi_2^*(t))$, such that

(i) $\psi_*(t)$ is a nontrivial solution of the adjoint system

$$\begin{cases} \dot{\psi}_1^*(t) = -(q_1(1 - x_*(t) - a_{12}y_*(t)) - q_1x_*(t) - m_1u_*(t))\psi_1^*(t) + \\ \qquad +q_2a_{21}y_*(t)\psi_2^*(t), \\ \dot{\psi}_2^*(t) = q_1a_{12}x_*(t)\psi_1^*(t) - (q_2(1 - y_*(t) - a_{21}x_*(t)) - q_2y_*(t) - \\ \qquad -m_2u_*(t))\psi_2^*(t), \\ \psi_1^*(T) = \alpha, \ \psi_2^*(T) = -1; \end{cases} \tag{7}$$

and

(ii) the control $u_*(t)$ maximizes the Hamiltonian $H(x_*(t), y_*(t), u, \psi_1^*(t), \psi_2^*(t))$ with respect to $u \in [0, u_{\max}]$ for almost all $t \in [0, T]$, and, therefore, the following relationship holds:

$$u_*(t) = \begin{cases} u_{\max} & \text{if } L_u(t) > 0, \\ \text{any } u \in [0, u_{\max}] & \text{if } L_u(t) = 0, \\ 0 & \text{if } L_u(t) < 0. \end{cases} \tag{8}$$

Here, function $L_u(t) = -m_1x_*(t)\psi_1^*(t) - m_2y_*(t)\psi_2^*(t)$ is the switching function, which defines the optimal control $u_*(t)$ via formula (8). Introducing auxiliary adjoint variables $\phi_1(t) = -x_*(t)\psi_1^*(t)$ and $\phi_2(t) = -y_*(t)\psi_2^*(t)$, we can rewrite adjoint system (7) and the switching function as

$$\begin{cases} \dot{\phi}_1(t) = q_1x_*(t)\phi_1(t) + q_2a_{21}x_*(t)\phi_2(t), \\ \dot{\phi}_2(t) = q_1a_{12}y_*(t)\phi_1(t) + q_2y_*(t)\phi_2(t), \\ \phi_1(T) = -\alpha x_*(T) < 0, \ \phi_2(T) = y_*(T) > 0, \end{cases} \tag{9}$$

and

$$L_u(t) = m_1\phi_1(t) + m_2\phi_2(t).$$

Systems (3) and (9) allows to formulate the Cauchy problem

$$\begin{cases} \dot{L}_u(t) = m_1^{-1}q_1(m_1x_*(t) + m_2a_{12}y_*(t))L_u(t) + \\ \qquad +m_1^{-1}\Big(m_1(m_1q_2a_{21} - m_2q_1)x_*(t) + m_2(m_1q_2 - m_2q_1a_{12})y_*(t)\Big)\phi_2(t), \\ L_u(T) = -m_1\alpha x_*(T) + m_2y_*(T). \end{cases} \tag{10}$$

for function $L_u(t)$.

An important property of functions $\phi_1(t)$, $\phi_2(t)$ is established by the following lemma.

Lemma 2 *The auxiliary adjoint variables $\phi_1(t)$, $\phi_2(t)$ are sign definite on the entire interval $[0, T]$: $\phi_1(t) < 0, \ \phi_2(t) > 0, \ t \in [0, T]$.*

Our task is to estimate the number of zeros of the switching function $L_u(t)$ and investigate the existence of singular arcs; see [4]. Analysis of the Cauchy problem (10) together with inequalities (4) leads us to the following conclusions:

(i) Let $m_1q_2a_{21} - m_2q_1 \geqslant 0, m_1q_2 - m_2q_1a_{12} \geqslant 0$ hold. If there is $t_0 \in [0, T]$ such that $L_u(t_0) = 0$, then $\dot{L}_u(t_0) > 0$. Then, by (8), the optimal control $u_*(t)$ is a piecewise constant function with one switching of the type

$$u_*(t) = \begin{cases} 0 & t \in [0, \theta_*], \\ u_{\max} & t \in (\theta_*, T], \end{cases}$$

where $\theta_* \in (0, T)$ is the moment of switching.

(ii) Let $m_1q_2a_{21} - m_2q_1 \leqslant 0, m_1q_2 - m_2q_1a_{12} \leqslant 0$ hold. If there is $t_0 \in [0, T]$ such that $L_u(t_0) = 0$, then $\dot{L}_u(t_0) < 0$. Hence, by (8), the optimal control $u_*(t)$ is a piecewise constant function with one switching of the type

$$u_*(t) = \begin{cases} u_{\max} & t \in [0, \theta_*], \\ 0 & t \in (\theta_*, T], \end{cases}$$

where $\theta_* \in (0, T)$ is the moment of switching.

(iii) Let either $m_1q_2a_{21} - m_2q_1 \geqslant 0$ and $m_1q_2 - m_2q_1a_{12} \leqslant 0$, or $m_1q_2a_{21} - m_2q_1 \leqslant 0$ and $m_1q_2 - m_2q_1a_{12} \geqslant 0$ hold. Then switching function $L_u(t)$ can become zero on some interval $\Delta \subset [0, T]$. This means that the optimal control $u_*(t)$ can have a singular arc on this interval. Then, on the interval Δ equalities $L_u(t) = 0$ and $\dot{L}_u(t) = 0$ hold. Therefore,

$$m_2(m_1q_2 - m_2q_1a_{12})y + m_1(m_1q_2a_{21} - m_2q_1)x = 0. \tag{11}$$

By equalities $L_u(t) = 0$ and $\dot{L}_u(t) = 0$, and assumption (2), the necessary condition of the optimality of a singular arc (the Kelly condition, see [8]) in a strengthened form

$$\frac{\partial}{\partial u}\ddot{L}_u(t) = -m_1^{-1}m_2(m_2 - m_1)(m_1q_2 - m_2q_1a_{12})y_*(t)\phi_2(t) > 0. \tag{12}$$

By Lemmas 1 and 2 and formula (11), one can immediately conclude that the Kelly condition (12) holds if $m_1q_2a_{21} - m_2q_1 > 0$ and $m_1q_2 - m_2q_1a_{12} < 0$ hold. This implies that the necessary condition of the optimality of a singular arc is valid in the strengthened form. Hence, on interval Δ, the optimal control $u_*(t)$ is

$$u^*_{\text{sing}}(t) = \frac{q_2 - q_1}{m_2 - m_1} + \frac{(m_1 + m_2)q_1q_2a_{12}a_{21} - (m_1q_2^2a_{21} + m_2q_1^2a_{12})}{m_2(m_1q_2 - m_2q_1a_{12})}x_*(t).$$

That is, the optimal control has the form of a feedback that depends only on the optimal solution $x_*(t)$.

If the inclusion $u^*_{\text{sing}}(t) \in (0, u_{\max})$ holds for all $t \in \Delta$ (we are only interested in such controls), then it is possible to concatenate the singular arc $u^*_{\text{sing}}(t)$ with bang–bang control portions $u_*(t)$.

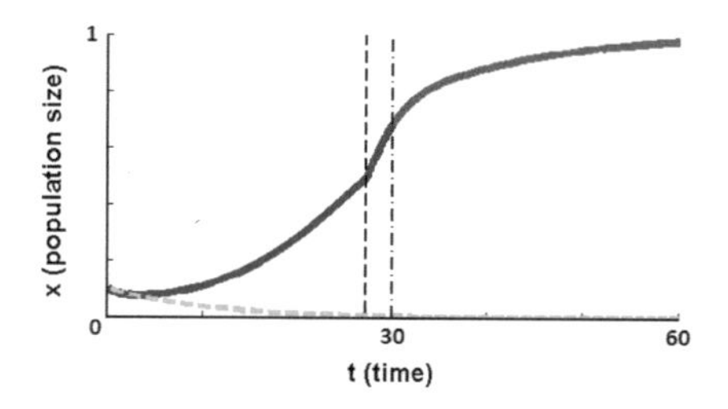

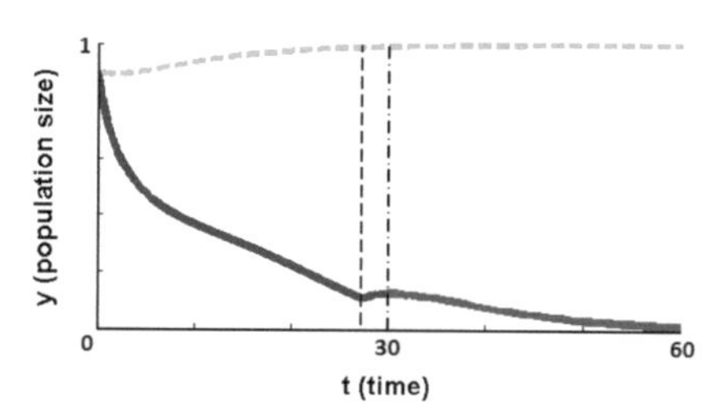

Fig. 2 Optimal solutions $x^*(t)$ and $y^*(t)$. Here, $x_0 = 0.1$, $y_0 = 0.9$, $\alpha = 1$, $u_{max} = 0.6$, $r = 0.6$, $a_{12} = 1.25$, $a_{21} = 1.25$, $\kappa_1 = 0.2$, $\kappa_2 = 0.6$, $m_1 = 0.2$, $m_2 = 0.4$, $T = 30$, and $w = 1$

If $m_1 q_2 a_{21} - m_2 q_1 < 0$ and $m_1 q_2 - m_2 q_1 a_{12} > 0$, then the Kelly condition (12) is not hold and, hence, the necessary condition of the optimality of a singular arc is not valid. Therefore, in this case the optimal control $u_*(t)$ does not have a singular arc on the interval Δ, and the optimal control on entire interval $[0, T]$ is a bang–bang control taking the values 0 or $u_{\max}$ with a finite number of switchings.

4 Numerical Results

To illustrate possible outcomes of the optimal controls, we run calculations using software package BOCOP 2.1.0; see [1]. Some results of these are given in Figs. 2, 3, and 4.

The optimal solutions in Figs. 2 and 3 correspond to the optimal control of the type

$$u_*(t) = \begin{cases} u_{\max} & t \in [0, \theta_*], \\ 0 & t \in (\theta_*, T], \end{cases}$$

where the moment of switching is $\theta_* = 27.3$ in Fig. 2 and $\theta_* = 27.9$ in Fig. 3. In Figs. 2, 3 and 4 the blue lines corresponds to the optimal solutions. The red lines are the continuations of the optimal solutions for a longer time interval (in this case, for

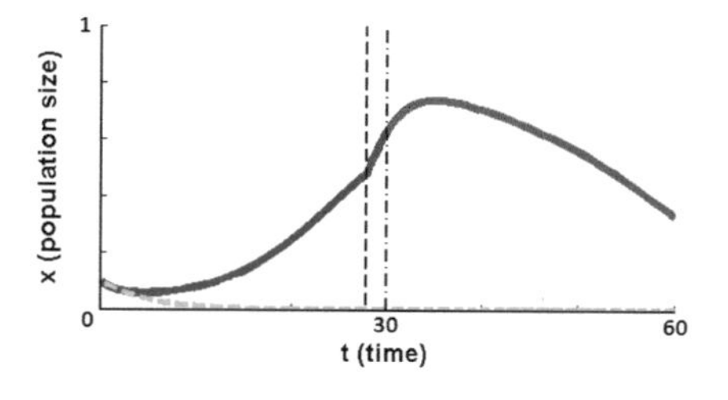

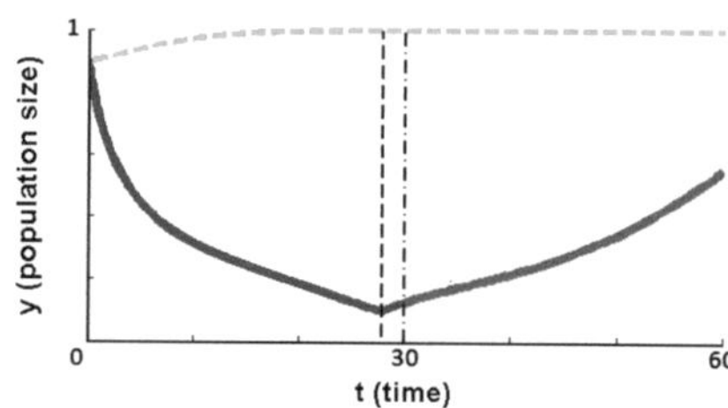

Fig. 3 Optimal solutions $x^*(t)$ and $y^*(t)$. Here, $x_0 = 0.1$, $y_0 = 0.9$, $\alpha = 1$, $u_{max} = 0.6$, $r = 0.6$, $a_{12} = 1.5$, $a_{21} = 0.9$, $\kappa_1 = 0.2$, $\kappa_2 = 0.7$, $m_1 = 0.2$, $m_2 = 0.4$, $T = 30$, and $w = 1$

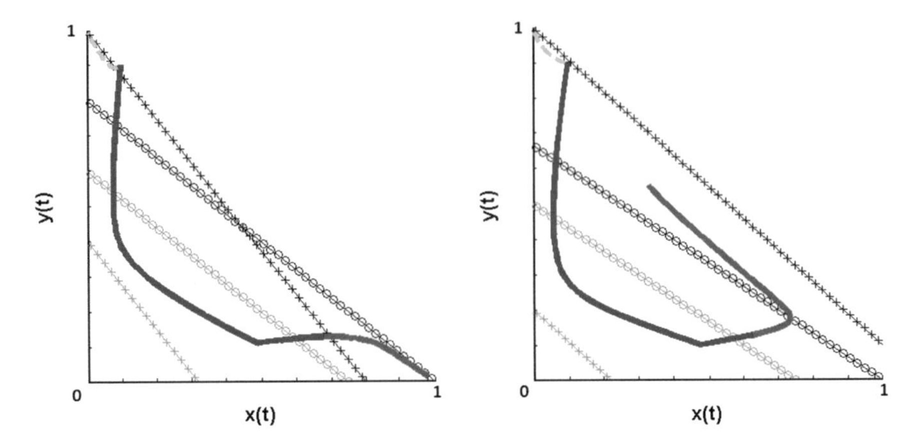

Fig. 4 Phase portraits corresponding to the optimal solutions in Figs. 2 and 3, respectively

$t \in [T, 2T]$). The green dashed curves represent the solutions in the absence of the control ($u(t) = 0$). The vertical dashed lines correspond to the switching moments, whereas the vertical dot-dashed lines correspond to $t = T$ (the end of the control interval). In Fig. 4, black lines are the nullclines of uncontrolled system ($u = 0$), while orange lines are the nullclines of the controlled system (in this case, $u = 0.6$).

Please note that in both these examples, the initial conditions are located in the domain of attraction of the point (0,1), which corresponds to the extinction of the normal cells. The phase portraits show that the behavior of the uncontrolled system match, respectively, Scenario 3 (the first example) and Scenario 2 (the second example) as shown in Fig. 1. The optimal control, when it is active (i.e., $u = u_{\max},\ t \in [0, \theta_*]$), transfers the system to Scenario 1. Thereafter, when the optimal control is passive (i.e., $u = 0,\ t \in (\theta_*, T]$), the system returns to its original scenario. In the first example, the optimal control is able to move the state of the system into the domain of attraction of the point (1,0), where cancer cell population goes to extinction. For this case, further treatment is not required. In the second example, while the therapy appears to be successful, and after the treatment, the system returns to the scenario where the cancer cell population continues to grow.

References

1. F. Bonnans, P. Martinon, D. Giorgi, V. Grélard, S. Maindrault, O. Tissot, J. Liu, BOCOP 2.1.0—User Guide (2017), http://www.bocop.org/
2. E.B. Lee, L. Marcus, *Foundations of Optimal Control Theory* (Wiley, New York, 1967)
3. L.S. Pontryagin, V.G. Boltyanskii, R.V. Gamkrelidze, E.F. Mishchenko, *Mathematical Theory of Optimal Processes* (Wiley, New York, 1962)
4. H. Schättler, U. Ledzewicz, *Optimal Control For Mathematical Models of Cancer Therapies. An Application of Geometric Methods* (Springer, New York-Heidelberg-Dordrecht-London, 2015)
5. R.V. Solé, T.S. Deisboeck, An error catastrophe in cancer? J. Theor. Biol. **228**, 47–54 (2004)

6. R.V. Solé, I.G. García, J. Costa, Spatial dynamics in cancer, in *Complex systems science in biomedicine*, ed by T.S. Deisboeck, J.Y. Kresh (Springer, New York, 2006), pp. 557–572
7. H. Weiss *A Mathematical Introduction to Population Dynamics* (2010)
8. M.I. Zelikin, V.F. Borisov, *Theory of Chattering Control with Applications to Astronautics, Robotics, Economics and Engineering* (Birkhäuser, Boston, 1994)

The Two-Scale Periodic Unfolding Technique

Anna Zubkova

Abstract In this paper, the definitions of the periodic unfolding and averaging operators are extended to the case of two sub-domains separated by a thin interface. Their properties are introduced and illustrative examples of these operators are given.

1 Introduction

This paper is devoted to a useful tool in the homogenization procedure for models defined in a two-phase domain, which is the two-phase periodic unfolding technique.

The periodic unfolding technique was introduced in continuous and perforated domains, see, e.g., [1, 3] and it is based on the periodic unfolding and the averaging operators. The paper [2] suggests an extension of the definition of the unfolding operator on a boundary. Our specific interest concerns the Poisson–Nernst–Planck system in a two-phase domain with an interface, see our works [4, 5]. For this reason, we extend the definitions to the case of two phases and their interface. We describe a two-phase medium with a microstructure consisting of solid and pore phases, which are separated by a thin interface. The corresponding geometry is represented by a disconnected domain. A special interest of our consideration is the interface between the two phases because of electrochemical reactions which occur here.

The paper has the following structure. In Sect. 2, the definitions of the periodic unfolding and the averaging operators in the two-phase domain are introduced. Section 3 is devoted to clarifying examples of these operators.

A. Zubkova (✉)
KFU, Institute for Mathematics and Scientific Computing,
Mozartgasse 14, 8010 Graz, Austria
e-mail: anna.zubkova@uni-graz.at

A. Korobeinikov et al. (eds.), *Extended Abstracts Spring 2018*,
Trends in Mathematics 11, https://doi.org/10.1007/978-3-030-25261-8_7

2 Definitions

We start with the two-phase geometry. A unit cell $Y = (0, 1)^d$, $d \in \mathbb{N}$, consists of two open, connected sub-domains: a solid part ω and a pore part Π, separated by a thin interface $\partial\omega$ which is assumed to be Lipschitz continuous, see Fig. 1. By scaling a unit cell Y with a small parameter $\varepsilon > 0$, we introduce a local cell Y_ε^l with some index l. Its solid part is denoted by ω_ε^l and its pore part by Π_ε^l.

Every spacial point $x \in \mathbb{R}^d$ can be decomposed as the following sum:

$$x = \varepsilon\lfloor\frac{x}{\varepsilon}\rfloor + \varepsilon\{\frac{x}{\varepsilon}\}, \tag{1}$$

where $\lfloor x/\varepsilon\rfloor \in \mathbb{Z}^d$ is the floor part and $\{x/\varepsilon\} \in Y = (0, 1)^d$ is the fractional part of x/ε.

We consider a domain $\Omega \subset \mathbb{R}^d$ with a Lipschitz boundary $\partial\Omega$. Based on the decomposition (1), it is covered by repeating periodically local cells Y_ε^l in such a way that all local cells lay inside of Ω. The union of these periodic local cells is denoted by $\Omega_\varepsilon := \bigcup_{l\in I^\varepsilon} Y_\varepsilon^l$ with the solid part $\omega_\varepsilon = \bigcup_{l\in I^\varepsilon} \omega_\varepsilon^l$, and the pore part $\Pi_\varepsilon = \bigcup_{l\in I^\varepsilon} \Pi_\varepsilon^l$. The interface $\partial\omega_\varepsilon := \bigcup_{l\in I^\varepsilon} \partial\omega_\varepsilon^l$ is the union of local interfaces in each local cell. A thin boundary layer attaching the external boundary $\partial\Omega$ is called $\Omega \setminus \Omega_\varepsilon$, see Fig. 2. Summarizing, the two-phase domain Ω consists of the pore phase $Q_\varepsilon = (\Omega \setminus \Omega_\varepsilon) \cup \Pi_\varepsilon$, the solid phase ω_ε, and the interface $\partial\omega_\varepsilon$.

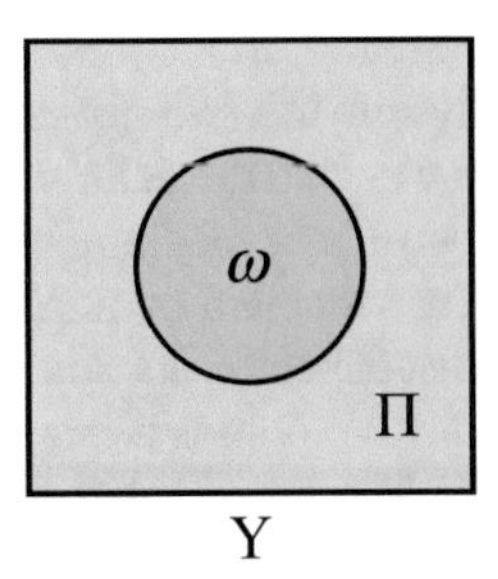

Fig. 1 A unit cell Y

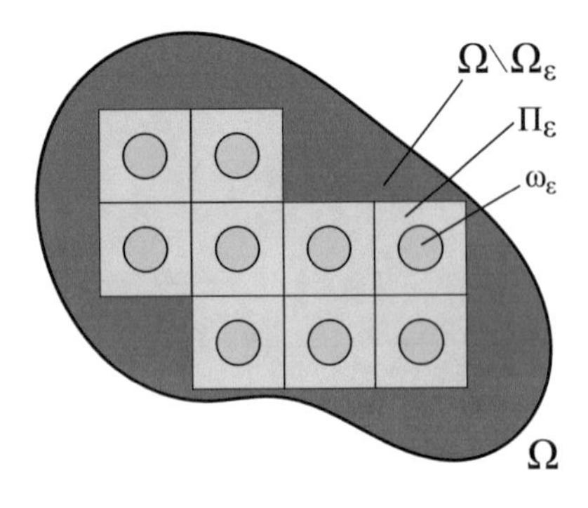

Fig. 2 The domain $\Omega = Q_\varepsilon \cup \omega_\varepsilon$

In the prescribed geometry, we define the periodic unfolding and the averaging operators in the two-phase domain and at the interface.

Definition 1 (*two-phase unfolding operator*) The linear continuous operator $f(x) \mapsto T_\varepsilon\colon H^1(Q_\varepsilon) \times H^1(\omega_\varepsilon) \mapsto L^2(\Omega; H^1(\Pi) \times H^1(\omega))$ is defined as

$$(T_\varepsilon f)(x, y) = \begin{cases} f\big(\varepsilon\lfloor \frac{x}{\varepsilon}\rfloor + \varepsilon y\big), & \text{for a.e. } x \in \Omega_\varepsilon \text{ and } y \in \Pi \cup \omega, \\ f(x), & \text{for a.e. } x \in \Omega \setminus \Omega_\varepsilon \text{ and } y \in \Pi \cup \omega. \end{cases} \tag{2}$$

Definition 2 (*two-phase averaging operator*) The left-inverse operator to T_ε (linear and continuous) is defined as $u(x, y) \mapsto T_\varepsilon^{-1}\colon L^2(\Omega; H^1(\Pi) \times H^1(\omega)) \mapsto H^1(\bigcup_{l \in I^\varepsilon} \Pi_\varepsilon^l) \times H^1(\Omega \setminus \Omega_\varepsilon) \times H^1(\omega_\varepsilon)$:

$$(T_\varepsilon^{-1} u)(x) = \begin{cases} \dfrac{1}{|Y|}\displaystyle\int_{\Pi\cup\omega} u\big(\varepsilon\lfloor \frac{x}{\varepsilon}\rfloor + \varepsilon z, \{\frac{x}{\varepsilon}\}\big)\, dz, & \text{for a.e. } x \in \Pi_\varepsilon \cup \omega_\varepsilon, \\ \dfrac{1}{|Y|}\displaystyle\int_{\Pi\cup\omega} u(x, y)\, dy, & \text{for a.e. } x \in \Omega \setminus \Omega_\varepsilon. \end{cases} \tag{3}$$

The operators satisfy the following properties:

Lemma 3 (Properties of the operators T_ε and T_ε^{-1} in the domain) *For arbitrary $f, g, h \in H^1(Q_\varepsilon) \times H^1(\omega_\varepsilon)$, the following equalities hold:*

(i) $(T_\varepsilon^{-1} T_\varepsilon) f(x) = f(x)$, *and* $(T_\varepsilon T_\varepsilon^{-1} u)(x, y) = u(y)$, *when u is constant for $x \in Q_\varepsilon \cup \omega_\varepsilon$, or a periodic function $u(y)$ of $y \in \Pi \cup \omega$ for $x \in \Pi_\varepsilon \cup \omega_\varepsilon$;*
(ii) *composition rule:* $T_\varepsilon(\mathcal{F}(f)) = \mathcal{F}(T_\varepsilon f)$ *for any elementary function $\mathcal{F}$;*
(iii) *integration rules:*

$$\int_{\Pi_\varepsilon\cup\omega_\varepsilon} f(x) g(x)\, dx = \frac{1}{|Y|}\int_{\Omega_\varepsilon}\int_{\Pi\cup\omega} (T_\varepsilon f)(x, y) \cdot (T_\varepsilon g)(x, y)\, dy\, dx,$$

$$\int_{\Omega\setminus\Omega_\varepsilon} f(x) g(x)\, dx = \frac{1}{|Y|}\int_{\Omega\setminus\Omega_\varepsilon}\int_{\Pi\cup\omega} (T_\varepsilon f)(x, y) \cdot (T_\varepsilon g)(x, y)\, dy\, dx;$$

(iv) *boundedness of T_ε:*

$$\int_{Q_\varepsilon\cup\omega_\varepsilon} h^2(x)\, dx = \frac{1}{|Y|}\int_{\Omega}\int_{\Pi\cup\omega} (T_\varepsilon h)^2(x, y)\, dy\, dx,$$

$$\int_{Q_\varepsilon\cup\omega_\varepsilon} |\nabla_x h|^2(x)\, dx = \frac{1}{\varepsilon^2|Y|}\int_{\Omega}\int_{\Pi\cup\omega} |\nabla_y (T_\varepsilon h)|^2(x, y)\, dy\, dx.$$

2.1 Restriction of the Operators to the Interface

The definitions (2) and (3) are extended to the interface in a natural way:

Definition 4 The restriction of the two-phase unfolding operator T_ε to the interface $\partial\omega_\varepsilon$ is well defined as follows: $f(x) \mapsto T_\varepsilon \colon L^2(\partial\omega_\varepsilon) \mapsto L^2(\Omega_\varepsilon) \times L^2(\partial\omega)$,

$$(T_\varepsilon f)(x, y) = f\big(\varepsilon\lfloor\frac{x}{\varepsilon}\rfloor + \varepsilon y\big), \quad \text{for a.e. } x \in \Omega_\varepsilon \text{ and } y \in \partial\omega. \tag{4}$$

The corresponding averaging operator $u(x, y) \mapsto T_\varepsilon^{-1} \colon L^2(\Omega_\varepsilon) \times L^2(\partial\omega) \mapsto L^2(\partial\omega_\varepsilon)$,

$$(T_\varepsilon^{-1} u)(x) = \frac{1}{|Y|} \int_{\Pi\cup\omega} u\big(\varepsilon\lfloor\frac{x}{\varepsilon}\rfloor + \varepsilon z, \{\frac{x}{\varepsilon}\}\big)\, dz, \quad \text{for a.e. } x \in \Omega_\varepsilon. \tag{5}$$

Analogously, the following properties at the interface hold:

Lemma 5 (Properties of the operators T_ε and T_ε^{-1} at the interface) *For arbitrary $f, g \in L^2(\partial\omega_\varepsilon)$, the following equalities hold.*

(i) $(T_\varepsilon^{-1} T_\varepsilon) f(x) = f(x)$;
(ii) composition rule: $T_\varepsilon(\mathcal{F}(f)) = \mathcal{F}(T_\varepsilon f)$ *for any elementary function* $\mathcal{F}$;
(iii) integration rule:

$$\int_{\partial\omega_\varepsilon} f(x) g(x)\, dS_x = \frac{1}{\varepsilon|Y|} \int_{\Omega_\varepsilon} \int_{\partial\omega} (T_\varepsilon f)(x, y) \cdot (T_\varepsilon g)(x, y)\, dS_y\, dx;$$

(iv) boundedness of T_ε:

$$\int_{\partial\omega_\varepsilon} f^2(x)\, dS_x = \frac{1}{\varepsilon|Y|} \int_{\Omega_\varepsilon} \int_{\partial\omega} (T_\varepsilon f)^2(x, y)\, dS_y\, dx.$$

3 Examples

In this section, two examples representing the behavior of the periodic unfolding operator is given.

Example 6 In the one-dimensional domain $\Omega = (-2\pi, 2\pi)$, we consider the function $f(x) = \sin x$ and the small parameter $\varepsilon = 2\pi$, which coincides with the period of the function f. We consider a unit cell $Y = (0, 1)$, therefore, the number of local cells is 2 which are the intervals $Y_\varepsilon^1 = (-2\pi, 0)$ and $Y_\varepsilon^2 = (0, 2\pi)$. On the first graph in Fig. 3, the red curve is the function f and the blue line represents the projection of the mapping $T_\varepsilon f$ for $y = 0$. On the second graph, both functions are presented on the two-dimensional (x, y)-plane, where the interval $Y = (0, 1)$ in y-axis is a unit cell and the intervals $(-2\pi, 0)$, $(0, 2\pi)$ in x-axis are local cells Y_ε^1 and Y_ε^2, and the

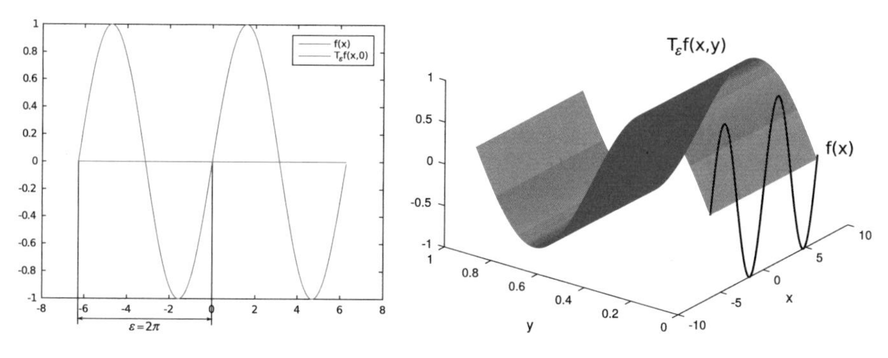

Fig. 3 $\varepsilon = 2\pi$

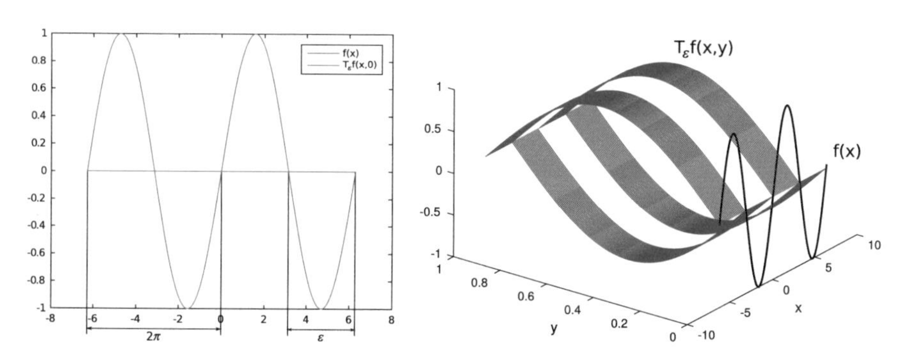

Fig. 4 $\varepsilon = \pi$

point $\{x = 0\}$ is a part of the cell boundary $\partial Y_{\varepsilon}^{1}$. We note that in the case of periodic functions $f \in H^1(\Omega_\varepsilon)$ with respect to a small parameter ε, the mapping $T_\varepsilon f$ is continuous across the boundary $\partial Y_{\varepsilon}^{l}$.

Example 7 We consider the same setting as in Example 6 but with $\varepsilon = \pi$, which does not coincide now with the period of the function f or, in other words, the function f is not periodic with respect to ε. Comparing with the Example 6, the mapping $T_\varepsilon f(x, y)$ is discontinuous along the x-variable, see Fig. 4. This example illustrates that for nonperiodic functions, the averaging mapping $(T_{\varepsilon}^{-1} T_\varepsilon) f$ does not belong to the space $H^1(\Omega_\varepsilon)$ but only to $L^2(\Omega_\varepsilon)$ even for continuous functions f from the space $H^1(\Omega_\varepsilon)$. Such functions can be smoothed by the gradient folding operator, see, e.g., [6].

References

1. D. Cioranescu, A. Damlamian, P. Donato, G. Griso, R. Zaki, The periodic unfolding method in domains with holes. SIAM J. Math. Anal. **44**, 718–760 (2012)
2. J. Franců, Modification of unfolding approach to two-scale convergence. Math. Bohem. **135**, 403–412 (2010)
3. M. Gahn, M. Neuss-Radu, P. Knabner, Homogenization of reaction-diffusion processes in a two-component porous medium with nonlinear flux conditions at the interface. SIAM J. App. Math. **76**, 1819–1843 (2016)
4. V.A. Kovtunenko, A.V. Zubkova, Mathematical modeling of a discontinuous solution of the generalized Poisson-Nernst-Planck problem in a two-phase medium. Kinet. Relat. Mod. **11**, 119–135 (2018)
5. V.A. Kovtunenko, A.V. Zubkova, Homogenization of the generalized Poisson–Nernst–Planck problem in a two-phase medium: correctors and residual error estimates. Appl. Anal., submitted
6. A. Mielke, S. Reichelt, M. Thomas, Two-scale homogenization of nonlinear reaction-diffusion systems with slow diffusion. J. Netw. Heterog. Media **9**, 353–382 (2014)

Limit Cycles for Piecewise Linear Differential Systems via Poincaré–Miranda Theorem

Armengol Gasull and Víctor Mañosa

Abstract In Gasull and Mañosa (Periodic orbits of discrete and continuous dynamical systems via Poincaré–Miranda theorem, Preprint 2018 [2]), we develop an effective procedure to prove the existence, determine the number, and locate periodic orbits of dynamical systems of both discrete and continuous nature. It is based on the use of the Poincaré–Miranda theorem. This note presents one of the results obtained in that paper: a new example of piecewise linear differential system with three limit cycles.

1 Introduction and Main Result

The study of the number of limit cycles for planar differential systems is a classical topic in the theory of dynamical systems. In the past years, many attention has been devoted to the study of nested limit cycles of piecewise linear systems, steered by the applicability of these systems in the modelling of biological and mechanical applications. In 2012, S.M. Huan and X.S. Yang gave numerical evidences of a piecewise linear system with two zones and a discontinuity straight line, having three nested limit cycles [3]. A proof based on the Newton–Kantorovich theorem of the existence

The authors are supported by Ministry of Economy, Industry and Competitiveness, State Research Agency of the Spanish Government through grants MTM2016-77278-P (MINECO/AEI/FEDER, UE, first author) and DPI2016-77407-P (MINECO/AEI/FEDER, UE, second author). The first author is also supported by the grant 2017-SGR-1617 from AGAUR, Generalitat de Catalunya. The second author acknowledges the group's research recognition 2017-SGR-388 from AGAUR, Generalitat de Catalunya.

A. Gasull (✉)
Departament de Matemàtiques, Universitat Autònoma de Barcelona, 08193 Bellaterra, Barcelona, Spain
e-mail: gasull@mat.uab.cat

V. Mañosa
Departament de Matemàtiques, Universitat Politècnica de Catalunya, Colom 11, 08222 Terrassa, Spain
e-mail: victor.manosa@upc.edu

A. Korobeinikov et al. (eds.), *Extended Abstracts Spring 2018*,
Trends in Mathematics 11, https://doi.org/10.1007/978-3-030-25261-8_8

of these limit cycles for this example and a nearby one, was given by Llibre and Ponce [5]. A different proof, from a bifurcation viewpoint, was presented by Freire, Ponce and Torres in [1]. Until now, as far as we know, three is the maximum observed number of limit cycles in a piecewise linear systems with two zones and a discontinuity straight line, but it is not known if this is the maximum number that such type of systems can have.

In this work, we present a new example, again with three limit cycles, inspired on the ones given in [3, 5]. The main contribution is that our proof relies on the so-called Poincaré–Miranda theorem and it is very simple. This theorem is essentially the extension of the intermediate value theorem (or more precisely, the Bolzano's theorem) to higher dimensions. It was stated by H. Poincaré in 1883 and 1884, and proved by himself in 1886 [7, 8]. In 1940, C. Miranda re-obtained the result as an equivalent formulation of Brouwer fixed point theorem [6]. Recent proofs are presented in [4, 10]. For completeness, we recall it. As usual, $\overline{S}$ and ∂S denote, respectively, the closure and the boundary of a set $S \subset \mathbb{R}^n$.

Theorem 1 (Poincaré–Miranda) *Set* $\mathcal{B} = \{\mathbf{x} = (x_1, \ldots, x_n) \in \mathbb{R}^n : L_i < x_i < U_i, 1 \le i \le n\}$. *Suppose that* $f = (f_1, f_2, \ldots, f_n)\colon \overline{\mathcal{B}} \to R^n$ *is continuous,* $f(\mathbf{x}) \neq \mathbf{0}$ *for all* $\mathbf{x} \in \partial\mathcal{B}$, *and* $f_i(x_1, \ldots, x_{i-1}, L_i, x_{i+1}, \ldots, x_n) \le 0$ *and* $f_i(x_1, \ldots, x_{i-1}, U_i, x_{i+1}, \ldots, x_n) \ge 0$, *for* $1 \le i \le n$. *Then, there exists* $\mathbf{s} \in \mathcal{B}$ *such that* $f(\mathbf{s}) = \mathbf{0}$.

We prove:

Theorem 2 *The two zones piecewise linear differential system is*

$$\dot{\mathbf{x}} = \begin{cases} A^+\mathbf{x} \text{ if } x \ge 1, \\ A^-\mathbf{x} \text{ if } x < 1, \end{cases} \tag{1}$$

where $\mathbf{x} = (x, y)^t$,

$$A^- := \begin{pmatrix} \frac{67}{50} & -\frac{833}{125} \\ \frac{1}{2} & -\frac{87}{50} \end{pmatrix} \text{ and } A^+ := \begin{pmatrix} \frac{3}{8} & -1 \\ 1 & \frac{3}{8} \end{pmatrix},$$

has at least three nested hyperbolic limit cycles surrounding the origin.

2 Proof of Theorem 2

Let $\varphi^{\pm}(t; p) = (x^{\pm}(t; p), x^{\pm}(t; p))$ denote the flow associated to the linear systems $\dot{\mathbf{x}} = A^{\pm}\mathbf{x}$. Observe that if there exists a limit cycle then it must lie on both sides of the line $x = 1$, so let $t^- > 0$ be the smaller time such that $x^-(t^-; (1, y)) = 1$ for a point $(1, y)$ with $y > 0$, and let $t^+ > 0$ be the first positive time such that $x^+(-t^+; (1, y)) = 1$. Then any limit cycle must satisfy both conditions and $y^+(-t^+; (1, y)) - y^-(t^-; (1, y)) = 0$, or equivalently,

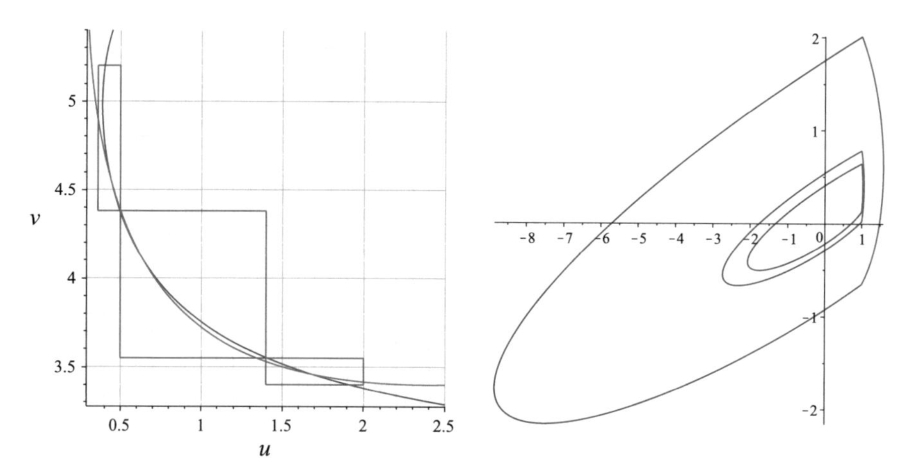

Fig. 1 Left part: Intersection points between $g_1(u, v) = 0$ (in blue) and $g_2(u, v) = 0$ (in magenta) and some boxes containing them. Right part: the three limit cycles of system (1)

$$e^{-\frac{3}{8}u}\left(\cos(u) + y\sin(u)\right) - 1 = 0, \tag{2}$$

$$\left(35\cos\left(\frac{49}{50}v\right) + (-238y + 55)\sin\left(\frac{49}{50}v\right)\right)\frac{e^{-\frac{1}{5}v}}{35} - 1 = 0, \tag{3}$$

$$\left(-49\cos\left(\frac{49}{50}v\right)y + (77y - 25)\sin\left(\frac{49}{50}v\right)\right)\frac{e^{-\frac{v}{5}}}{49} + e^{-\frac{3}{8}u}\left(\cos(u)\,y - \sin(u)\right) = 0, \tag{4}$$

where $u = t^+ > 0$ and $v = t^- > 0$. By solving Eq. (2) we get $y = (e^{-3u/8} - \cos(u))/\sin(u)$. By substituting this expression in Eqs. (3) and (4), we obtain

$$\begin{aligned} g_1(u, v) &:= a(v)\cos(u) + b(v)\sin(u) - a(v)e^{\frac{3}{8}u} = 0, \\ g_2(u, v) &:= c(v)\cos(u) + d(v)\sin(u) + e(v)e^{\frac{3}{8}u} + f(v)e^{-\frac{3}{8}u} = 0, \end{aligned} \tag{5}$$

where

$$a(v) = 238\,e^{-\frac{v}{5}}\sin\left(\frac{49}{50}v\right), \quad b(v) = 55\,e^{-\frac{v}{5}}\sin\left(\frac{49}{50}v\right) + 35\,e^{-\frac{v}{5}}\cos\left(\frac{49}{50}v\right) - 35,$$

$$c(v) = 49\,e^{-\frac{v}{5}}\cos\left(\frac{49}{50}v\right) - 77\,e^{-\frac{v}{5}}\sin\left(\frac{49}{50}v\right) + 49, \quad d(v) = -25\,e^{-\frac{v}{5}}\sin\left(\frac{49}{50}v\right),$$

$$e(v) = 77\,e^{-\frac{v}{5}}\sin\left(\frac{49}{50}v\right) - 49\,e^{-\frac{v}{5}}\cos\left(\frac{49}{50}v\right), \quad f(v) = -49.$$

Numerically, it is easy to guess that there are three different solutions of system (5), see Fig. 1. Their approximate values in (u, v) variables are $(0.441441, 4.554696)$, $(0.639391, 4.105752)$ and $(1.686596, 3.458345)$. Once we prove that near these values there are actual solutions of system (5), each one of them will correspond to a solution of the system of Eqs. (2)–(4) and, consequently, all them will give rise to 3 limit cycles of (1), see again the Fig. 1.

To prove the existence of three solutions of system (5), we consider the three boxes:

$$\mathcal{B}_1 := \left[\frac{9}{25}, \frac{1}{2}\right] \times \left[\frac{219}{50}, \frac{26}{5}\right], \quad \mathcal{B}_2 := \left[\frac{1}{2}, \frac{7}{5}\right] \times \left[\frac{71}{20}, \frac{219}{50}\right], \quad \text{and } \mathcal{B}_3 := \left[\frac{7}{5}, 2\right] \times \left[\frac{17}{5}, \frac{71}{20}\right]$$

which are also shown in Fig. 1 and we apply the Poincaré–Miranda theorem to each of them. In short, we only give some details for $\mathcal{B}_1$. We write $[\underline{u}, \overline{u}] := [9/25, 1/2]$ and $[\underline{v}, \overline{v}] := [219/50, 26/5]$.

The existence of a solution in $\mathcal{B}_1$ will follow by applying the Poincaré–Miranda theorem to this box if we prove the following two claims:

(i) it holds that $g_2(u, \underline{v}) > 0$ and $g_2(u, \overline{v}) < 0$ for all $u \in [\underline{u}, \overline{u}]$;
(ii) it holds that $g_1(\underline{u}, v) < 0$ and $g_1(\overline{u}, v) > 0$ for all $v \in [\underline{v}, \overline{v}]$.

To control the sign of g_j on the sides of each box, we use next lemma:

Lemma 3 *Set* $h(x) = A\cos(\alpha x) + B\sin(\alpha x) + C\mathrm{e}^{\beta x} + D\mathrm{e}^{-\beta x}$, *with* $A, B, C, D \in \mathbb{R}$, $\alpha \neq 0$, $\beta > 0$, *and* $x \in [\underline{x}, \overline{x}] \subset \mathbb{R}^+$. *Then for each* $n \geq 0$ *we have* $h(x) = \sum_{j=0}^{n} a_j x^j + m_n(x)x^{n+1}$, *where*

$$a_j = \frac{1}{j!}\left(\alpha^j\left[A\cos\left(j\frac{\pi}{2}\right) + B\sin\left(j\frac{\pi}{2}\right)\right] + \beta^j\left[C + (-1)^j D\right]\right), \tag{6}$$

$$|m_n(x)| \leq \overline{m}_n = \frac{|\alpha|^{n+1}(|A| + |B|) + |\beta|^{n+1}\left(|C|\mathrm{e}^{\beta\overline{x}} + |D|\mathrm{e}^{-\beta\underline{x}}\right)}{(n+1)!}. \tag{7}$$

In fact, we only give the details to prove in item (i) that $g_2(u, \underline{v}) > 0$ for all $u \in [\underline{u}, \overline{u}]$. All the other sides of the box and the study of the other two boxes can be done by adapting the same procedure. We have that

$$g_2(u, \underline{v}) = c\left(\frac{219}{50}\right)\cos(u) + d\left(\frac{219}{50}\right)\sin(u) + e\left(\frac{219}{50}\right)\mathrm{e}^{\frac{3}{8}u} + f\left(\frac{219}{50}\right)\mathrm{e}^{-\frac{3}{8}u},$$

with

$$A = c\left(\tfrac{219}{50}\right) = 49\,\mathrm{e}^{-\frac{219}{250}}\cos\left(\tfrac{10731}{2500}\right) - 77\,\mathrm{e}^{-\frac{219}{250}}\sin\left(\tfrac{10731}{2500}\right) + 49,$$
$$B = d\left(\tfrac{219}{50}\right) = -25\,\mathrm{e}^{-\frac{219}{250}}\sin\left(\tfrac{10731}{2500}\right),$$
$$C = e\left(\tfrac{219}{50}\right) = \left(-49\cos\left(\tfrac{10731}{2500}\right) + 77\sin\left(\tfrac{10731}{2500}\right)\right)\mathrm{e}^{-\frac{219}{250}}, \quad D = f\left(\tfrac{219}{50}\right) = -49.$$

By applying Lemma 3 with $n = 4$, $\alpha = 1$ and $\beta = 3/8$, we have that $g_2(u, \underline{v}) = \sum_{j=0}^{4} a_j u^j + m_4(u)u^5$, with a_j given in (6) and $|m_4(u)| < \overline{m}_4 \simeq 0.66642 < 0.7 =$

M, see (7). Taking $a_j^- := \text{Trunc}(a_j \cdot 10^k) \cdot 10^{-k} - 10^{-k}$ with $k = 3$, for each $j = 0, \ldots, 4$ we obtain that $\sum_{j=0}^4 a_j u^j > \sum_{j=0}^4 a_j^- u^j$ in $[\underline{u}, \overline{u}]$, where

$$\sum_{j=0}^{4} a_j^- u^j = -\frac{1}{1000} + \frac{1001}{50} u - \frac{39899}{1000} u^2 - \frac{669}{500} u^3 + \frac{357}{125} u^4.$$

Putting all the inequalities together we get that in $[\underline{u}, \overline{u}]$,

$$g_2(u, \underline{v}) = \sum_{j=0}^{4} a_j u^j + m(u) u^5 > \sum_{j=0}^{4} a_j^- u^j - \frac{7}{10} u^5 := Q_5(u).$$

Finally, Q_5 is a polynomial with rational coefficients. Computing its Sturm sequence [9] we get that it has no zeroes $[\underline{u}, \overline{u}]$ and it is positive, as we wanted to prove.

On the other two sides of $\mathcal{B}_1$ or on the boundaries of the other two boxes we can use the same approach. The only changes are that n and k vary from one to another, the corresponding upper bound M must be computed and sometimes instead of g_j is is convenient to consider $e^{v/5} g_j$, see [2] for more details.

To prove the hyperbolicity of the limit cycles we can follow the same ideas that in [5].

References

1. E. Freire, E. Ponce, F. Torres, The discontinuous matching of two planar linear foci can have three nested crossing limit cycles. Publ. Mat. 221–253 (2014)
2. A. Gasull, V. Mañosa, Periodic orbits of discrete and continuous dynamical systems via Poincaré–Miranda theorem. Discrete Contin. Dyn. Syst. Ser. B (submitted). See arXiv:1809.06208 [math.DS]
3. S.M. Huan, X.S. Yang, On the number of limit cycles in general planar piecewise linear systems. Discrete Contin. Dyn. Syst. A **32**, 2147–2164 (2012)
4. W. Kulpa, The Poincaré-Miranda theorem. Amer. Math. Month. **104**, 545–550 (1997)
5. J. Llibre, E. Ponce, Three nested limit cycles in discontinuous piecewise linear differential systems with two zones. Dynam. Contin. Discrete Impuls. Syst. **19**, 325–335 (2012)
6. C. Miranda, Una osservazione su una teorema di Brouwer. Boll. Unione Mat. Ital. **3**, 5–7 (1940)
7. H. Poincaré, Sur certaines solutions particulieres du probléme des trois corps. C.R. Acad. Sci. Paris **97**, 251–252; and Bull. Astronomique **1**(1884), 63–74 (1883)
8. H. Poincaré, Sur les courbes définies par une équation différentielle IV. J. Math. Pures Appl. **85**, 151–217 (1886)
9. J. Stoer, R. Bulirsch, *Introduction to Numerical Analysis* (Springer, New York, 2002)
10. M.N. Vrahatis, A short proof and a generalization of Miranda's existence theorem. Proc. AMS **107**, 701–703 (1989)

Resonance: The Effect of Nonlinearity, Geometry and Frequency Dispersion

Michael P. Mortell and Brian R. Seymour

Abstract Three basic experiments define nonlinear resonant oscillations in continuous media: a gas in a straight closed tube, in a closed tube of variable cross section, and shallow water in a tank. These experiments and the associated mathematical techniques used to explain them are described.

1 Introduction

We deal with one-dimensional waves in a region of finite extent, so that reflections from boundaries are intrinsic to the problems considered and waves pass through each other. For linear theory, solutions can be superposed and hyperbolic waves pass through each other without interaction or distortion. In contrast, the fundamental difficulty for nonlinear hyperbolic waves is that such waves interact and distort, see Riemann [10]. Then a wave traveling in one direction is affected by that traveling in the opposite direction and the characteristic equations cannot, in general, be integrated. There is, however, a class of nonlinear waves where, to first order in the amplitude, the waves do not interact while the signal distorts. Resonant oscillations in tubes and tanks belong to such a class.

Three basic experiments define nonlinear resonant oscillations in continuous media. These experiments and the associated mathematical techniques used to explain them are described here. The first involves small amplitude resonant acoustic oscillations in a closed tube and the appearance of shocks in the flow. Saenger and Hudson [11] observed that the shocks travel with the linear sound speed and do not interact after reflection. The second experiment involves resonant acoustic oscillations in a closed tube of varying cross section, e.g., a cone or bulb shape, see Lawrenson et al. [6]. Due to the interaction of nonlinearity and geometry, the motion

M. P. Mortell (✉)
Department of Applied Mathematics, University College Cork, Cork, Ireland
e-mail: m.mortell@ucc.ie

B. R. Seymour
Department of Mathematics, University of British Columbia, Vancouver, Canada
e-mail: seymour@math.ubc.ca

A. Korobeinikov et al. (eds.), *Extended Abstracts Spring 2018*,
Trends in Mathematics 11, https://doi.org/10.1007/978-3-030-25261-8_9

of the gas can have a large amplitude and yet remain continuous. The final experiment, reported on by Chester and Bones [2], concerns resonant sloshing of shallow water in a tank. The presence of frequency dispersion, which spreads the wave and counteracts the steepening due to nonlinearity, ensures the fluid motion is continuous. The outcome is a series of solitary waves and is governed by a periodically forced KdV equation, [4].

2 Experiment 1: Resonance in a Straight Closed Tube

The basic experiment is that described in Saenger and Hudson [11]. They observed that oppositely traveling waves contain shocks, obey the rule of linear superposition and travel at constant adiabatic sound speed. The amplitude of the flow at resonance is significantly greater than that of the piston amplitude.

In dimensionless variables, the fluid velocity, $u(x, t)$, and the condensation $e = \rho - 1$ ($\rho(x, t)$ is the fluid density) both satisfy the linear wave equation

$$\frac{\partial^2 u}{\partial t^2} - \frac{\partial^2 u}{\partial x^2} = 0, \ 0 \leq x \leq 1, \ t > 0. \tag{1}$$

The boundary conditions are that the tube is closed at $x = 0$ and a piston oscillates periodically at $x = 1$, i.e.,

$$u(0, t) = 0, \quad u(1, t) = M \sin(2\pi\omega t), \tag{2}$$

where M is the applied Mach number and ω the dimensionless frequency. The general solution to (1) is

$$u(\alpha, \beta) = f(\alpha) + g(\beta), \quad e = f(\alpha) - g(\beta), \tag{3}$$

where f and g are arbitrary functions and the linear characteristics are

$$\alpha = t - x, \quad \beta = t + x - 1. \tag{4}$$

Applying the boundary conditions (2) to the solution (3) implies that $g(y)$, $y = \omega t$, satisfies the *linear difference equation*

$$g(y) - g(s) = M \sin(2\pi y), \ y = s + 2\omega. \tag{5}$$

In this notation, 2ω is the linear round-trip travel time in the tube. For $\omega = 1/2$, the period of the input oscillation is unity, the same as the linear travel time. Then $y = s + 1$ and there is no solution of (5) with unit period; $\omega = 1/2$ is the fundamental resonant frequency. The basic conservation laws together with the boundary conditions imply the zero mean condition that $\int_0^1 g(y)dy = 0$.

We will now correct the linear difference equation (5) by including a contribution from the nonlinear travel time. We refer to this as *nonlinearization,* see Mortell and Seymour [9]. Then 2ω is replaced by $2\omega + \omega(\gamma + 1)g(s)$, where γ is the polytropic gas constant and g then satisfies the *nonlinear difference equation*

$$g(y) - g(s) = M \sin(2\pi y), \quad y = s + 2\omega + \omega(\gamma + 1)g(s). \tag{6}$$

Exactly at the fundamental resonant frequency, $\omega = 1/2$, (6) becomes

$$g(y) - g(s) = M \sin(2\pi y), \quad y = s + \frac{\gamma + 1}{2} g(s), \tag{7}$$

on using the unit periodicity of $g(y)$.

In the small rate limit, when $|g| \ll 1$ and $|g'| \ll 1$, (7) are approximated using a Taylor expansion by a nonlinear ODE with solutions

$$g(y) = \pm\left(\frac{4M}{\pi(\gamma + 1)}\right)^{1/2} \sin(\pi y). \tag{8}$$

To satisfy the zero mean condition, these are combined with a discontinuity at $y = 0.5$, hence inserting a shock.

The required solution is a linear standing wave given by (3) with $f(y) = -g(y - 1)$ where $y = t/2$ and $g(y)$ is given by the nonlinear solution (8). Thus, $u(x, t)$ is the superposition of linear waves traveling at the sound speed and the signal $|g| = O(M^{1/2})$. These agree with the results of the Saenger and Hudson [11] experiment.

The full small rate solution is described in [12], while the evolution of these oscillations is given in [3]. The finite rate evolution is given in [13].

3 Experiment 2: Resonant Macrosonic Synthesis

The purpose of the experiments described in Lawrenson et al. [6] was to examine the effect of a variable cross section on resonant oscillations. Of particular interest was whether shocks were present. Shocks are a dissipative mechanism, converting kinetic energy into heat, so eliminating shocks results in higher pressures. They show that shocks (acoustic saturation) can be avoided by shaping the resonator; with no shocks the same energy input can produce peak acoustic overpressure exceeding 340% of ambient pressure. The analytic underpinning for these results is given in Mortell and Seymour [8], and numerical solutions are given in Ilinskii et al. [5].

3.1 Nonlinear Equations for Variable Tube Area

The dimensionless equations of conservation of mass and momentum relating the velocity $u(x,t)$ and density $\rho(x,t)$ for a polytropic gas in Eulerian coordinates are

$$\frac{\partial(s\rho)}{\partial t}+\frac{\partial}{\partial x}(su\rho)=0, \qquad \frac{\partial u}{\partial t}+u\frac{\partial u}{\partial x}+\rho^{\gamma-2}\frac{\partial \rho}{\partial x}=-a(t), \tag{9}$$

where $e(x,t)=\rho/\rho_0-1$ is the condensation, γ is the adiabatic constant (1.2 for air), $c_0=\sqrt{\gamma p_0/\rho_0}$ is the associated linear sound speed and $a(t)$ is the acceleration applied along the axis of the tube that is closed at both ends.

A new variable is defined by $v(x,t)=s(x)u(x,t)$ and $a(t)=\varepsilon^3\cos\theta$, $\theta=\omega t$, $0<\varepsilon\ll 1$. We seek solutions with period $2\pi/\omega$ in t.

3.2 Dominant First Harmonic Approximation

A perturbation expansion of the form

$$\begin{aligned} e(x,t)&=\varepsilon e_1(x,t)+\varepsilon^2 e_2(x,t)+\cdots,\\ v(x,t)&=\varepsilon v_1(x,t)+\varepsilon^2 v_2(x,t)+\cdots,\\ \omega(\varepsilon)&=\lambda_1+\varepsilon^2\delta+\cdots, \end{aligned}$$

where $|e_i|, |v_i|=O(1), i=1,2,3,\ldots$, and λ_1 is the fundamental eigenvalue, yields the Webster–Horn equation for $v_1(x,t)$

$$\frac{\partial^2 v_1}{\partial t^2}-s(x)\frac{\partial}{\partial x}(\frac{1}{s}\frac{\partial v_1}{\partial x})=0. \tag{10}$$

The eigenfunction corresponding to λ_1 is $\varphi(x)$, where $v_1(x,t)=A\varphi(x)\sin\theta$ and $A(\delta)$ is a constant. There is an infinity of eigenvalues, $\lambda=\lambda_n$, and the critical assumption is that $\lambda_n\neq n\lambda_1, n=2,3,\ldots$, for the given $s(x)$.

$$\text{At } O(\varepsilon^3): \qquad v_3(x,\theta)=P(x)\sin\theta+Q(x)\sin 3\theta.$$

Since $3\lambda_1$ is not an eigenvalue, $Q(x)$ exists with no restriction on A, but $P(x)$ is determined by

$$\frac{d}{dx}(\frac{1}{s}\frac{dP}{dx})+\frac{\lambda_1^2}{s}P=A^3G(x),\ \ P(0)=0,\ \ P(1)=0, \tag{11}$$

where $G(x)$ is a combination of lower order solutions. The orthogonality condition to ensure a solution $P(x)$ yields the *amplitude–frequency relation*

$$NA^3 - 2\delta\lambda_1 A = M, \tag{12}$$

where M and N are constants. The experimental and theoretical curves for this relation are qualitatively similar, and show a "hard" or "soft" spring response depending on the cylinder shape. The fundamental solution is a linear standing wave with a signal determined by a nonlinear equation. The evolution of such solutions is given in [7].

4 Experiment 3: Resonant Sloshing in a Shallow Tank

Near resonance waves in the tank have high peaks and low troughs and abrupt changes in amplitude occur at certain discrete frequencies, see [2]. The theory for periodic resonant oscillations is given in [1] and the evolution in [4]. The derivation of the basic equations is given in Chap. 13 of Whitham [14], but here, we follow the notation in [4].

$\varphi(x, z, t)$ is the velocity potential, where the particle velocity $u = \frac{\partial\varphi}{\partial x}$ and $z = \eta(x, t)$ is the free surface. The wavemaker at $x_w = 1 - \varepsilon\cos(\pi t)$ implies the boundary condition $\frac{\partial\varphi}{\partial x}(x_w, z, t) = 2\pi\varepsilon\omega\sin(\pi t)$, while the tank is closed at $x = 0$, so that $\frac{\partial\varphi}{\partial x}(0, z, t) = 0$.

We assume an expansion of the form

$$\varphi(x, z, t) = \varepsilon\varphi_0(x, z, t) + \varepsilon^{3/2}\varphi_1(x, z, t) + \cdots,$$
$$\eta(x, z, t) = \varepsilon\eta_0(x, z, t) + \varepsilon^{3/2}\eta_1(x, z, t) + \cdots,$$

where the detuning Δ is given by $2\omega = 1 + \varepsilon^{1/2}\Delta + \cdots$, and ε is the amplitude of the periodic input. Then at $O(\varepsilon^{3/2})$ the linear differential-difference equation is

$$2\pi\omega\varepsilon\sin(\pi t) = \varepsilon[h(t) - h(t-2)] + \varepsilon^{3/2}[-\frac{\kappa}{3}h'''(t) + 2\Delta h'(t)],$$

where the dispersion and detuning are included.

The nonlinear terms from the free surface conditions are inserted by nonlinearization, i.e., the linear travel time 2 is replaced by the nonlinear travel time $2 + (\gamma + 1)\varepsilon h(t)$, with $\gamma = 2$, which reflects the underlying hydraulic flow.

Then we get

$$\pi\omega\sin(\pi t) = \frac{3}{2}\varepsilon h(t)h'(t) + \varepsilon^{1/2}[-\frac{\kappa}{6}h'''(t) + \Delta h'(t)] + O(\varepsilon^{3/2}).$$

The substitution $R = \varepsilon^{1/2}h(t)$ reduces this equation to

$$\pi\omega\sin(\pi t) = \frac{3}{2}R(t)R'(t) + \Delta R'(t) - \frac{\kappa}{6}R'''(t),$$

which is a periodically forced, steady state, KdV equation. The solution then is

$$u = \frac{\partial \varphi}{\partial x} = \varepsilon[h(t + x - 1) - h(t - x - 1)]$$

with $h(t)$given by the equation for $R(t)$. Thus the solution is the superposition of linear waves traveling in opposite directions at the linear wave speed, while the signal is determined by a nonlinear ODE.

References

1. W. Chester, Resonant oscillations of water waves I. Theory. Proc. Roy. Soc. London A **306**, 5–22 (1968)
2. W. Chester, J.A. Bones, Resonant oscillations of water waves II. Experiment. Proc. Roy. Soc. Lond. A **306**, 23–39 (1968)
3. E.A. Cox, M.P. Mortell, The evolution of resonant oscillations in closed tubes. ZAMP **34**(6), 845–866 (1983)
4. E.A. Cox, M.P. Mortell, The evolution of resonant water-wave oscillations. J. Fluid Mech. **162**, 99–116 (1986)
5. Y.A. Ilinskii, B. Lipkens, T.S. Lucas, T.W. Van Doren, E.A. Zabolotskaya, Nonlinear standing waves in an acoustical resonator. J. Acoust. Soc. Am. **104**(5), 2664–2674 (1998)
6. C.C. Lawrenson, B. Lipkens, T.S. Lucas, D.K. Perkins, T.W. Van Doren, Measurements of macrosonic standing waves in oscillating closed cavities. J. Acoust. Soc. Am. **104**(2), 623–636 (1998)
7. M.P. Mortell, K.F. Mulchrone, B.R. Seymour, The evolution of macrosonic standing waves in a resonator. Int. J. Eng. Sci. **47**(11), 1305–1314 (2009)
8. M.P. Mortell, B.R. Seymour, Nonlinear resonant oscillations in closed tubes of variable cross-section. J. Fluid Mech. **519**, 183–199 (2004)
9. M.P. Mortell, B.R. Seymour, Nonlinearization and waves in bounded media: old wine in a new bottle. J. Phys. Conf. Ser. **811**, 1–9 (2017)
10. R. Riemann, Uber die fortpflanzung ebener luftwellen von endlicher schwingungsweite, Gottingun Abhandlungen, vol. viii (1858), p. 43
11. R.A. Saenger, G.E. Hudson, Periodic shock waves in resonating gas columns. J. Acoust. Soc. Am. **32**(8), 961–970 (1960)
12. B.R. Seymour, M.P. Mortell, Resonant acoustic oscillations with damping: small rate theory. J. Fluid Mech. **58**, 353–373 (1973)
13. B.R. Seymour, M.P. Mortell, The evolution of a finite rate periodic oscillation. Wave Motion **7**(5), 399–409 (1985)
14. G.B. Whitham, *Linear and Nonlinear Waves* (Wiley, New York, 1974)

On the Null Controllability of the Heat Equation with Hysteresis in Phase Transition Modeling

Chiara Gavioli and Pavel Krejčí

Abstract We prove the null controllability of the relaxed Stefan problem, which models phase transitions in two-phase systems. The technique relies on the penalty approximation of the differential inclusion describing the phase dynamics, solving a constrained minimization problem, and passing to the limit.

Introduction

The null controllability problem for various kinds of linear and semilinear parabolic equations has been an intensively studied subject in the recent decades and a nice survey can be found in the monograph [2]. Here, we propose to discuss the null controllability problem for the parabolic equation with hysteresis of the form

$$u_t(x,t) - \Delta u(x,t) + \mathcal{F}[u](x,t) = v(x,t), \quad x \in \Omega \subset \mathbb{R}^n,\ t \in (0,T) \tag{1}$$

with a hysteresis operator $\mathcal{F}$, a right-hand side v called the *control*, and initial and boundary conditions specified below. Existence, uniqueness, and regularity results for Eq. (1) with a given right-hand side v, can be found in the monograph [7]. The null controllability problem for Eq. (1) consists in proving that for an arbitrary initial condition and arbitrary final time T, it is possible to choose the control v in a suitable class of functions of x and t in such a way that the solution satisfies $u(\cdot, T) = 0$, a.e., in Ω.

First results about the null controllability of Eq. (1) were obtained by F. Bagagiolo in [1]: His technique relies on a linearization followed by a fixed-point procedure, and we briefly comment on it in Sect. 2. We will see that hysteresis operators arising from

C. Gavioli (✉)
Dipartimento di Scienze Fisiche, Informatiche E Matematiche,
Università Degli Studi di Modena E Reggio Emilia, via Campi 213/b, I-41125 Modena, Italy
e-mail: chiara.gavioli@unimore.it

P. Krejčí
Faculty of Civil Engineering, Czech Technical University, Thákurova 7, CZ-16629 Praha 6, Czech Republic
e-mail: krejci@math.cas.cz

A. Korobeinikov et al. (eds.), *Extended Abstracts Spring 2018*,
Trends in Mathematics 11, https://doi.org/10.1007/978-3-030-25261-8_10

phase transition modeling cannot be linearized. To establish the null controllability of the system, new techniques based on M. Brokate's previous works [3, 4] on optimal control of ODEs with hysteresis need to be developed, and this will be done in Sect. 3.

1 The Physical Problem

Consider a bounded connected Lipschitzian domain $\Omega \subset \mathbb{R}^3$, fix an arbitrary $T > 0$ and define $Q = \Omega \times (0, T), \Gamma = \partial\Omega \times (0, T)$. The unknown functions of the space variable $x \in \Omega$ and time $t \in [0, T]$ are $s(x, t) \in [-1, 1]$ for the phase parameter ($s = -1$ solid, $s = 1$ liquid, $s \in (-1, 1)$ mixture), and $\theta(x, t) > 0$ for the absolute temperature.

The system we consider is the following:

$$\begin{cases} c\theta_t + Ls_t - \kappa\Delta\theta = h & \text{in } Q, \\ \rho s_t + \partial I(s) \ni L(\theta - \theta_c) & \text{in } Q, \\ \text{initial and boundary conditions,} \end{cases} \tag{2}$$

where I is the indicator function of the interval $[-1, 1]$, ∂I is its subdifferential, $h = h(x, t)$ is the heat source density, and c specific heat, L latent heat, κ heat conductivity, ρ phase relaxation parameter and θ_c critical temperature are given positive constants. In the literature this is known as the *relaxed Stefan problem* (see, e. g., A. Visintin's monograph [8]), and it models the phase transition in solid–liquid systems: the first equation is the energy balance, whereas the second one describes the phase dynamics. In particular:

(i) the smaller ρ is, the faster the transition takes place. When $\rho = 0$ we get the classical Stefan problem, in which the phase transition is assumed to be instantaneous;
(ii) when $\theta > \theta_c$ then $s_t \geq 0$, which means that the substance is melting; when $\theta < \theta_c$ then $s_t \leq 0$, which means that the substance is freezing.

We now show that system (2) can be transformed into the form (1). Indeed, we define a new unknown u by the formula

$$u_t = \frac{L}{\rho}(\theta - \theta_c).$$

Then the phase dynamics equation in (2) is of the form $s_t + \partial I(s) \ni u_t$. This is nothing but the definition of the stop operator with threshold 1, $s = \mathfrak{s}[u]$; see Fig. 1.

The first equation in (2) thus reads

$$\frac{c\rho}{L}u_{tt} + L\mathfrak{s}[u]_t - \frac{\kappa\rho}{L}\Delta u_t = h.$$

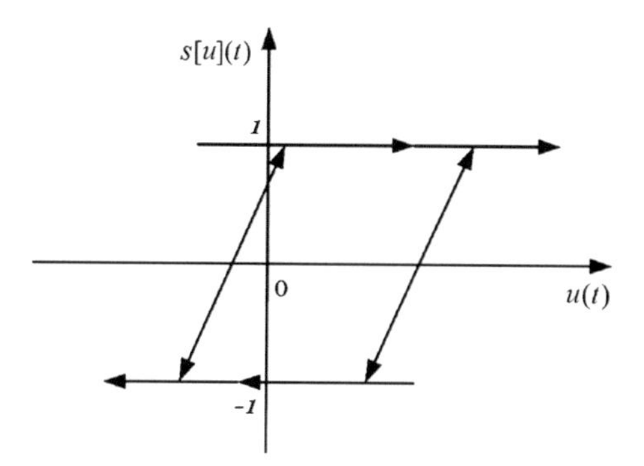

Fig. 1 Hysteresis loop of the stop operator

Integrating the above equation in time leads, up to constants, to an equation of the form (1), more specifically,

$$\frac{c\rho}{L}u_t + L\mathfrak{s}[u] - \frac{\kappa\rho}{L}\Delta u = v \tag{3}$$

with $\mathcal{F}[u] = \mathfrak{s}[u]$, and with v containing the time integral of h and additional terms coming from the initial conditions.

2 Null Controllability by Linearization

The system considered by F. Bagagiolo in [1] is the following:

$$\begin{cases} u_t(x,t) - \Delta u(x,t) + \mathcal{F}[u](x,t) = m(x)v(x,t) & \text{in } Q, \\ u(x,t) = 0 & \text{on } \Gamma, \\ u(x,0) = u_0(x) & \text{in } \Omega, \end{cases} \tag{4}$$

where m is the characteristic function of a set $\omega \subset\subset \Omega$ and $\mathcal{F}\colon L^2(\Omega; C^0([0,T])) \longrightarrow L^2(\Omega; C^0([0,T]))$ is a hysteresis operator satisfying the following condition: there exist two constants $L > 0$ and $m \in \mathbb{R}$ such that, for all $z \in L^2(\Omega; C^0([0,T]))$, for all $t \in [0,T]$ and for a.e. $x \in \Omega$

$$|\mathcal{F}[z](x,t)| \leq L|z(x,t)|, \tag{5}$$

$$\text{if } z(x,t) = 0 \text{ then } \lim_{\tau \to t,\, z(x,\tau) \neq 0} \frac{\mathcal{F}[z](x,\tau)}{z(x,\tau)} = m \text{ uniformly in } [0,T]. \tag{6}$$

Similarly as above, $v : Q := \Omega \times (0,T) \to \mathbb{R}$ is the control function which, being multiplied by m, acts only on a compact subregion of the original domain. The null controllability of system (4) strongly relies on the following result from V. Barbu's paper [2].

Theorem 1 (Null controllability in the linear case) *Let $\Omega \subset \mathbb{R}^n$ be an open and bounded domain with boundary of class C^2, let $\omega \subset \Omega$ be a compactly embedded subset, and let $a \in L^\infty(Q)$ be given. Then for every initial datum $u_0 \in L^2(\Omega)$ there is a control function $v \in L^2(Q)$ such that the (unique) corresponding solution $u^v \in C^0([0,T]; L^2(\Omega)) \cap L^2(0,T; W_0^{1,2}(\Omega))$ of*

$$\begin{cases} u_t(x,t) - \Delta u(x,t) + a(x,t)u(x,t) = m(x)v(x,t) & \text{in } Q, \\ u(x,t) = 0 & \text{on } \Gamma, \\ u(x,0) = u_0(x) & \text{in } \Omega, \end{cases} \tag{7}$$

satisfies $u^v(x,T) = 0$ a. e. $x \in \Omega$. Moreover, the control v can be taken in such a way that

$$\|v\|_{L^2(Q)} \leq C\|u_0\|_{L^2(\Omega)},$$

where the constant C only depends on $\|a\|_{L^\infty(Q)}$.

Note that V. Barbu's proof of this result relies on the *Pontryagin's Maximum Principle*, and the *Carleman estimates*.

In particular, Pontryagin's Maximum Principle requires the study of the dual system associated with (7), whereas Carleman estimates allow us to bound the L^2-norm of the dual variable in terms of the L^2-norm of the control on the subregion $\omega \times (0,T)$.

F. Bagagiolo's condition (5)–(6) implies, in particular, that all hysteresis branches pass through the origin. System (4) thus can be reduced to the form (7) with a factor $a(x,t)$ depending on the unknown function u. The null controllability result is then obtained by a fixed-point argument.

In the case that the operator $\mathcal{F}$ is the stop operator given by Eq. (3), the assumption (5) is not satisfied. It is well known (cf. [7]) that typical hysteresis branches of the stop operator do not pass through the origin as required by condition (5), see Fig. 1.

3 A New Approach: Null Controllability by Penalization

The exact values of the physical constants c, ρ, L, κ are irrelevant for our analysis. We can therefore represent Eq. (3) as a system of the form

$$\begin{cases} u_t - \Delta u + s = v & \text{in } Q, \\ s_t + \partial I(s) \ni u_t & \text{in } Q, \\ \nabla u \cdot n = 0 & \text{on } \Gamma, \\ u(x,0) = u_0(x) & \text{in } \Omega, \\ s(x,0) = s_0(x) & \text{in } \Omega. \end{cases} \tag{8}$$

The homogeneous Neumann boundary condition for u has the physical meaning of a thermally insulated body in the original setting (2). The main result for the system (8) reads as follows.

Theorem 2 *Let* $u_0 \in W^{1,2}(\Omega) \cap L^\infty(\Omega)$ *and* $s_0 \in L^\infty(\Omega)$ *be given,* $|s_0(x)| \le 1$, *a.e. Then the system* (8) *is null controllable, that is, there exists* $v \in L^2(Q)$ *such that the corresponding solution* $u^v \in W^{1,2}((0,T); L^2(\Omega)) \cap L^\infty(0,T; W^{1,2}(\Omega))$ *of* (8) *satisfies* $u^v(\cdot, T) = 0$, *a.e., in* Ω.

Note that controls with support restricted to a subdomain $\omega \subset \Omega$ as in Theorem 1 are not admissible in Theorem 2. This is related to the problem whether Carleman estimates are compatible with the penalty approximation. This question will be given appropriate attention in future work.

Proof The argument consists in penalizing the subdifferential ∂I and replacing the differential inclusion with an ODE. In particular, we choose the penalty function

$$\Psi(s) = \begin{cases} \phi(s-1) & \text{for } s > 1, \\ 0 & \text{for } s \in [-1,1], \\ \phi(-s-1) & \text{for } s < -1, \end{cases} \tag{9}$$

with a convex C^2-function $\phi\colon [0,\infty) \to [0,\infty)$ with quadratic growth, for example

$$\phi(r) = \begin{cases} \frac{1}{6} r^3 & \text{for } r \in [0,1], \\ \frac{1}{2} r^2 - \frac{1}{2} r + \frac{1}{6} & \text{for } r > 1. \end{cases}$$

Choosing a small parameter $\gamma > 0$, we replace (8) with a system of one PDE and one ODE for unknown functions $(u, s) = (u^\gamma, s^\gamma)$

$$\begin{cases} u_t - \Delta u + s = v & \text{in } Q, \\ s_t + \frac{1}{\gamma} \Psi'(s) = u_t & \text{in } Q, \end{cases} \tag{10}$$

with the same initial and boundary conditions, and with the intention to let γ tend to 0. We choose another small parameter $\varepsilon > 0$ independent of γ and define the cost functional

$$J(u,s,v) = \frac{1}{2} \iint_Q v^2 \mathrm{d}x\mathrm{d}t + \frac{1}{2\varepsilon} \int_\Omega u^2(x,T)\mathrm{d}x,$$

where the two summands represent the cost to implement the control and to reach the desired null final state. Then, for each $\gamma > 0$ we solve the following optimal control problem:

$$\text{minimize } J(u, s, v) \text{ subject to (10).} \tag{11}$$

It is not difficult to see (see, e. g., Tröltzsch [6]) that for each $\varepsilon > 0$ problem (11) has a unique solution $(u_\varepsilon^\gamma, s_\varepsilon^\gamma, v_\varepsilon^\gamma)$. It is found as a critical point of the Lagrangian

$$L(u, s, v) = J(u, s, v) + \langle p, G_1(u, s, v)\rangle + \langle q, G_2(u, s, v)\rangle$$

where p, q are Lagrange multipliers, the brackets denote the canonical scalar product in $L^2(\Omega)$, and the constraints are

$$G_1(u, s, v) = u_t - \Delta u + s - v, \qquad G_2(u, s, v) = s_t + \frac{1}{\gamma}\Psi'(s) - u_t.$$

The first-order necessary optimality condition for $(u, s, v, p, q) = (u_\varepsilon^\gamma, s_\varepsilon^\gamma, v_\varepsilon^\gamma, p_\varepsilon^\gamma, q_\varepsilon^\gamma)$ reads

$$v = p \text{ a. e. in } Q, \tag{12}$$

and $p \in W^{1,2}(0, T; L^2(\Omega)) \cap L^\infty(0, T; W^{1,2}(\Omega))$, $q \in W^{1,2}(0, T; L^2(\Omega))$ are the solutions to the backward dual problem

$$\begin{cases} p_t + \Delta p - q_t = 0 & \text{in } Q, \\ q_t - \frac{1}{\gamma}\Psi''(s)q - p = 0 & \text{in } Q, \\ \nabla p \cdot n = 0 & \text{on } \Gamma, \\ p(x, T) = -\frac{1}{\varepsilon}u(x, T) & \text{in } \Omega, \\ q(x, T) = 0 & \text{in } \Omega. \end{cases} \tag{13}$$

3.1 Estimates

In order to pass to the limits $\varepsilon \to 0$, $\gamma \to 0$, we first derive a series of estimates for $(u, s, v, p, q) = (u_\varepsilon^\gamma, s_\varepsilon^\gamma, v_\varepsilon^\gamma, p_\varepsilon^\gamma, q_\varepsilon^\gamma)$ satisfying the system (10), (12), and (13). In what follows, we denote by C any constant independent of γ and ε.

We first multiply the second equation of (13) by $-\text{sign}(q)$ and integrate from an arbitrary $t \in [0, T)$ to T to obtain

$$|q(x, t)| + \int_t^T \frac{1}{\gamma}\Psi''(s(x, \tau))|q(x, \tau)|\,\mathrm{d}\tau \le \int_t^T |p(x, \tau)|\,\mathrm{d}\tau \quad \text{for a. e. } x \in \Omega. \tag{14}$$

In the next step, we combine the first and the second equation of (13) to get

$$p_t + \Delta p - \frac{1}{\gamma}\Psi''(s)q - p = 0,$$

and multiply the resulting equation by an approximation $S_n(p)$ of $-\text{sign}(p)$, say, $S_n(p) = -\text{sign}(p)$ for $|p| \geq 1/n$, $S_n(p) = -np$ for $|p| < 1/n$. Integrating over Ω and letting n tend to infinity we obtain

$$-\frac{\mathrm{d}}{\mathrm{d}t}\int_{\Omega}|p(x,t)|\mathrm{d}x + \int_{\Omega}|p(x,t)|\mathrm{d}x \leq \int_{\Omega}\frac{1}{\gamma}\Psi''(s(x,t))|q(x,t)|\mathrm{d}x \quad \text{for a. e. } t \in (0,T). \tag{15}$$

Integrating (15) consecutively $\int_0^{\tau} \mathrm{d}t$ and then $\int_0^{T} \mathrm{d}\tau$ and using the estimate (14) gives a bound for $p(x,0)$, namely

$$\int_{\Omega}|p(x,0)|\mathrm{d}x \leq C\int_0^T\int_{\Omega}|p(x,t)|\mathrm{d}x\mathrm{d}t. \tag{16}$$

Finally, we multiply the first equation in (10) by p, the second equation in (10) by q, the first equation in (13) by u, the second equation in (13) by s, integrate in space and time, and sum up (note that $p = v$ by virtue of (12)):

$$\int_0^T\int_{\Omega}p^2\mathrm{d}x\mathrm{d}t + \frac{1}{\varepsilon}\int_{\Omega}u^2(x,T)\mathrm{d}x = -\int_{\Omega}u_0(x)p(x,0)\mathrm{d}x + \int_{\Omega}u_0(x)q(x,0)\mathrm{d}x$$
$$-\int_{\Omega}s_0(x)q(x,0)\mathrm{d}x + \frac{1}{\gamma}\int_0^T\int_{\Omega}q\left(\Psi'(s) - s\Psi''(s)\right)\mathrm{d}x\mathrm{d}t. \tag{17}$$

The choice (9) of Ψ guarantees that

$$|\Psi'(s) - s\Psi''(s)| \leq \frac{3}{2}\Psi''(s).$$

Hence, by virtue of (14)–(16), we infer from (17) the estimate

$$\int_0^T\int_{\Omega}p^2(x,t)\mathrm{d}x\mathrm{d}t + \frac{1}{\varepsilon}\int_{\Omega}u^2(x,T)\mathrm{d}x \leq C\int_0^T\int_{\Omega}|p(x,t)|\mathrm{d}x\mathrm{d}t \tag{18}$$

with a constant C depending on the L^{∞}-norm of u_0, which, together with Hölder's inequality, implies in turn that

$$\int_0^T\int_{\Omega}(v_{\varepsilon}^{\gamma})^2(x,t)\mathrm{d}x\mathrm{d}t + \frac{1}{\varepsilon}\int_{\Omega}(u_{\varepsilon}^{\gamma})^2(x,T)\mathrm{d}x \leq C \tag{19}$$

with a constant C independent of γ and ε.

3.2 Passage to the Limit

As a consequence of (19), we see by a standard result on parabolic PDEs that the solutions u_ε^γ, s_ε^γ of (10) are for each fixed $\gamma > 0$ uniformly bounded in $W^{1,2}(0,T;L^2(\Omega)) \cap L^\infty(0,T;W^{1,2}(\Omega))$. Keeping thus γ fixed for the moment, letting $\varepsilon \to 0$ and using the compact embedding of $W^{1,2}(0,T;L^2(\Omega)) \cap L^\infty(0,T;W^{1,2}(\Omega))$ into $L^2(\Omega;C([0,T]))$, we conclude that along a subsequence for each fixed γ we have

$$v_\varepsilon^\gamma \rightharpoonup v_*^\gamma,\ s_\varepsilon^\gamma \to s_*^\gamma,\ (s_\varepsilon^\gamma)_t \rightharpoonup (s_*^\gamma)_t \text{ in } L^2(Q), \qquad \|u_\varepsilon^\gamma(x,T)\|_{L^2(\Omega)}^2 \to 0,$$
$$u_\varepsilon^\gamma \to u_*^\gamma \text{ in } L^2(\Omega;C([0,T])) \text{ and } u_*^\gamma(x,T) = 0 \text{ a. e.}$$

The convergence $\gamma \to 0$ is more delicate. By (19), the controls contain a weakly convergent subsequence

$$v_*^\gamma \rightharpoonup v_* \text{ in } L^2(Q).$$

The same parabolic PDE argument as above yields

$$(u_*^\gamma)_t \rightharpoonup (u_*)_t \text{ in } L^2(Q), \qquad u_*^\gamma \to u_* \text{ in } L^2(\Omega;C([0,T])) \text{ with } u_*(x,T) = 0. \tag{20}$$

It remains to prove that the solutions s_*^γ to the equation

$$(s_*^\gamma)_t + \frac{1}{\gamma}\Psi'(s_*^\gamma) = (u_*^\gamma)_t$$

converge weakly to $\mathfrak{s}[u_*]$. To this end, we denote by y^γ the solution of the ODE

$$y_t^\gamma + \frac{1}{\gamma}\Psi'(y^\gamma) = (u_*)_t, \quad y^\gamma(x,0) = s_0(x). \tag{21}$$

By [5, Theorem 1.12], we have for a. e. $(x,t) \in Q$

$$|s_*^\gamma(x,t) - y^\gamma(x,t)| \le 2 \max_{\tau\in[0,t]} |u_*^\gamma(x,\tau) - u_*(x,\tau)|. \tag{22}$$

Multiplying (21) by y_t^γ and integrating over Q we see that the $L^2(Q)$-norm of y_t^γ is bounded independently of γ. Hence, up to a subsequence,

$$y_t^\gamma \rightharpoonup y_t,\ y^\gamma \rightharpoonup y,\ \frac{1}{\gamma}\Psi'(y^\gamma) \rightharpoonup w \text{ in } L^2(Q), \tag{23}$$

and it suffices to prove that $y = \mathfrak{s}[u_*]$. To this end note that y and w satisfy the equation

$$y_t + w = (u_*)_t, \quad y(x,0) = s_0(x). \tag{24}$$

Furthermore, for every function $z \in L^\infty(Q)$ we have

$$\iint_Q y^\gamma z \, dx dt \le \iint_Q |y^\gamma| \, |z| \, dx dt \le \iint_Q (|y^\gamma| - 1)^+ \, |z| \, dx dt + \iint_Q |z| \, dx dt,$$

hence, choosing z such that $\iint_Q |z| dx dt \le 1$, by (23) we have $\iint_Q yz \, dx dt \le 1$ which in turn implies that $|y(x,t)| \le 1$ a.e.

We now multiply (21) by y^γ and (24) by y and integrate over Q. By virtue of the weak convergence we have $\int_\Omega y^2(x,T) \, dx \le \liminf_{\gamma \to 0} \int_\Omega (y^\gamma)^2(x,T) \, dx$, hence

$$\liminf_{\gamma \to 0} \frac{1}{\gamma} \iint_Q \Psi'(y^\gamma) y^\gamma \, dx dt \le \iint_Q wy \, dx dt.$$

Since Ψ' is monotone and vanishes in $[-1, 1]$, it follows that

$$\iint_Q \Psi'(y^\gamma)(y^\gamma - \rho) \, dx dt \ge 0$$

for every measurable function ρ such that $|\rho(x,t)| \le 1$ a.e. Hence, for every ρ we have

$$\iint_Q w(y - \rho) \, dx dt \ge 0,$$

which implies that $y = \mathfrak{s}[u_*]$, and the proof is complete.

References

1. F. Bagagiolo, On the controllability of the semilinear heat equation with hysteresis. Phys. B Condens. Matter **407**, 1401–1403 (2012)
2. V. Barbu, Controllability of parabolic and Navier-Stokes equations. Sci. Math. Jpn. **56**(1), 143–211 (2002)
3. M. Brokate, *Optimale Steuerung von gewöhnlichen Differentialgleichungen mit Nichtlinearitäten vom Hysteresis-Typ* (Verlag Peter D, Lang, Frankfurt a. M., 1987)
4. M. Brokate, ODE control problems including the Preisach hysteresis operator: necessary optimality conditions, in *Dynamic Economic Models and Optimal Control* (Vienna, North-Holland). Amsterdam **1992**, 51–68 (1991)
5. P. Krejčí, Hysteresis operators—a new approach to evolution differential inequalities. Comment. Math. Univ. Carolinae **33**(3), 525–536 (1989)
6. F. Tröltzsch, Optimal control of partial differential equations, theory, methods and applications. Graduate Studies in Mathematics (American Mathematical Society, 2010)
7. A. Visintin, *Differential Models of Hysteresis* (Springer, Berlin Heidelberg, Applied Mathematical Sciences, 1994)
8. A. Visintin, *Models of Phase Transitions* (Progress in nonlinear differential equations and their applications, Birkhäuser Boston, 1996)

Evidence of Critical Transitions and Coexistence of Alternative States in Nature: The Case of Malaria Transmission

David Alonso, Andy Dobson and Mercedes Pascual

Abstract Sometimes abrupt changes occur in nature. Examples of these phenomena exist in lakes, oceans, terrestrial ecosystems, climate, evolution, and human societies. Dynamical systems theory has provided useful tools to understand the nature of these changes. When certain non-linearities underlie system dynamics, rapid transitions may happen when critical thresholds for certain parameter values are overcome. Here we describe a malaria dynamical model that couples vector and human disease dynamics through mosquito infectious bites, with the possibility of super-infection, this is, the reinfection of asymptomatic hosts before they have cleared a prior infection. This key feature creates the potential for sudden transitions in the prevalence of infected hosts that seem to characterize malaria's response to environmental conditions. This dynamic behavior may challenge control strategies in different locations. We argue that the potential for critical transitions is a general and overlooked feature of any model for vector borne diseases with incomplete, complex immunity.

This work was funded by the Spanish government through the Ramón y Cajal Fellowship program (DA).

D. Alonso (✉)
Theoretical and Computational Ecology, Center for Advanced Studies (CEAB–CSIC), Blanes, Spain
e-mail: dalonso@ceab.csic.es

A. Dobson
Ecology and Evolutionary Biology, Eno Hall, Princeton University, Princeton, NJ 08540, USA
e-mail: dobber@princeton.edu

M. Pascual
Ecology and Evolutionary Biology, University of Chicago, Chicago, IL 60637, USA

Santa Fe Institute, Hyde Park Road, Santa Fe, NM 87501, USA
e-mail: pascualmm@uchicago.edu

A. Korobeinikov et al. (eds.), *Extended Abstracts Spring 2018*,
Trends in Mathematics 11, https://doi.org/10.1007/978-3-030-25261-8_11

1 Introduction

Critical transitions have received considerable attention in ecology, geophysics, hydrology and economics for the last decade [12]. They occur when natural systems drastically shift from one state to another. Comparatively less attention has been given to carefully characterize the underlying dynamic structure of the system under study. We believe the focus should change from describing, and understanding single transitions to characterizing the full dynamic behavior of the systems along with the environmental conditions in which these transitions occur. In epidemiology, critical transitions may underlie and potentially enhance (or undermine) attempts to control and eliminate infectious pathogens. Following an intervention, the trajectory of the host-pathogen system may cross a critical transition where pathogen prevalence drops to apparent eradication. However, the final success of eradication efforts depends strongly on dynamic underlying structure of the transition. Critical transitions are often associated to the coexistence of alternative equilibria. In that case, small changes in a driving parameter can lead to large shifts from low to high levels of prevalence (or *vice versa*). Continuous external pressure on critical transmission parameters, or seasonal variation in vector abundance, can also lead to hysteresis, whereby the inertial response of the system would effectively keep it trapped longer in either the endemic or disease-free state.

There is some (theoretical) evidence for the existence of alternative steady-states in infectious disease dynamics [3, 6, 7]. Here, we decribe one potentially important pre-condition for the existence of alternative steady states in malaria that stems from the complex immune response of the host to a highly diverse pathogen, the *Plasmodium* parasite. Humans are infected by concurrent multiple strains of the pathogen (*superinfection*). As as consequence, malaria infections are not fully immunizing, and multiplicity of infection is common in endemic regions. Under these conditions, rates of full recovery slow down. As a consequence, significant levels of superinfection create a positive feedback between infecting mosquitoes, which increase as humans remain infected longer, and disease prevalence, which also increases at the exposure to infecting mosquitoes increases. This loop has the potential to generate multiple alternative equilibria and associated tipping points.

We provide a model formulation of superinfection that explicitly allows infections to occur concurrently without interfering with each other. In addition, we present a semi-analytical, but general approach to identify alternative equilibria in models for vector-transmitted diseases. We then apply these methods to a vector-borne disease model (SECIR-LXVW) that has been successfully used to understand the origins of environmentally driven fluctuations of malaria, and the potential impact of increasing temperatures, in epidemic regions [1]. We demonstrate that irrespective of the details, superinfection consistently creates tipping points that can generate hysteresis in responses to control efforts (as well as seasonal variation in vector abundance). We argue that complex malaria immunity underlies abrupt transitions in response to control strategies or slight environmental variation. Models that fail to consider the complexity of malaria-induced immunity response may be misleading, and, therefore, their utility in practice is very limited when used to examine transitions towards low prevalence levels in response to different control strategies affecting the vectors as well as the pathogen.

2 The Model

The model can be considering an extension of the standard Ross–McDonald model [8–11]. Details on model formulation and parameter definitions (including biologically reasonable parameter value ranges) are found in Alonso, Bouma and Pascual [1]). The model considers the dynamics of both humans and mosquitoes populations by means of two sub-models (Eqs. 1 and 2) that are coupled through mosquito bites; see Fig. 1b. The full ODE system can be written as:

$$\begin{aligned}
\frac{dS}{dt} &= f_H\, N - \beta\, S + \sigma R - \delta\, S + \rho\, C \\
\frac{dE}{dt} &= \beta\, S - \delta\, E - \gamma\, E \\
\frac{dI}{dt} &= (1-\chi)\,\gamma\, E - \eta\,\beta\, I + \nu\, C - r\, I - \Psi\, I - \delta\, I \\
\frac{dR}{dt} &= -\sigma R + r\, I - \delta\, R \\
\frac{dC}{dt} &= \chi\,\gamma\, E + \eta\,\beta\, I - \nu\, C - \rho\, C - \alpha\, C - \delta\, C
\end{aligned} \tag{1}$$

$$\begin{aligned}
\frac{dL}{dt} &= f\,(X + V + W)\left(\frac{K_0 - L}{K_0}\right) - \delta_L\, L - d_L\, L \\
\frac{dX}{dt} &= -c\, a\, y\, X - \delta_M\, X + d_L\, L \\
\frac{dV}{dt} &= +c\, a\, y\, X - \gamma_V\, V - \delta_M\, V \\
\frac{dW}{dt} &= \gamma_V\, V - \delta_M\, W,
\end{aligned} \tag{2}$$

where N is the total human population, which is assumed constant ($f_H = \delta$), y is the fraction of infectious humans ($y = (C + I)/N$) or disease prevalence, β is the *per capita* rate of disease acquisition by humans through infectious bites ($\beta = a\, b\, W/N$), and r is a function of the number of infectious mosquitoes—see Eq. (3). This is a key point of our formulation: the way effective *per capita* recovery rates, r, behave as transmission intensity (the rate of infectious bites per human) change. Under the assumptions that (1) infectious bites arrive at a constant Poissonian rate, (2) the individual infections within a host progress independently, and (3) last a constant

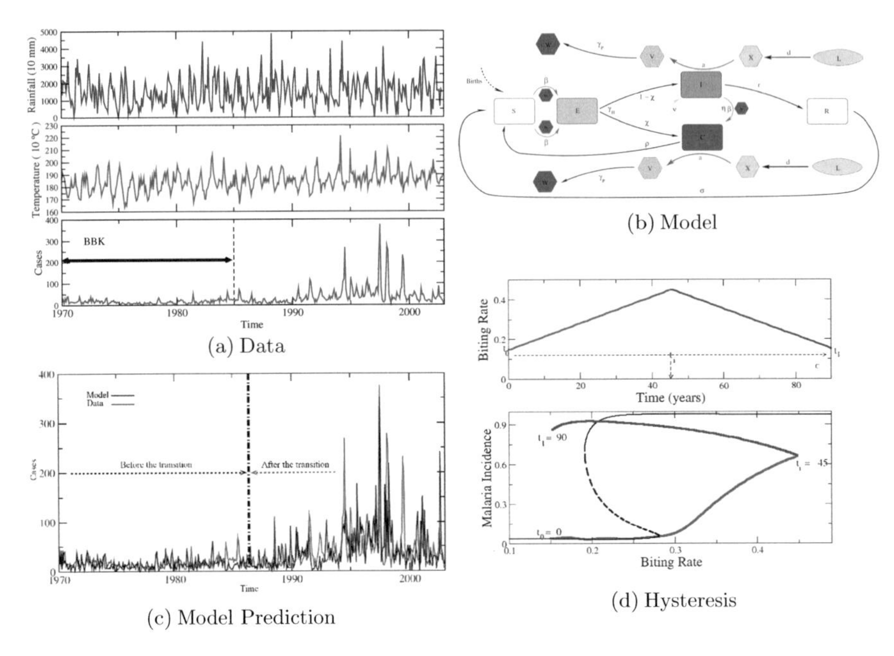

Fig. 1 The human-mosquito SECIR-LXVW coupled model. Fluctuations in rainfall and temperature induce variability in malaria cases **a** through the dynamics of disease transmission represented in the model **b** including response functions that map temperature and rainfall onto certain mosquito model parameters. Model predictions capture real variability in cases including an abrupt shift to higher variability in the 2nd half of the time series (**c**) [1]. The model also predicts hysteresis (**d**)

period, $1/r_0$, Dietz, Molineax, and Thomas [4] derived the following expression for the effective *per capita* recovery rate,

$$r(\Lambda) = \frac{\Lambda}{\exp(\Lambda/r_0) - 1}, \tag{3}$$

where Λ denotes the rate of total infectious bites per human ($\Lambda = a\,W/N$), and r_0, is the basal recovery rate when disease transmission is very low (more precisely, in the limit of the infectious mosquito population tending to zero). Thus, the higher the rate of infectious bites per human host, Λ, the slower the disease clearance rate, and, therefore, the longer humans remain infectious. In vector-borne diesase models, the Λ parameter is usually measured per year, and called the entomological inoculation rate (EIR).

3 Results

3.1 Saddle-Point Bifurcation

Stationary points of the coupled system are identifed by a semi-analytical method that consists of first finding the equilibria of the two submodels separately, this is, first, finding the expression for the number of infectious mosquitoes as a function of a given fraction of infectious humans (according to the mosquito submodel), and, second, the expression for the fraction of infectious humans for a given number of infecting mosquitoes (according to the human submodel). The fixed points should be, therefore, defined by the intersection of these two curves; see Fig. 2. The generality and feasibility of this method relies on the linearity of the human and mosquito submodels when considered separately. This means that both the human submodel (for a given number of infectious mosquitoes, $W^{\star}$), and the mosquito submodel (for a given fraction of infectious humans, y) are linear ODE systems.

Figure 2 shows that the intersection of the curves can produce more that one fixed point. As the biting rate a increases, the system undergoes two bifurcations. The first one corresponds to a transcritical bifurcation [5], and represents the transition from a free-disease situation ($R_0 < 1$) to an endemic stable equilibrium ($R_0 > 0$). The second one corresponds to a saddle node bifurcation (also called a tangential or fold bifurcation). The tangential intersection of the two curves defines a critical biting rate ($a_C = 0.19089$). For $a > a_C$, there is the sudden appearance of a pair of resting points, a saddle node and a second stable point with a higher fraction

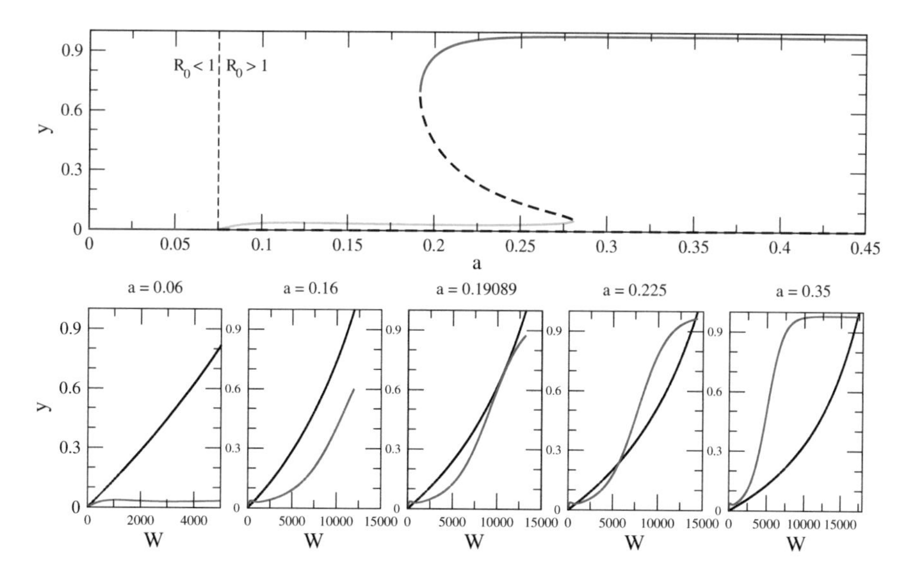

Fig. 2 Saddle-node bifurcation and mosquito biting rates

of total infectious humans. The first stable point corresponding to a lower disease prevalence remains. As a result, two basins of attraction coexist, each consisting of initial conditions that lead to one of the two alternative stable states, separated by the existence, of an intermediate unstable state.

3.2 Hysteresis

Coexistence of stable equilibria give rise to a hysteretic behavior. When an external perturbation is applied through a gradual increase of a model parameter (for instance, the biting rate, a), the system responds with an abrupt, non-linear increase in disease prevalence. However, the symmetric gradual decrease of the same parameter is unable to drive the system back to the initial disease incidence levels. This involves an asymmetry in the temporal trajectories from endemicity to elimination, and from elimination to re-emergence. These hysteresis effects are illustrated in Fig. 1d. Although decreasing a back to its initial low values would eventually lead the system to settle down at the initial low incidence equilibrium, the transient trajectory to this state can take very long.

4 Conclusion

Our work demonstrates that inclusion of superinfection in malaria models, not only determines the lengthening of infectious periods [4], but is a key factor responsible for the coexistence of multiple stationary states, and the possibility of nonlinear regime shifts, including hysteresis. This has important implications [2]. Small changes in parameters (for instance, biting rate a or mosquitoes' carrying capacity K) can give rise to large changes in disease incidence. Control efforts may see no progressive decrease of incidence until a sudden effect finally occurs. Conversely, the progressive relaxation of control efforts in endemic regions could generate sudden transitions from low to high incidence. Finally, concerning variability, as it is conjectured in Fig. 1c, since mosquito vital rates critically respond to temperature, sudden shifts from low to large fluctuations in incidence may follow in epidemic regions as average temperatures slowly increase due to global warming.

References

1. D. Alonso, M.J. Bouma, M. Pascual, Epidemic malaria and warmer temperatures in recent decades in an east african highland. Proc. R. Soc. Biol. Sci. **278**, 1661–1669 (2011)
2. D. Alonso, A. Dobson, M. Pascual, Critical transitions in malaria transmission models are consistently generated by superinfection. Philos. Trans. R. Soc. Lond. B Biol. Sci. **374**, 20180275 (2019)

3. N. Chitnis, J.M. Cushing, J.M. Hyman, Bifurcation analysis of a mathematical model for malaria transmission. SIAM J. Appl. Math. **67**, 24–45 (2006)
4. K. Dietz, L. Molineax, A. Thomas, Malaria model tested in african savannah. Bull. World Health Organ. **50**, 347–357 (1974)
5. G. Iooss, D.D. Joseph, *Elementary Stability and Bifurcation Theory* (Springer, New York, 1980)
6. J.S. Lavine, A.A. King, V. Andreasen, O.N. Bjørnstad, Immune boosting explains regime-shifts in prevaccine-era pertussis dynamics. PLoS ONE **8**, e72086 (2013)
7. J.S. Lavine, A.A. King, O.N. Bjørnstad, Natural immune boosting in pertussis dynamics and the potential for long-term vaccine failure. Proc. Natl. Acad. Sci. USA. **17**, 7259–7264 (2011)
8. G. MacDonald, Ross's a priori Pathometry—a Perspective. Trop. Dis. Bull. **49**, 813–829 (1952)
9. G. MacDonald, *The Epidemiology and Control of Malaria* (Oxford University Press, 1957)
10. R. Ross, *The Prevention of Malaria*, ed. by J. Murray (1910)
11. R. Ross, An application of the theory of probabilities to the study of a priori pathometry. Proc. R. Soc. Lond. Ser. A **92**, 204–230 (1916)
12. M. Scheffer, *Critical Transition in Nature and Society* (Princeton University Press, 2009)

A Mathematical Model of Cancer Evolution

David Moreno Martos and Andrei Korobeinikov

Abstract Cancer appears as a result of mutations and, due to a very high mutability of cancer cells, undergoes a continuous evolution through its development. In this note, we introduce a mathematical model of the cancer initial development. The model describes the cancer evolution in a continuous variant space. It is based upon the logistic population growth model (the Lotka–Volterra model of competing populations) and includes a possibility for cell mutations and competition of different genotypes for limited resources. Numerical simulations are performed to demonstrate the model suitability.

1 Model

Cancer initially appears as a result of a mutation of a normal cell. Subsequently, this mutant genotype undergoes through a series of mutations. However, in either case, to form into a cancer, the mutant genotype has to fix and successfully reproduce in the cell population. Thus, the mutant genotypes have to successfully compete against normal cells for the available resources, such as oxygen, nutrients or, simply, space. The majority of the cancer development models in the literature reflect this fact [7]. Due to a very high mutagenicity of cancer cells, mutations of cancer genotypes continue further leading to the appearance of a very wide diversity of genotypes simultaneously present in a tumor.

We start with an assumption that a multitude of genotypes exist and can be present in a system. Each genotype forms a subpopulation, whose behavior is governed

A. Korobeinikov is supported by Ministerio de Economía y Competitividad of Spain via grant MTM2015-71509-C2-1-R.

D. M. Martos (✉)
Departament de Matemàtiques, Universitat Autònoma de Barcelona, Barcelona, Spain
e-mail: damormar95@gmail.com

A. Korobeinikov
Centre de Recerca Matemàtica, Campus de Bellaterra Edifici C, 08193 Barcelona, Spain
e-mail: akorobeinikov@crm.cat

A. Korobeinikov et al. (eds.), *Extended Abstracts Spring 2018*,
Trends in Mathematics 11, https://doi.org/10.1007/978-3-030-25261-8_12

by the logistic population growth model (the Lotka–Volterra model of competing populations). However, in order to allow the possibility of random mutations, we have to make some modifications in this model.

Let us assume the existence of n different malicious cell genotypes. We are interested in finding the current concentration of each of these genotypes. The concentration of the i-th genotype can be defined as the number of the cells of this genotype per a given volume and denoted by $c_i(t)$. (Here, we assume that the cell distribution is spatially homogeneous.) For a population that comprises a single genotype, the current concentration of this genotype $c(t)$ can be described by the logistic population growth model.

$$\frac{dc(t)}{dt} = Bc(t)\,(1 - h\,bc_1(t)) - Dc(t)\,(1 + g\,bc(t)) = Ac(t)(1 - bc(t)). \quad (1)$$

Here, A, B and D are the per capita growth, birth, and death rates, respectively, in the most favorable conditions, when the concentration $c(t)$ is small; $b = 1/K$, where K is the carrying capacity of the environment; and h and g are positive constants (the weights), which reflect the impacts of the shortage of the resources on the birth and death rates. It is obvious that the equalities $B + D = A$ and $Bh + Dg = A$ hold.

Likewise, for a system composed of n different genotypes, assuming the logistic growth of these genotypes and the competition among them for the limited resources, we can write the following ODE system for the concentrations $c_i(t)$, where $i = 1, 2, \ldots, n$,

$$\frac{dc_i}{dt} = B_i c_i \left(1 - h_i \sum_{k-1}^{n} b_{ik} c_k\right) - D_i c_i \left(1 + g_i \sum_{k=1}^{n} b_{ik} c_k\right). \quad (2)$$

Here, $b_{ii} = 1/K_i$ and b_{ij} reflect the comparative competitive abilities of the genotypes; see [7]. All the parameters in Eq. (2) are nonnegative.

Model (2) does not include a possibility of mutations and, therefore, cannot be used to describe evolution. Following the concept suggested in [5], we assume that random mutations occur in the process of cell division and that the death rates are not affected by mutations. That is, with probability p_{ji}, a cell of the j-th genotype produces a daughter cell of the i-th genotype. Then, the system of ordinary differential equations can be rewritten as the following:

$$\frac{dc_i}{dt} = \sum_{j=1}^{n} B_j p_{ji} c_j \left(1 - h_j \sum_{k=1}^{n} b_{jk} c_k\right) - D_i c_i \left(1 + g_i \sum_{k=1}^{n} b_{ik} c_k\right), \quad (3)$$

$i = 1, \ldots, n$.

The diversity of genotypes in a tumor is usually very large. Moreover, phenotypes of the cells of the same genotype are never identical. Therefore, it is reasonable to assume that in a population (in a tumor, in this case) the phenotypic traits (which are described by the constants in the model) follow a continuous distribution. Then one can replace the n ordinary differential equations in model (3) by a single integro-partial differential equation. In order to do this, following [1–6, 8], one has to assume that function $c(s, t)$, where $s \in \Omega$ is a variable related to are genotype and Ω is the variant space, describes density distribution of the cell by their variants (or "types" in a more general sense). The model parameters are now functions of s as well. Then, the cell density distribution can be represented by the following integro-partial differential equation:

$$\frac{\partial c(s,t)}{\partial t} = \int_{r\in\Omega} B(r)p(r,s)c(r,t)\left(1 - h(r)\int_{q\in\Omega} b(r,q)c(q,t)dq\right)dr - \\ -D(s)c(s,t)\left(1 + g(s)\int_{q\in\Omega} b(s,q)c(q,t)dq\right). \quad (4)$$

The dimension of variant space Ω is to be determined. Typically (see [1–6, 8]), the dimension of the variant space is equal to the number of independent phenotypical traits, but it can be reduced assuming that some of these are variant independent.

2 Numerical Simulations

For simplicity, for numerical simulations with model (4), we choose $\Omega = [0, 1]$. Also for simplicity, we set $h(s) = 0$: that is, we assume that the competition does not affect the birth rate. We set g to be variant independent. Then, $g(s) = (B - D)/D$, where $B - D > 0$. For the competing capabilities we postulate $b(s, r) = 1 + \beta(r - s)$, where β is constant. The density probability $p(r, s)$ is defined by the triangle function

$$p(r,s) = \begin{cases} -\frac{1}{\sigma^2}|r - s| + \frac{1}{\sigma} & \text{if } r \in [s - \sigma, s + \sigma], \\ 0 & \text{otherwise,} \end{cases}$$

where the parameter σ is to be defined.

We have to define initial and boundary conditions. For simplicity, we impose the no-flux boundary conditions at the both ends of the interval [0, 1]. As initial conditions, we usually used a small-magnitude narrow triangular distribution near a point of Ω.

2.1 *Changing the Competition Coefficient*

Figure 1 illustrates results of numerical simulations for $\tilde{\beta} = 0.5$ and 0.9 and for variant independent B and D. (Here $\tilde{\beta}$ is the nondimensional analogous of parameter β.) In this figure, one can see contour plots of the nondimensional concentration $x(s, \tau)$. The colors are related to the levels of $x(s, \tau)$ as is indicated in the color bar next to the graphics.

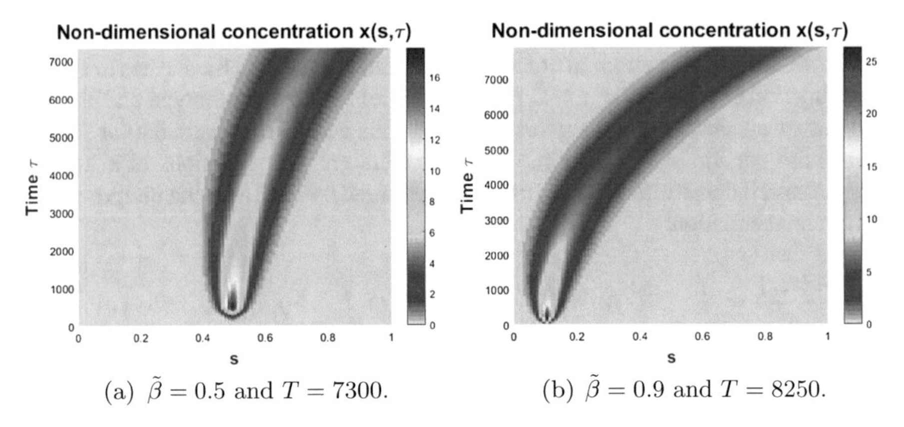

(a) $\tilde{\beta} = 0.5$ and $T = 7300$.

(b) $\tilde{\beta} = 0.9$ and $T = 8250$.

Fig. 1 Density distribution in the variant space Ω for $\tilde{\beta} = 0.5$ and 0.9. The density of variants is represented by colors (see the color bar at the right hand side of the graphic). Here, $\sigma = 0.01$, $B = 1$, and $D = 0.1$. Please note formation of a traveling wave of evolution in the space Ω

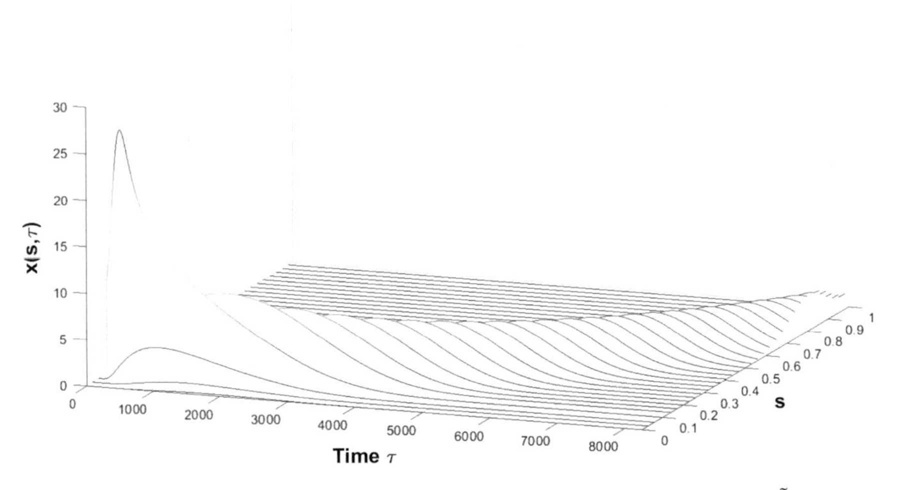

Fig. 2 Traveling wave of evolution in space Ω. Here, $\sigma = 0.01$, $B = 1$, $D = 0.1$ and $\tilde{\beta} = 0.9$

It is easy to see that the higher value $\tilde{\beta}$ is, the faster the system evolves. However, the qualitative behavior in the situations is the same: namely, the Darwinian fitness of the population grows.

Figure 2 is a waterfall plot that illustrates the formation of a traveling wave of evolution in Ω. For this figure we used initial conditions in the form of the triangular function with base $s = 0.02$ and height 100 centered at $s = 0.1$. This corresponds to the initial predominance of the normal cells in the population and ensures that the initial total population is equal to 1.

We also would like to note that the simulation demonstrates that variant dependence of the nondimensional birth and death rates does not make any change in the system dynamics. This is hardly surprising because B and D are mostly responsible for the time scales.

References

1. A.A. Archibasov, A. Korobeinikov, V.A. Sobolev, Asymptotic expansions of solutions for a singularly perturbed model of viral evolution. Comput. Math. Math. Phys. **55**(2), 240–250 (2015)
2. A.A. Archibasov, A. Korobeinikov, V.A. Sobolev, Passage to the limits in a singularly perturbed partial integro-differential system. Differ. Equ. **52**(9), 1115–1122 (2016)
3. A. Korobeinikov, Immune response and within-host viral evolution: immune response can accelerate evolution. J. Theor. Biol. **456**, 74–83 (2018)
4. A. Korobeinikov, A. Archibasov, V. Sobolev, Multi-scale problem in the model of RNA virus evolution. J. Phys. Conf. Ser. **727**, 012007 (2016)
5. A. Korobeinikov, K.E. Starkov, P.A. Valle, Modeling cancer evolution. J. Phys. Conf. Ser. **811**, 012004 (2017)
6. D. Masip, A. Korobeinikov, A continuous phenotype space model of cancer evolution. J. Phys. Conf. Ser. **811**, 012005 (2017)
7. J.D. Murray, *Mathematical Biology: I. An Introduction*, 3rd edn. (Springer, NY, 2002)
8. S. Pagliarini, A. Korobeinikov, A mathematical model of marine bacteriophage evolution. R. Soc. Open Sci. **5**, 171661 (2018)

Resonance of Isochronous Oscillators

David Rojas

Abstract An oscillator such that all motions have the same minimal period is called isochronous. When the isochronous is forced by a time-dependent perturbation with the same natural frequency as the oscillator the phenomenon of resonance can appear. This fact is well understood for the harmonic oscillator and we extend it to the nonlinear scenario.

1 Introduction

In this communication, we present some results from [4] that aim to characterize the class of periodic forcings producing resonances in nonlinear isochronous oscillators.

A well-known fact from physics and mathematics is that the harmonic oscillator with period 2π perturbed by a periodic forcing

$$\ddot{x} + n^2 x = p(t),$$

$n = 1, 2, \ldots$, exhibits resonance whenever the Fourier coefficient

$$\hat{p}_n := \frac{1}{2\pi} \int_0^{2\pi} p(t) e^{-int} \, dt$$

does not vanish. In this context, resonance means that all solutions of the perturbed equation are unbounded. After this example, the question that naturally arises is if there exists an equivalent condition for general nonlinear isochronous oscillators. As far as we know, this question was first raised by Prof. Roussarie in the Open Problems Session of the II Symposium on Planar Vector Fields (Lleida, 2000).

In this direction, Ortega [3] proved that if the nonlinear isochronous oscillator satisfies a Lipschitz condition then there exist functions $p(t)$ producing resonance.

D. Rojas (✉)
Departament D'Enginyeria Informàtica I Matemàtiques, Universitat Rovira I Virgili, Tarragona, Spain
e-mail: david.rojas@urv.cat

A. Korobeinikov et al. (eds.), *Extended Abstracts Spring 2018*, Trends in Mathematics 11, https://doi.org/10.1007/978-3-030-25261-8_13

Also, Bonheure et al. [2] give concrete examples of perturbations. Our contribution in [4] may be interpreted as a nonlinear version of condition $\hat{p}_n \neq 0$.

2 Statement of the Results

Consider the oscillator

$$\ddot{x} + V'(x) = 0, \tag{1}$$

$x \in \mathbb{R}$, where $V \in C^2(\mathbb{R})$ is a potential defined on the whole real line satisfying $V(0) = 0$, $xV'(x) > 0$ if $x \neq 0$, and such that all its solutions are 2π-periodic. The purpose of the following results is to identify the class of 2π-periodic perturbations $p(t)$ such that all the solutions of the nonautonomous equation

$$\ddot{x} + V'(x) = \varepsilon p(t) \tag{2}$$

are unbounded for $\varepsilon \neq 0$ small. More precisely, we say that the equation is resonant if every solution $x(t)$ of (2) satisfies

$$\lim_{|t| \to +\infty} (|x(t)| + |\dot{x}(t)|) = +\infty.$$

Let us denote by $\mathcal{C} := (\mathbb{R}/2\pi\mathbb{Z}) \times [0, \infty)$ the cylinder with coordinates (θ, r). The analogous function that plays the role of the Fourier coefficient $\hat{p}_n$ in the nonlinear case is given by the function $\Phi_p : \mathcal{C} \to \mathbb{C}$ defined by

$$\Phi_p(\theta, r) := \frac{1}{2\pi} \int_0^{2\pi} p(t - \theta)\psi(t, r)dt,$$

where $\psi(t, r)$ is the complex-valued solution of the variational equation

$$\ddot{y} + V''(\varphi(t, r))y = 0, \ \ y(0) = 1, \ \ \dot{y}(0) = i,$$

and $\varphi(t, r)$ denotes the solution of system (1) with initial data $x(0) = r$ and $\dot{x}(0) = 0$.

According to [4, Theorem A], if V satisfies the previous conditions and V'' is bounded over the reals then Eq. (2) is resonant for small $\epsilon \neq 0$ for any $p \in L^1(\mathbb{T})$ satisfying the condition

$$\inf_{\mathcal{C}} |\Phi_p(\theta, r)| > 0. \tag{3}$$

This result is a sufficient condition for resonance, but in fact condition (3) is not too far from being also necessary. Under the same assumptions on V, [4, Proposition 2.2] shows that if Φ_p has a nondegenerate zero (θ_*, r_*) with $r_* > 0$ then system (2) has a 2π-periodic solution for small $\epsilon \neq 0$. In particular, resonance is excluded in this situation.

These two results motivate the choice of condition (3) as the nonlinear version of $\hat{p}_n \neq 0$ for oscillators defined in the whole real line. Indeed, for the linear oscillator $V(x) = \frac{1}{2}n^2x^2$, $n = 1, 2, \ldots$, elementary computations lead to the estimates

$$\frac{1}{2\pi n}|\hat{p}_n| \leq |\Phi_p(\theta, r)| \leq \frac{1}{2\pi}|\hat{p}_n|,$$

which show the equivalence between the condition $\hat{p}_n \neq 0$ and (3). However, there are also isochronous oscillators having a singularity. This is the case of the well-known isochronous center

$$\ddot{x} + \frac{1}{4}\left(x + 1 - \frac{1}{(x+1)^3}\right) = 0,$$

defined for all $x \in (-1, +\infty)$, solved explicitly by Pinney [5]. Bonheure et al. [2] considered the perturbed equation

$$\ddot{x} + \frac{1}{4}\left(x + 1 - \frac{1}{(x+1)^3}\right) = \varepsilon \sin t \tag{4}$$

and proved that all solutions are unbounded for $\varepsilon \neq 0$ small enough. Our contribution in this scenario is an analogous version of the sufficient condition theorem for resonance. In this case, [4, Theorem B] proves that if $p \in L^1(\mathbb{T})$ satisfies condition (3) then all solutions of the equation

$$\ddot{x} + \frac{1}{4}\left(x + 1 - \frac{1}{(x+1)^3}\right) = \varepsilon p(t) \tag{5}$$

are unbounded for $\varepsilon \neq 0$ small enough.

Although this result is stated for Pinney equation, the same proof can be extended to a larger class of potentials V having a singularity. Indeed, Bonheure et al. [2] observed that usually the existence of a singularity of the potential at $x = a$, $a < 0$, determines the behavior of V at infinity. This behavior is precisely the key on the proof of the Theorem. We refer to [4, App.] for more details.

The computation of the resonance condition (3) may be difficult in general. In this case, thanks to the contribution of Pinney [5], we are able to compute Φ_p explicitly for the class of linear trigonometric forcings $p(t) = a_0 + a_1 \cos t + b_1 \sin t$. Applying [4, Theorem B] we obtain that Eq. (5) is resonant if $a_1^2 + b_1^2 > 9a_0^2$. In particular, we recover the result for (4) in [2].

Motivated by mechanical oscillators, all the perturbations taken into account up to now have been of additive type but in general other kind of perturbations may appear. Inspired by a problem from geometry, Ai–Chou–Wei [1] studied the equation

$$\ddot{x} + x = \frac{R(t)}{x^3}, \tag{6}$$

$x > 0$, where $R(t)$ is a T-periodic function, and proved existence of T-periodic solutions when R is a positive C^2-function and $T < \pi$. The previous equation with $R \equiv 1$ turns out to be isochronous with minimal period π. This fact suggests that condition $T < \pi$ seems to be sharp due to the presence of resonance. [4, Theorem C] is in the direction of proving this fact, showing that the π-periodic function

$$R(t) = \begin{cases} 1 & \text{if } t \in [0, \frac{\pi}{2}), \\ c & \text{if } t \in [\frac{\pi}{2}, \pi), \end{cases}$$

with $c > 0$ produces that all solutions of (6) are unbounded if $c \neq 1$.

3 Open Problems

We end this contribution with some related problems that remain unsolved.

First, both the results concerning the identification of the forcings producing resonance for nonlinear isochronous oscillators defined in the whole plane, we have presented and the construction of examples by Ortega [3] require the oscillator to be Lipschitz continuous. However, this requirement seems to be a technicality not intrinsically linked to the problem itself but to the proof. We expect that no specific regularity properties of the potential are needed to produce resonance or at least weaker properties than Lipschitz continuity.

Second, we give a sufficient condition of resonance for the Eq. (5) perturbed by a linear trigonometric function. Based on the fact that the condition (3) seems also close to be necessary for resonance, it would be interesting to study if Eq. (5) have periodic orbits for linear trigonometric forcings satisfying $a_1^2 + b_1^2 \leq 9a_0^2$.

Third, the example $R(t)$ we have given subscribes the idea that Eq. (6) exhibits resonance if R is π-periodic, but it is discontinuous. We think that smooth examples can also be constructed but the approach in [4] do not apply in this situation.

Finally, the results presented deal with nonlinear isochoronous oscillators with one degree of freedom. In more degrees of freedom, the notion of isochronicity is strongly related with the notion of superintegrability, at least in the Hamiltonian framework. It would be interesting to relate properly superintegrable Hamiltonian systems with isochronicity and to construct resonance of such systems.

References

1. J. Ai, K.S. Chou, J. Wei, Self-similar solutions for the anisotropic affine curve shortening problem. Calc. Var. Partial Differ. Equations **13**, 311–337 (2001)
2. D. Bonheure, C. Fabry, D. Smets, Periodic solutions of forced isochronous oscillators at resonance. Discrete Contin. Dyn. Syst. **8**, 907–930 (2002)

3. R. Ortega, Periodic perturbations of an isochronous center. Qual. Theory Dyn. Syst. **3**, 83–91 (2002)
4. R. Ortega, D. Rojas, Periodic oscillators, isochronous centers and resonance. Nonlinearity **32**, 800–832 (2019)
5. E. Pinney, The nonlinear differential equation $y'' + p(x)y + cy^{-3} = 0$. Proc. Amer. Math. Soc. **1**, 681 (1950)

Modeling of N-Methyl-D-Aspartate Receptors

Denis Shchepakin, Leonid Kalachev and Michael Kavanaugh

Abstract Several reduced kinetic models for N-methyl-D-aspartate receptors were derived in order to suit different experimental protocols. Their simultaneous application allows for a step-wise estimation of parameters of a conventional model that is otherwise overparameterized with respect to the existing data.

1 Introduction

One of the major subtypes of glutamate receptors on neurons is the N-methyl-D-aspartate receptor (NMDAR). The receptor plays critical role in neural plasticity, development, learning, and memory. Disrupted function is associated with disorders including epilepsy, depression, schizophrenia, ischemic brain injury, and others. NMDARs have been targets of numerous studies, and several models have been proposed and published over the last two decades to explain the dynamics of the currents mediated by the NMDAR ion channel, e.g., [1, 2]. However, conclusions about receptor kinetics based on these Markov models are typically limited by model overparameterization with respect to the available data. Such obstacles cannot be resolved by switching fitting methods; rather the model and the experiments must be adjusted to be in line with each other. In this work, we design the experiments alongside with model development to resolve this issue of overparameterization.

D. Shchepakin (✉) · L. Kalachev
Department of Mathematical Sciences, University of Montana, Missoula, MT, USA
e-mail: denis.shchepakin@umconnect.umt.edu

L. Kalachev
e-mail: kalachevl@mso.umt.edu

D. Shchepakin · M. Kavanaugh
Department of Biomedical Pharmaceutical Sciences, University of Montana, Missoula, MT, USA
e-mail: michael.kavanaugh@mso.umt.edu

A. Korobeinikov et al. (eds.), *Extended Abstracts Spring 2018*,
Trends in Mathematics 11, https://doi.org/10.1007/978-3-030-25261-8_14

2 NMDAR Desensitization

For common NMDARs are commonly heterotetramers composed of two NR1- and two NR2-subunits [2]. For NMDARs to signal by ion channel opening, they must bind glutamate at each of two NR2 subunits as well as co-agonist (either D-serine or glycine) at each of two NR1-type subunits. In response to prolonged agonist pulses, NMDARs desensitize: a process in which the response amplitude decays over time. The desensitization effect increases and speeds up in the presence of limiting co-agonist [1]. This phenomenon could potentially be explained by different mechanisms or a combination of mechanisms. One possibility is that co-agonist already bound to the NMDAR could experience a reduction in affinity following glutamate binding ("glycine (or D-serine)–dependent desensitization"; see [1]). Alternatively, the effect of co-agonist concentration on desensitization may not depend on agonist–co-agonist site interactions: upon binding all four molecules, some fraction of the receptors transfers into a long-lived nonconductive state instead [3]. The general chemical kinetic model of the process is shown in Fig. 1. Estimating the reaction rates for state transitions reflecting these processes will answer the question of the nature of NMDAR desensitization.

3 Modeling and Experiments

The activation of NMDAR receptors by agonist and co-agonist binding facilitates flow of ions across the cell membrane and this can be recorded as a current in a patch clamp experiment. Piezoelectric switching of solutions bathing an excised outside-out patch allows for fast agonist application. During the experiment, one substrate is chronically present while other one is supplied in a short pulse manner. Varying

$$R \underset{K_1}{\overset{S}{\rightleftharpoons}} RS \underset{K_3}{\overset{S}{\rightleftharpoons}} RS_2$$

$$R \overset{G}{\underset{K_2}{\rightleftharpoons}} RG, \quad RS \overset{G}{\underset{K_4}{\rightleftharpoons}} GRS, \quad RS_2 \overset{G}{\underset{K_6}{\rightleftharpoons}} GRS_2$$

$$RG \underset{K_5}{\overset{S}{\rightleftharpoons}} GRS \underset{K_7}{\overset{S}{\rightleftharpoons}} GRS_2$$

$$RG \overset{G}{\underset{K_8}{\rightleftharpoons}} RG_2, \quad GRS \overset{G}{\underset{K_{10}}{\rightleftharpoons}} G_2RS, \quad GRS_2 \overset{G}{\underset{K_{12}}{\rightleftharpoons}} G_2RS_2$$

$$RG_2 \underset{K_9}{\overset{S}{\rightleftharpoons}} G_2RS \underset{K_{11}}{\overset{S}{\rightleftharpoons}} G_2RS_2 \underset{K_{14}}{\rightleftharpoons} G_2R^*S_2$$

$$G_2RS_2 \underset{K_{13}}{\rightleftharpoons} G_2R'S_2$$

Fig. 1 General model of NMDA receptor with two binding sites for L-glutamate and D-serine agonists. R denotes the receptor, S denotes D-serine, and G denotes L-glutamate. $G_2R'S_2$ is a long-lived nonconductive state and $G_2R^*S_2$ is a conductive state. Each K_i is an equilibrium constant for the corresponding reaction: $K_i = k_i^+/k_i^-$, where k_i^+ and k_i^- are the forward and reverse reaction rate constants, respectively

the concentrations of D-serine and L-glutamate from saturating to relatively low, we can accelerate or slow down reactions in particular directions. Applying Boundary Function Method [4] to the model depicted in the Fig. 1 will yield different results for different scenarios. Here, we show that the appropriate choice of an experimental design allows for a reliable step-wise parameter estimation: the resulting models have different subsets of parameters of the original model, and some parameter estimates obtained in one experiment can be used in the other ones.

The model that correspond to the chemical kinetic scheme depicted in Fig. 1 is a system of differential equations with 11 variables, each corresponding to one state. Let us denote these variables using states notations, i.e., the variable $R(t)$ is a fraction of all receptors that are in R state, etc. Therefore, the sum of all variables is 1. For all reduced models, we denote the leading order approximations in the following manner:

$$\begin{aligned} R(t) &= \alpha(t) + O(\varepsilon_s), & G_2RS(t) &= y(t) + O(\varepsilon_s), \\ RG(t) &= \beta(t) + O(\varepsilon_s), & GRS_2(t) &= \zeta(t) + O(\varepsilon_s), \\ RG_2(t) &= \gamma(t) + O(\varepsilon_s), & G_2RS_2(t) &= z(t) + O(\varepsilon_s), \\ RS(t) &= \eta(t) + O(\varepsilon_s), & G_2R'S_2(t) &= z'(t) + O(\varepsilon_s), \\ RS_2(t) &= x(t) + O(\varepsilon_s), & G_2R^*S_2(t) &= z^*(t) + O(\varepsilon_s), \\ GRS(t) &= \theta(t) + O(\varepsilon_s), \end{aligned} \tag{1}$$

where ε is a small parameter, which is different for each experiment and is defined individually in each corresponding section. The current recorded during the experiments is directly proportional to the only conducting state $G_2R^*S_2$.

3.1 *Saturating Concentration of D-Serine*

In this experiment, we apply a saturating concentration of D-Serine and let the system reach the steady state before a short pulse of L-glutamate at a low concentration. The presence of saturating D-serine allows us to introduce a small parameter $0 < \varepsilon_s \ll 1$:

$$S \cdot k_i^+ = \frac{\widetilde{k_i^+}}{\varepsilon_s}, \quad \widetilde{k_i^+} \sim O(1),$$

$i = 1, 3, 5, 7, 9, 11$, where S is a D-Serine saturating concentration. For the leading order approximations of the functions (1), we obtain

$$\alpha(t) \equiv 0, \quad \eta(t) \equiv 0, \quad \beta(t) \equiv 0, \quad \theta(t) \equiv 0, \quad \gamma(t) \equiv 0, \quad \zeta(t) \equiv 0,$$

$$\begin{aligned}
\frac{dx}{dt} &= -k_6^+ Gx + k_6^- y,\\
\frac{dy}{dt} &= k_6^+ Gx - k_6^- y - k_{12}^+ Gy + k_{12}^- z,\\
\frac{dz}{dt} &= k_{12}^+ Gy - k_{12}^- z - k_{13}^+ z + k_{13}^- z' - k_{14}^+ z + k_{14}^- z^*,\\
\frac{dz_0'}{dt} &= k_{13}^+ z - k_{13}^- z',\\
\frac{dz_0^*}{dt} &= k_{14}^+ z - k_{14}^- z^*,
\end{aligned}$$

with initial conditions $x(0) = 1$, $y(0) = 0$, $z(0) = 0$, $z'(0) = 0$, and $z^*(0) = 0$.

3.2 Saturating Concentration of L-Glutamate

This case mirrors the experiment from the Sect. 3.1: we apply saturating concentration of L-glutamate and let the system reach the steady state before a short pulse of D-serine at a limiting concentration. We introduce a small parameter $0 < \varepsilon_g \ll 1$ in a similar manner:

$$G \cdot k_i^+ = \frac{\widetilde{k_i^+}}{\varepsilon_g}, \quad \widetilde{k_i^+} \sim O(1),$$

$i = 2, 4, 5, 8, 10, 12$, where G is a L-Glutamate saturating concentration. For the leading order approximations of the functions (1), we obtain

$$\alpha(t) \equiv 0, \quad \eta(t) \equiv 0, \quad x(t) \equiv 0, \quad \beta(t) \equiv 0, \quad \theta(t) \equiv 0, \quad y(t) \equiv 0,$$

$$\begin{aligned}
\frac{d\gamma}{dt} &= -k_9^+ S\gamma + k_9^- \zeta,\\
\frac{d\zeta}{dt} &= k_9^+ S\gamma - k_9^- \zeta_0 - k_{11}^+ S\zeta + k_{11}^- z,\\
\frac{dz}{dt} &= k_{11}^+ S\zeta - k_{11}^- z - k_{13}^+ z + k_{13}^- z' - k_{14}^+ z + k_{14}^- z^*,\\
\frac{dz'}{dt} &= k_{13}^+ z - k_{13}^- z',\\
\frac{dz^*}{dt} &= k_{14}^+ z - k_{14}^- z^*,
\end{aligned}$$

with initial conditions $\gamma(0) = 1$, $\zeta(0) = 0$, $z(0) = 0$, $z'(0) = 0$, and $z^*(0) = 0$.

3.3 Saturating Concentrations of both Substrates

In this experiment, the concentrations of both agonist and co-agonist are high, therefore, a small parameter $0 < \varepsilon_b \ll 1$ can be defined as

$$S \cdot k_i^+ = \frac{\widetilde{k_i^+}}{\varepsilon_b}, \quad \widetilde{k_i^+} \sim O(1),$$

$$G \cdot k_j^+ = \frac{\widetilde{k_j^+}}{\varepsilon_b}, \quad \widetilde{k_j^+} \sim O(1),$$

$i = 1, 3, 5, 7, 9, 11$, $j = 2, 4, 5, 8, 10, 12$. The functions (1) depend on which substance is always present and which is given in a short pulse manner. Let us consider the case when D-serine is applied continuously throughout the whole experiment with steps of L-glutamate. Then we have

$$\alpha(t) \equiv 0, \quad \eta(t) \equiv 0, \quad \beta(t) \equiv 0, \quad \theta(t) \equiv 0, \quad \gamma(t) \equiv 0, \quad \zeta(t) \equiv 0,$$

$$\begin{aligned}
\frac{dz_0}{dt} &= -k_{13}^+ z_0 + k_{13}^- z' - k_{14}^+ z_0 + k_{14}^- z^* \\
\frac{dz'}{dt} &= k_{13}^+ z_0 - k_{13}^- z', \\
\frac{dz^*}{dt} &= k_{14}^+ z_0 - k_{14}^- z^*,
\end{aligned}$$

with initial conditions $z_0(0) = 1$, $z'(0) = 0$, and $z^*(0) = 0$. And

$$\begin{aligned}
x(t) &= e^{-k_6^+ G t}, \\
y(t) &= \frac{k_6^+}{k_{12}^+ - k_6^+} \left(e^{-k_6^+ G t} - e^{-k_{12}^+ G t} \right), \\
z(t) &= z_0(t) + \frac{k_6^+ k_{12}^+}{k_{12}^+ - k_6^+} \left(\frac{1}{k_{12}^+} e^{-k_{12}^+ G t} - \frac{1}{k_6^+} e^{-k_6^+ G t} \right).
\end{aligned}$$

Let us also notice that after the L-Glutamate pulse stops, one should use the model from Sect. 3.1 with $G = 0$.

4 Conclusion

The experiments and the corresponding reduced models described in Sect. 3 can be used for the estimation of parameters of the full model depicted in Fig. 1 in a step-wise manner. The low number of parameters at each step helps to resolve the overparameterization issue. The estimates of reaction rates constants will give us the answer about the nature of NMDAR desensitization.

References

1. M. Benveniste, J. Clements, L. Vyklický, M.L. Mayer, A kinetic analysis of the modulation of N-methyl-D-aspartic acid receptors by glycine in mouse cultured hippocampal neurones. J Physiol. **428**, 333–357 (1990)
2. K.A. Cummings, G.K. Popescu, Glycine-dependent activation of NMDA receptors. J. Gen Physiol. **145**(6), 513–527 (2015)
3. R. Nahum-Levy, D. Lipinski, S. Shavit, M. Benveniste, Desensitization of NMDA receptor channels is modulated by glutamate agonists. Biophys J. **80**(5), 2152–2166 (2001)
4. A.B. Vasileva, V.F. Butuzov, L.V. Kalachev, The boundary function method for singularly perturbed problems. SIAM (1995)

Some Lessons from Two Simple Approaches to Model the Impact of Harvest Timing on Seasonal Populations

Eduardo Liz

Abstract Using two different discrete-time models for seasonal populations, we review the potential effects of harvesting in regard to population abundance, stability, and extinction, and we emphasize how these effects strongly depend on harvest timing. We also stress the fact that census timing is crucial, and populations should be measured as many times as a discrete event occurs during the year cycle.

1 Introduction

An important challenge in harvesting and management theory is predicting population responses to the removal of individuals. In seasonal populations, for which population growth is controlled by a combination of density-dependent processes during different periods of the annual cycle, the timing of harvesting not only influences the impact of captures on population abundance, but also can alter stability properties and extinction risk.

Recently, we have addressed this problem for discrete-time population models, assuming proportional harvesting and using two different approaches. Both lead to one-dimensional discrete dynamical systems, which makes the mathematical analysis simpler than with other approaches. The first model was introduced by Seno in 2008 [12], and assumes that there is a specific season during which population accumulates energy for reproduction. Harvesting is assumed to be a discrete event that can occur at any moment within the season. The second model follows the ideas introduced by Jonzén and Lundberg in 1999 [6] (see also [10]), and assumes that there are a breeding and a nonbreeding season in the annual cycle, and harvesting is considered as a new event that can take place after or before the breeding season.

In this note, we review some potential effects of harvesting, emphasizing how these effects strongly depend on harvest timing. Of particular interest is the *hydra effect,* defined as a paradoxical increase in population size in response to an increasing mor-

E. Liz (✉)
Departamento de Matemática Aplicada II, Universidad de Vigo, 36310 Vigo, Spain
e-mail: eliz@dma.uvigo.es

A. Korobeinikov et al. (eds.), *Extended Abstracts Spring 2018*,
Trends in Mathematics 11, https://doi.org/10.1007/978-3-030-25261-8_15

tality [1]. In a recent paper, McIntire–Juliano [9] found strong evidence that timing of mortality contributes to overcompensation and the hydra effect in mosquitoes.

We also stress the fact that census timing is crucial, and populations should be measured as many times as a discrete event occurs during the year cycle.

The paper is based on Refs. [3, 7], although it contains new insight concerning analogies and differences between the two mentioned models; the role of harvest timing in models with Allee effects; and how to choose harvest time based on different criteria, such as sustainability or maximizing the yield. Paper [3] has recently won the Bellman Prize, awarded every two years for the best paper published in the journal Mathematical Biosciences over the preceding two years [2].

2 Seno's Model

For many species, births occur in well-defined breeding seasons in the annual cycle, and their life stories are usually described by discrete-time models. We consider a seasonal population subject to some form of harvesting (e.g., fishing, hunting, control). The simplest models assume only two processes in each generation: reproduction and harvesting. If harvesting is proportional to population stock, so that a percentage γx of the population size x is harvested every year, the between-year dynamics is defined by the simple model

$$x_{n+1} = (1 - \gamma) f(x_n), \tag{1}$$

$n = 0, 1, 2, \ldots$, starting at an initial population size $x_0 > 0$. Here, x_n is the population size after n generations and f is the production function. For a review of some potential effects of harvesting in this simple model, we refer to [8] and references therein. One important remark is that even in this situation, census time is relevant for management decisions. Indeed, if population is censused after reproduction, the model reads

$$x_{n+1} = f((1 - \gamma)x_n), \tag{2}$$

$n = 0, 1, 2, \ldots$. Although Eqs. (1) and (2) are topologically conjugated, and therefore they exhibit the same dynamics, in case of overcompensatory dynamics (usually represented by a unimodal map f), Eq. (2) can exhibit hydra effects.

A manager who only measures population after reproduction and before harvesting, may have the impression that increasing harvesting intensity can be beneficial for population abundance, while if census takes place after harvesting, the message can be the opposite; see [5] for more details.

Since we are interested in the influence of harvest timing, we need to consider a more sophisticated model. Seno [12] suggested a model in which it is assumed that there is a prereproductive season of length T, and harvesting can take place at any moment $\tau = \theta T, 0 \leq \theta \leq 1$, during this season. The intraspecific density effect on reproduction is then divided in two parts: one depending on x_n, and the other on $(1-\gamma)x_n$. For a recruitment map f and proportional harvesting, Seno's model writes

$$x_{n+1} = \theta(1-\gamma)f(x_n) + (1-\theta)f((1-\gamma)x_n). \tag{3}$$

In mathematical terms, the right-hand side of (3) is a convex combination of the right-hand sides of models (1) and (2). Seno showed that, for stable overcompensatory populations, the earlier we harvest, the most the population size is increased. In [3], we extended the mathematical analysis to study how harvest time may determine the stability of the equilibrium, and to analyze models with Allee effect [4], where there is a higher risk of population extinction. In these models, overharvesting leads population to extinction when the equilibrium $x = 0$ becomes globally asymptotically stable after a tangent bifurcation occurs. The main conclusions from Seno's model are the following:

(i) In models without Allee effects:

(1) the critical value of γ leading to extinction does not depend on harvest time;
(2) later harvesting leads to a decrease in population size at the equilibrium, so early harvest is better for conservation purposes;
(3) in the presence of dynamical instabilities, intermediate harvest times can help to stabilize the positive equilibrium, sometimes preventing population extinction due to stochastic perturbations.

(ii) In the presence of Allee effects, mid-season harvesting can induce extinction.

In Fig. 1, we represent these effects for the Ricker map $f(x) = xe^{2.5(1-x)}$ (without Allee effects), and for the Ricker–Schreiber map $f(x) = 0.5x^2e^{4(1-x)}/(1+0.5x)$, which exhibits a strong Allee effect [11].

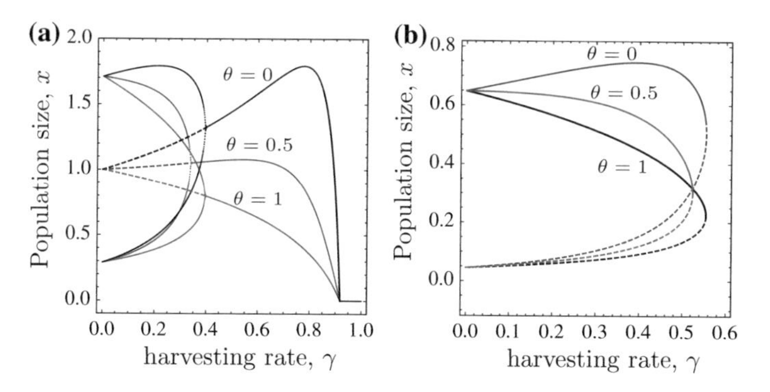

Fig. 1 Bifurcation diagrams for Seno's model (3) as harvesting rate γ is increased, with different harvest times. **a** For a Ricker map, the critical value of γ leading to extinction does not depend on θ, but intermediate values of θ help to stabilization; **b** For a Ricker–Schreiber map with Allee effect, intermediate values of θ may increase the risk of extinction. In both cases, a hydra effect is observed for early harvesting. Dashed lines correspond to unstable equilibria

3 The Seasonal Model of Jonzén and Lundberg

There are species subject to harvesting whose annual cycle is divided into a breeding season (typically, summer) and a nonbreeding season (typically, winter), and a harvesting season is only allowed either in spring or in autumn; see, e.g., [7]. For these populations, the main lesson that we could learn from Seno's model is that autumn harvest is preferable to increase average population abundance. Perhaps, the main drawback of Seno's model is that it predices similar qualitative behavior for very early and very late harvest within the season, because the cases $\theta = 0$ and $\theta = 1$ are equivalent in Eq. (3). Hence, in regard to stability and bifurcations, it would not show any difference between autumn and spring harvest; see Fig. 2.

The discrete-time Jonzén–Lundberg model [6, 7] consists of a composition of three discrete events during the annual cycle: density-dependent breeding, that we represent by a Ricker map $R(x) = xe^{r(1-x)}$, $r > 0$; density-dependent mortality represented by $M(x) = xe^{-ax}$, $a > 0$; and proportional harvesting, given by $H(x) = (1 - \gamma)x$, $\gamma \in (0, 1)$. In this way, spring harvest is represented by equation

$$x_{n+1} = R(H(M(x_n))), \tag{4}$$

$n = 0, 1, 2, \ldots$, while autumn harvest is governed by

$$x_{n+1} = R(M(H(x_n))), \tag{5}$$

$n = 0, 1, 2, \ldots$.

In both models, population is sampled after reproduction, but the relative order of the three discrete events is different, which can dramatically affect the size and dynamics of populations [10]. Regarding permanence, it is easy to prove that the

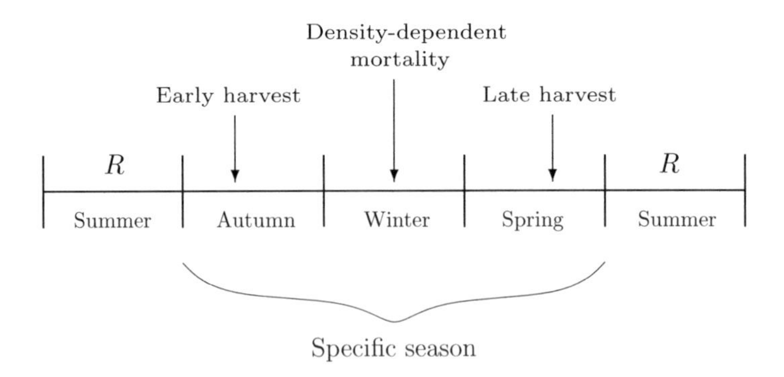

Fig. 2 Representation of the different seasons in the Jonzén–Lundberg model, where reproduction (R) occurs in summer, mortality during winter, and the harvest season is either autumn or spring. In Seno's model (3), autumn harvest would correspond to early harvest within the specific season (small θ), and spring harvest would correspond to late harvest within the specific season (θ close to 1)

critical value of γ leading to extinction does not depend on harvest time, and it is the same value predicted by Seno's model.

Proposition 1 (Extinction) *Assume that* $R(x) = xe^{r(1-x)}$, $M(x) = xe^{-ax}$, $H(x) = (1-\gamma)x$, *with* $r > 0$, $a > 0$, $0 < \gamma < 1$. *Then Eqs. (4) and (5) have at least one positive equilibrium if and only if* $\gamma < \gamma^* := 1 - e^{-r}$. *If* $\gamma \geq \gamma^*$, *then all solutions converge to zero.*

Regarding stability, Jonzén–Lundberg [6] argued that increasing harvesting is stabilizing, and the stability effect is stronger for autumn harvest. However, a further analysis shows that this effect depends strongly on the mortality rate a. Roughly speaking, the effect is stronger for autumn harvest only if a is large enough. Notice that mortality rates also influence which harvest season is better for population size, and again (5) is better for higher mortalities [7]; see Fig. 3a, b.

As far as we know, the Jonzén–Lundberg model has not been studied when the recruitment function R exhibits Allee effects. As in Seno's model, there is a critical value γ^* of the harvesting rate parameter for which a saddle–node bifurcation occurs. This value is different for spring harvest and autumn harvest. Numerical simulations suggest that autumn harvest can be beneficial to prevent sudden collapses; see Fig. 3c. This issue deserves further study.

An important feature of the seasonal models (4) and (5) that it is not present in Seno's model is that, for large growth rates r (a necessary condition is $r > 4$), the reproduction function $R(x)$ can have 3 critical points and 3 positive equilibria. This property opens the possibility for the coexistence of two nonzero attractors, and increasing harvesting can either promote or prevent bistability. Moreover, other phenomena such as non-smooth hydra effects, hysteresis and stability switches are possible; see [7] for more details.

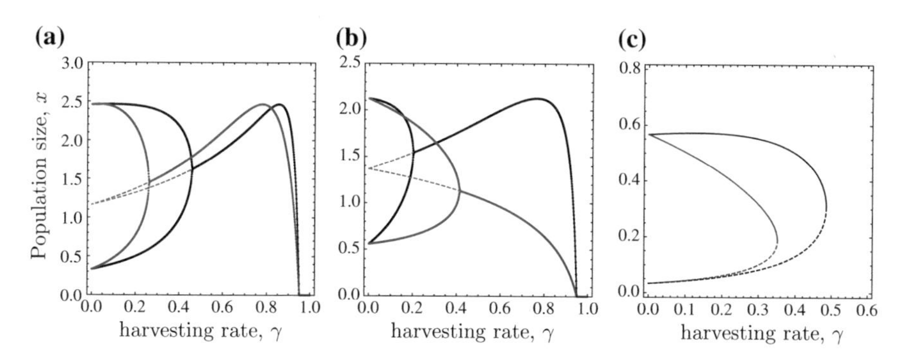

Fig. 3 Bifurcation diagrams for the Jonzén–Lundberg model as harvesting rate γ is increased, with blue color for spring harvest and black color for autumn harvest. Dashed lines correspond to unstable equilibria. **a** $R(x) = xe^{3(1-x)}$, $M(x) = xe^{-0.2x}$; **b** $R(x) = xe^{3(1-x)}$, $M(x) = xe^{-2x}$; **c** $R(x) = 2x^2e^{3(1-x)}/(1+2x)$, $M(x) = xe^{-2x}$

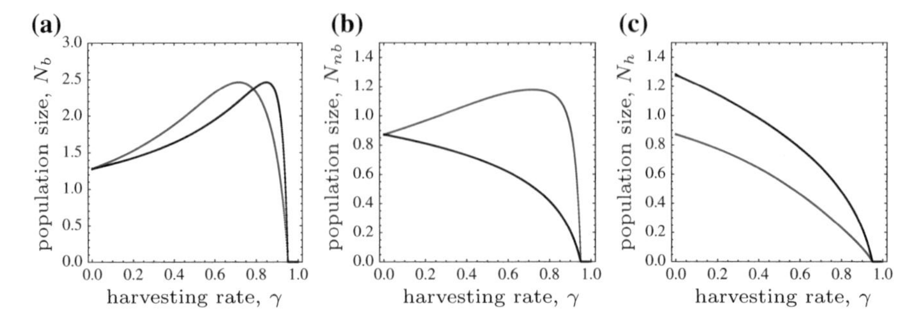

Fig. 4 Population size at equilibrium as harvesting intensity is increased, with blue color for spring harvest and black color for autumn harvest. **a** Census after reproduction; **b** Census after natural mortality; **c** Census after harvesting

4 The Importance of Census Time

Another difference between Seno's model and the approach of Jonzén–Lundberg is that in the latter census can take place in three different moments, namely, after every discrete event occurs in the annual cycle. As we have mentioned, even in simple models with only two discrete events, census time is important for management decisions. Although this aspect has been emphasized in previous work (e.g., [5, 7, 10]), we give an illustrative example and propose two criteria to choose between spring and autumn harvest.

In Fig. 4, we plot the positive equilibrium of models (4) and (5), with $R(x) = xe^{3(1-x)}$, $M(x) = xe^{-0.3x}$, $H(x) = (1-\gamma)x$. We denote by N_b, N_{nb}, and N_h the population sizes at the end of the breeding, the nonbreeding, and the harvesting seasons, respectively. A manager looking at panel (b) (census after winter mortality) would say that spring harvest is preferable for conservation purposes, but panel (c) sends the opposite message.

We suggest two criteria for choosing between autumn harvest and spring harvest. Having in mind the diagrams in Fig. 5, we arrive at the following conclusions:

(i) looking at census with lower population densities: if the target is sustainability, it seems autumn harvest is preferable;
(ii) looking at the yield: if the target is maximizing yield, again autumn harvest seems to be better.

5 Discussion

Theoretical and experimental results indicate that the size of seasonal populations depends strongly on harvest time. Using two different approaches based on discrete-time population models, we aimed to contribute to the understanding of the influence

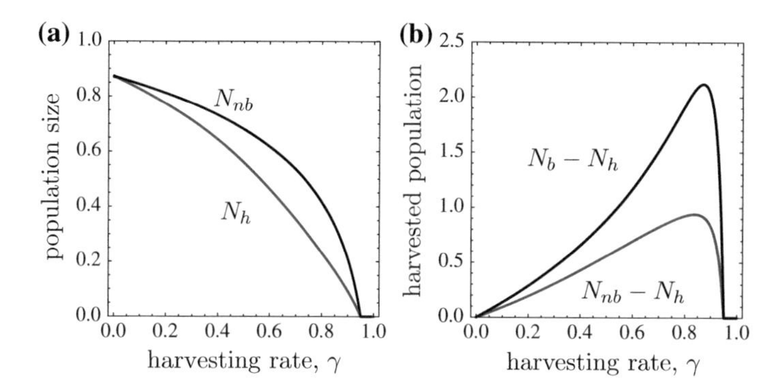

Fig. 5 Comparison between spring harvest (blue color) and autumn harvest (black color). **a** Low population densities are attained censusing after harvesting in the first case, and after natural mortality in the second one; **b** The yield is obtained as the difference between the number of individuals just after the harvesting season and just before it

of harvest time and census time in the management of exploited populations. In summary, we list the following main conclusions:

(i) *Extinction*: for globally persistent models, harvest timing does not change the critical harvest intensity leading to extinction. In models with Allee effects, a suitable harvest timing could prevent population collapses.
(ii) *Stability*: harvest timing has a strong influence on the stability properties of the population. However, choosing a suitable harvest season depends on other population parameters (birth rate, mortality rate).
(iii) *Hydra effects*: population can increase in response to an increasing mortality. Harvesting seasonal populations can lead to new forms of this paradoxical effect.
(iv) *Census time*: seasonal models emphasize the importance of sampling the population after every discrete event occurs during one cycle.

References

1. P.A. Abrams, When does greater mortality increase population size? the long history and diverse mechanisms underlying the hydra effect. Ecol. Lett. **12**, 462–474 (2009)
2. Announcement. Math. Biosci. **294**, I–II (2017)
3. B. Cid, F.M. Hilker, E. Liz, Harvest timing and its population dynamic consequences in a discrete single-species model. Math. Biosci. **248**, 78–87 (2014)
4. F. Courchamp, L. Berec, J. Gascoigne, *Allee Effects in Ecology and Conservation* (Oxford University Press, 2008)
5. F.M. Hilker, E. Liz, Harvesting, census timing and 'hidden' hydra effects. Ecol. Complex. **14**, 95–107 (2013)
6. N. Jonzén, P. Lundberg, Temporally structured density dependence and population management. Annal. Zool. Fenn. **36**, 39–44 (1999)

7. E. Liz, Effects of strength and timing of harvest on seasonal population models: stability switches and catastrophic shifts. Theor. Ecol. **10**, 235–244 (2017)
8. E. Liz, F.M. Hilker, Harvesting and dynamics in some one-dimensional population models, in *Theory and Applications of Difference Equations and Discrete Dynamical Systems*. Springer Proceedings in Mathematics & Statistics 102 (Springer, Berlin, Heidelberg, 2014), pp. 61–73
9. K.M. McIntire, S.A. Juliano, How can mortality increase population size? a test of two mechanistic hypotheses. Ecology (2018)
10. I.I. Ratikainen, J.A. Gill, T.G. Gunnarsson, W.J. Sutherland, H. Kokko, When density dependence is not instantaneous: theoretical developments and management implications. Ecol. Lett. **11**, 184–198 (2008)
11. S.J. Schreiber, Allee effects, extinctions, and chaotic transients in simple population models. Theor. Popul. Biol. **64**, 201–209 (2003)
12. H. Seno, A paradox in discrete single species population dynamics with harvesting/thinning. Math. Biosci. **214**, 63–69 (2008)

Global Stability Conditions of the Disease-Free Equilibrium for a Lymphatic *Filariasis* Model

Egberanmwen Barry Iyare, Daniel Okuonghae and Francis E. U. Osagiede

Abstract This work presents a mathematical model for the spread of lymphatic *filariasis* in a population. We use the Metzler Matrix Theory and the Kamgang–Sallet [2] (Math Biosci 213, 1–12, 2008) algorithm to compute the threshold conditions and global stability of the disease-free equilibrium. We showed that if $R_0 < 1$ the disease-free equilibrium (DFE) is globally asymptotically stable (GAS), and if $R_0 > 1$, the disease-free equilibrium is unstable.

1 Introduction

Lymphatic *filariasis* is one of the neglected tropical diseases. The disease is caused by a microscopic thread-like parasitic worm called *filariae*. It is transmitted by various species of mosquitoes [1]. The species Brugia malayi, Brugia timori, and Wuchereria bancrofti inhabit the lymphatic system, hence, the disease they cause is called lymphatic *filariasis* [3]. According to [4], over 120 million people are infected globally with about 40 million disfigured and incapacitated by the disease.

In this paper, we constructed a model of lymphatic *filariasis*. We use the Metzler matrix theory and the techniques in [2], to study the threshold conditions and global stability of the disease-free equilibrium.

E. B. Iyare (✉)
Department of Mathematical and Physical Sciences, Samuel Adegboyega University, Ogwa, Edo, Nigeria
e-mail: barryiyare@yahoo.com

D. Okuonghae · F. E. U. Osagiede
Department of Mathematics, University of Benin, Benin city, Nigeria
e-mail: daniel.okuonghae@uniben.edu

F. E. U. Osagiede
e-mail: francis.osagiede@uniben.edu

A. Korobeinikov et al. (eds.), *Extended Abstracts Spring 2018*,
Trends in Mathematics 11, https://doi.org/10.1007/978-3-030-25261-8_16

2 Model Formulation

The total human population at time t is denoted by $N_h(t)$, and subdivided into five mutually disjoint compartments of the susceptible humans $S_h(t)$, the latent individuals, $E_h(t)$, the asymptomatic individuals, $A_h(t)$, the symptomatic individuals, $I_h(t)$, and the treated individuals, $T_h(t)$. The total vector population at time t is denoted by$N_v(t)$ and subdivided into three mutually disjoint compartments of susceptible mosquitoe,s $S_v(t)$, the latent mosquitoes, $E_v(t)$, and the infectious mosquitoes, $I_v(t)$, that is

$$N_h(t) = S_h(t) + E_h(t) + A_h(t) + I_h(t) + T_h(t). \tag{1}$$

$$N_v(t) = S_v(t) + E_v(t) + I_v(t) \tag{2}$$

Based on the assumptions above, the model is given by the following system of nonlinear ordinary differential equations:

$$\begin{aligned}
\dot{S}_h &= \Lambda_h - \lambda_v S_h - \mu_h S_h,\\
\dot{E}_h &= \lambda_v S_h + \nu\lambda_v T_h - (k_1 + \mu_h)E_h,\\
\dot{A}_h &= k_1 E_h - (k_2 + \mu_h)A_h\\
\dot{I}_h &= k_2 A_h - (\tau_{h1} + \delta_L + \mu_h)I_h\\
\dot{T}_h &= \tau_{h1} I_h - \nu\lambda_v T_h - \mu_h T_h\\
\dot{S}_v &= \Lambda_v - \lambda_h S_v - \mu_v S_v,\\
\dot{E}_v &= \lambda_h S_v - k_v E_v - \mu_v E_v,\\
\dot{I}_v &= k_v E_v - \mu_v I_v,
\end{aligned} \tag{3}$$

where $\lambda_v = (\beta_v\sigma_v\sigma_h I_v)/(\sigma_v N_v + \sigma_h N_h)$ and $\lambda_h = (\beta_h\sigma_v\sigma_h(\eta_1 E_h + \eta_{h1} A_h + I_h))/(\sigma_v N_v + \sigma_h N_h)$. The variables of system (3) are summarize in Table 1.

2.1 Global Stability of the Model

The domain $\mathcal{D} = \mathcal{D}_h \times \mathcal{D}_v$, where $\mathcal{D}_h = \{(S_h, E_h, A_h, I_h, T_h) \in \mathbb{R}^5_+ : N_h \leq \Lambda_h/\mu_h\}$ and $\mathcal{D}_v = \{(S_v, E_v, I_v) \in \mathbb{R}^3_+ : N_v \leq \Lambda_v/\mu_v\}$, is positively invariant and attracts all the positive trajectories of model (3). Thus, the system (3) is dissipative on Ω, the trajectories of (3) are forward bounded. We shall study the system (3) in Ω. As in [2], we write system (3) in the form

$$\dot{x_1} = A_1(x_1, 0)(x_1 - x_1^*), \tag{4}$$

$$\dot{x_2} = A_2(x) \cdot x_2, \tag{5}$$

Table 1 Description of state variables of the model

Variable	Description
$S_h(t)$	Population of susceptible individuals
$E_h(t)$	Population of individuals with latent LF
$A_h(t)$	Population of individuals with asymptomatic LF
$I_h(t)$	Population of individuals with symptomatic LF
$T_h(t)$	Population of individuals treated of LF
$S_v(t)$	Population of susceptible mosquitoes
$E_v(t)$	Population of exposed mosquitoes
$I_v(t)$	Population of infectious mosquitoes

where $x_1 = (S_h, T_h, S_v)$ are the non-transmitting compartments, $x_2 = (E_h, A_h, I_h, E_v, I_v)$ are the transmitting compartments, and x_1^* is the disease-free equilibrium point.

Theorem 1 (Kamgang–Sallet, [2]) *The disease-free equilibrium is globally asymptotically stable if all eigenvalue of A_1 are real negative and A_2 is a Metzler matrix.*

We express the subsystem (4) as

$$\begin{aligned}\dot{S}_h &= \Lambda_h - \lambda_v S_h - \mu_h S_h,\\ \dot{T}_h &= \tau_{h1} I_h - \nu\lambda_v T_h - \mu_h T_h,\\ \dot{S}_v &= \Lambda_v - \lambda_h S_v - \mu_v S_v,\end{aligned} \tag{6}$$

where

$$A_1 = \begin{pmatrix} -(\lambda_v + \mu_h) & 0 & 0 \\ 0 & -(\nu\lambda_v + \mu_h) & 0 \\ 0 & 0 & -(\lambda_h + \mu_v) \end{pmatrix}. \tag{7}$$

Since all the diagonal entries are negative, the subsystem is globally asymptotically stable at the disease-free equilibrium $(\Lambda_h/\mu_h, 0, \Lambda_v/\mu_v)$. Hence, assumptions H_1 and H_2 in [2] are satisfied. The subsystem (5) is given as

$$\begin{aligned}\dot{E}_h &= \lambda_v S_h + \nu\lambda_v T_h - (k_1 + \mu_h)E_h,\\ \dot{A}_h &= k_1 E_h - (k_2 + \mu_h)A_h,\\ \dot{I}_h &= k_2 A_h - (\tau_{h1} + \delta_L + \mu_h)I_h,\\ \dot{E}_v &= \lambda_h S_v - k_v E_v - \mu_v E_v,\\ \dot{I}_v &= k_v E_v - \mu_v I_v,\end{aligned} \tag{8}$$

where $A_2(x)$ is

$$A_2 = \begin{pmatrix} -g_1 & 0 & 0 & 0 & G_1 \\ k_1 & -g_2 & 0 & 0 & 0 \\ 0 & k_2 & -g_3 & 0 & 0 \\ \eta_1 G_2 & \eta_{h1} G_2 & G_2 & -g_4 & 0 \\ 0 & 0 & 0 & k_v & -\mu_v \end{pmatrix},$$

$g_1 = k_1 + \mu_h$, $g_2 = k_2 + \mu_h$, $g_3 = \tau_{h1} + \delta_L + \mu_h$, $g_4 = k_v + \mu_v$, $G_1 = (\beta_v \mu_v \Lambda_h \sigma_v \sigma_h)/\ (\sigma_v \Lambda_v \mu_h + \sigma_h \Lambda_h \mu_v)$, and $G_2 = (\beta_h \mu_h \Lambda_v \sigma_v \sigma_h)/(\sigma_v \Lambda_v \mu_h + \sigma_h \Lambda_h \mu_v)$. Since all the off-diagonal entries of $A_2(x)$ are nonnegative, $A_2(x)$ is a Metzler matrix and irreducible for x in Ω, satisfying assumption H_3 in [2].

According to assumption H_4 in [2], there exists an upper bound matrix $\bar{A_2}$ for $\bar{\Omega} = \{A_2(x) \mid x \in \Omega\}$ with the property that either $\bar{A_2} \notin \bar{\Omega}$ or $\bar{A_2} \in \bar{\Omega}$ That is, $\bar{A_2} = max_{\Omega} \bar{\Omega}$. The points where these maximum is realized are contained in the disease-free of subsystem (5).

Proposition 2 (Kamgang–Sallet, [2]) *Let M be a Metzler matrix, which is block decomposed as follows:*

$$M = \begin{pmatrix} A & B \\ C & D \end{pmatrix},$$

where A and D are square matrices. Then M is Metzler stable if and only if A and $D - CA^{-1}B$ are Metzler stable.

We decompose the matrix A_2 in block form as follows:

$$A_2 = \begin{pmatrix} A & B \\ C & D \end{pmatrix},$$

where

$$A = \begin{pmatrix} -g_1 & 0 & 0 \\ k_1 & -g_2 & 0 \\ 0 & k_2 & -g_3 \end{pmatrix},\quad B = \begin{pmatrix} 0 & G_1 \\ 0 & 0 \\ 0 & 0 \end{pmatrix},\quad C = \begin{pmatrix} \eta_1 G_2 & \eta_{h1} G_2 & G_2 \\ 0 & 0 & 0 \end{pmatrix},\quad D = \begin{pmatrix} -g_4 & 0 \\ k_v & -\mu_v \end{pmatrix}.$$

Assumption H_5 in [2] requires $\alpha(A_2) \le 0$, where α is the largest eigenvalue of A_2.

Matrix A_2 is Metzler stable if A and $D - CA^{-1}B$ are Metzler stable. Obviously, A is Metzler stable. The matrix

$$D - CA^{-1}B = \begin{pmatrix} -g_4 & \hat{G} \\ k_v & -\mu_v \end{pmatrix},\quad CA^{-1}B = \begin{pmatrix} 0 & -\hat{G} \\ 0 & 0 \end{pmatrix},$$

where $\hat{G} = G_1 G_2 (\eta_1 g_2 g_3 + k_1 \eta_{h1} g_3 + k_1 k_2)/g_1 g_2 g_3$, is a regular splitting with a diagonal negative matrix.

Proposition 3 (Kamgang–Sallet, [2]) *Let $M = \Lambda + N$ be a regular splitting of a real Metzler matrix M. Then, M is Metzler stable if and only if the spectral radius $\rho(-N\Lambda^{-1}) < 1$.*

For subsystem (8), we have

$$D^{-1} = \begin{pmatrix} \frac{-1}{g_4} & 0 \\ \frac{k_v}{\mu_v g_4} & \frac{-1}{\mu_v} \end{pmatrix}, \quad -CA^{-1}BD^{-1} = \begin{pmatrix} \frac{\hat{G}k_v}{\mu_v g_4} & \frac{\hat{G}}{\mu_v} \\ 0 & 0 \end{pmatrix}.$$

The spectral radius $\rho(-CA^{-1}BD^{-1})$ is given by

$$R_0 = \frac{\sigma_h \sigma_v}{(\Lambda_h \mu_v \sigma_h + \Lambda_v \mu_h \sigma_v)} \sqrt{\frac{k_v \beta_h \beta_v \Lambda_h \Lambda_v \mu_h \hat{G}}{G_1 G_2 g_4}}. \tag{9}$$

Now, by [2, Corollary 4.4], we establish the following result.

Theorem 4 *The disease-free equilibrium of the model (3) is globally asymptotically stable in $\mathcal{D}$ if $\mathcal{R}_0 < 1$, and is unstable if $\mathcal{R}_0 > 1$.*

Here, R_0 is the threshold quantity called the basic reproduction number. It represent the average number of secondary cases that one individual (or mosquito) infected with lymphatic *filariasis* would generate over the duration of the average infectious period, if introduced into a susceptible human (or mosquito) population. Satisfying the condition $\mathcal{R}_0 < 1$ leads to eradication of the disease in the population while $\mathcal{R}_0 > 1$, leads to persistence of the disease.

References

1. Center for Disease Control (CDC). Accessed 23 Aug 2018
2. J.C. Kamgang, G. Sallet, Computation of threshold conditions for epidemiological models and glob al stability of the disease-free equilibrium (DFE). Math. Biosci. **213**, 1–12 (2008)
3. C.N. Ukaga, B.E.B. Nwoke, E.A. Nwoke, M.J. Nwachukwu, Epidemiological characteristics of Bancrofti filariasis and the nigerian environment. J. Public Health Epidemiol. **2**(6) (2010)
4. World Health organisation (WHO). Accessed 23 Aug 2018

Repulsive Invariant Manifolds in Modeling the Critical Phenomena

Elena Shchepakina

Abstract The paper shows how the repulsive invariant manifolds of multiscale dynamical systems are used for modeling the critical phenomena. A dynamical model of fuel spray ignition is considered to illustrate this approach.

1 Introduction

The paper outlines an approach to modeling the critical phenomena in multiscale dynamical models. Such models are usually described by singularly perturbed systems of differential equations to reflect the significant distinction in characteristic relaxation times of different processes. The approach is based on the geometric asymptotic method of invariant manifolds; see [4].

By a critical phenomenon, we mean a sharp change in a model's dynamics via a transition from a slow process to a self-accelerating mode. According to the theory of invariant manifolds, processes with self-acceleration correspond to trajectories that either has no common point with an attractive slow invariant manifold, or leave it after a while; see [3]. The last situation occurs when the trajectory reaches a boundary separating the attractive slow invariant manifold from repulsive one.

Using a fuel spray ignition model, we demonstrate how the repulsive slow invariant manifold can be used for modeling the critical phenomena.

This work was funded by RFBR and Samara Region (project 16-41-630529-p) and the Ministry of Education and Science of the Russian Federation under the Competitiveness Enhancement Program of Samara University (2013–2020).

E. Shchepakina (✉)
Department of Differential Equations and Control Theory, Samara National Research University, Moskovskoye shosse 34, Samara 443086, Russian Federation
e-mail: shchepakina@ssau.ru

A. Korobeinikov et al. (eds.), *Extended Abstracts Spring 2018*,
Trends in Mathematics 11, https://doi.org/10.1007/978-3-030-25261-8_17

2 Model

The model of ignition and combustion of a fuel spray is formulated using an adiabatic approach. The fuel spray is considered as a two-phase medium: the combustible gas mixture and the combustible liquid droplets. The effects of changes in pressure are neglected. The usual assumption is made that the thermal conductivity of the liquid phase is much greater than that in the gas phase. Thus, the heat transfer coefficient in the liquid gas mixture is supposed to be defined by the thermal properties of the gas phase. The droplets boundary is assumed to be on a saturation line, i.e., the liquid temperature is constant and is equal to the liquid saturation temperature. The combustion reaction is modeled as a first order, highly exothermic chemical reaction. The model is built with the usual assumptions of the theory of combustion processes in chemical homogeneity at each point of the reaction vessel. The dimensionless model has the following form (see [1])

$$\gamma \frac{d\theta}{d\tau} = \eta \exp\left(\frac{\theta}{1+\beta\theta}\right) - \varepsilon_1 r\theta(1+\beta\theta), \tag{1}$$

$$\frac{dr^3}{d\tau} = -\varepsilon_1\varepsilon_2 r\theta, \tag{2}$$

$$\frac{d\eta}{d\tau} = -\eta\frac{1}{1+\beta\theta}\exp\left(\frac{\theta}{1+\beta\theta}\right) + \varepsilon_1 r\psi\theta, \tag{3}$$

where θ is the dimensionless fuel gas temperature; r is the dimensionless radius of the droplets; η is the dimensionless concentration of the flammable gas; τ is the dimensionless time; γ is the dimensionless parameter equal to the final dimensionless adiabatic temperature thermally isolated system after the explosion; β gives the initial temperature; ε_1, ε_2 characterize the interaction between the gas and liquid phases; ψ is a parameter characterizing the ratio of the energy of combustion gas mixture to the liquid evaporation energy.

The initial conditions for the Eqs. (1)–(3) are

$$\theta(0) = 0, \quad \eta(0) = 1, \quad r(0) = 1.$$

Appropriate combination of Eqs. (1)–(3) and integration over time yields the following energy integral

$$\eta - 1 + \frac{\gamma}{\beta}\ln(1+\beta\theta) + \frac{\psi - 1}{\varepsilon_2}\left(r^3 - 1\right) = 0,$$

which allows to reduce the order of the system (1)–(3) to the form

$$\gamma \frac{d\theta}{d\tau} = \left(1 - \frac{\gamma}{\beta}\ln(1+\beta\theta) - \frac{\psi - 1}{\varepsilon_2}\left(r^3 - 1\right)\right)\exp\left(\frac{\theta}{1+\beta\theta}\right) - \varepsilon_1 r\theta(1+\beta\theta), \tag{4}$$

$$\frac{dr^3}{d\tau} = -\varepsilon_1\varepsilon_2 r\theta. \tag{5}$$

The degenerate equation, which follows from the fast subsystem (4) for $\gamma = 0$, describes a slow curve

$$\Lambda(\theta, r) = \left(1 - \frac{\gamma}{\beta}\ln(1+\beta\theta) - \frac{\psi - 1}{\varepsilon_2}\left(r^3 - 1\right)\right)\exp\left(\frac{\theta}{1+\beta\theta}\right) - \varepsilon_1 r\theta(1+\beta\theta) = 0$$

in the phase plane. The flow of system (4), (5) near the slow curve has a velocity of $O(1)$ as $\gamma \to 0$, while far from the slow curve the variable θ is changed very rapidly. In a γ neighborhood of the slow curve, there exists a slow invariant manifold of the system which is defined as an invariant surface of slow motions.

The set of points on the slow curve where $\partial\Lambda/\partial\theta < 0\,(\partial\Lambda/\partial\theta > 0)$ forms a stable (unstable) part of the slow curve. The stable and unstable parts of the slow curve are the zero-order approximations ($\gamma = 0$) of the stable (or attractive) and unstable (or repulsive) slow invariant manifolds of the system (4), (5), respectively.

For $0 < \psi < 1 - \varepsilon_2$, the slow curve is concave, see Fig. 1a. The part PT of the slow curve is stable while the part TQ is unstable. The ordinate of the point T depending on the parameters' values can be equal to, greater, or less than 1. If the point T has an ordinate greater than 1, a trajectory of the system starts from the initial point and tends to the stable part PT of the slow curve. Then it follows PT to the origin. This is the case of a slow combustion regime; see the trajectory $C''JP$ in Fig. 1a.

If the point T has an ordinate less than 1, then a trajectory of the system will pass beyond the basin of attraction of the PT. This case corresponds to an explosion regime; see the trajectory $C'D$ in Fig. 1a.

3 Critical Phenomenon

A critical phenomenon corresponds to the case when the trajectory of the system falls into a small vicinity of the point T and passes along the unstable part TQ of the slow curve (see the trajectory CTQ in Fig. 1a). This trajectory corresponds to the critical regime, which separates the safe regimes from explosive modes [3, 5].

The crucial result is that the repulsive slow manifold may be used to construct the critical trajectory CTQ and to calculate the corresponding value of a control parameter of the system, say, ε_1. To do this, we use the asymptotics proposed in [2]. To make this use eligible, we introduce the new "reverse" time $t = -\tau$ in (4), (5) that make the slow invariant manifold near TQ attractive; see [3].

The part CT of the critical trajectory can be represented in the form

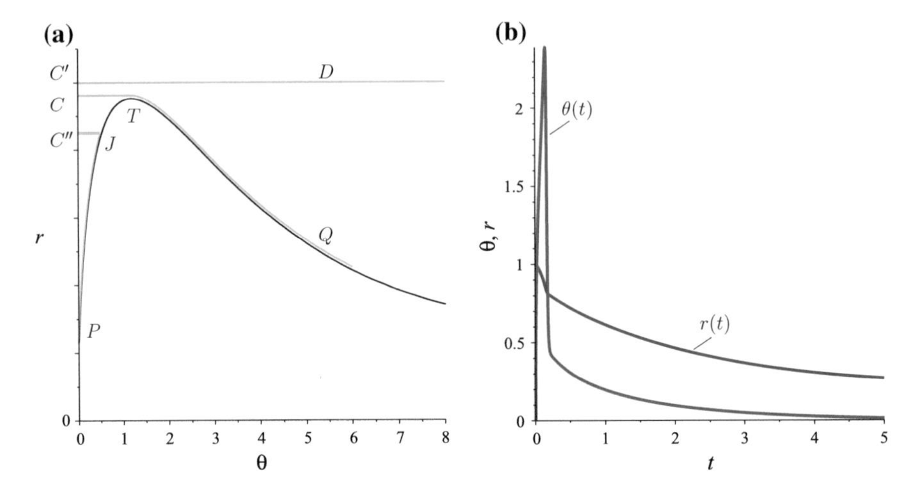

Fig. 1 **a** The slow curve (black) and the trajectories (green) of the system (4), (5) with $0 < \psi < 1 - \varepsilon_2$ in the limit case ($\gamma = 0$); **b** the solutions of (1)–(3) in the case of critical regime for $\gamma = 0.01$, $\varepsilon_1 = 2.2100108$, $\varepsilon_2 = 0.8$, $\beta = 0.05$, $\psi = 0.19$

$$r(\theta, \gamma) = r^* + \gamma^{\frac{2}{3}} \Gamma_0^{\frac{2}{3}} \Omega \operatorname{sign} f(\theta^*, r^*) + \frac{1}{3} \gamma \ln \frac{1}{\gamma} \Gamma_1 \operatorname{sign} f(\theta^*, r^*) + O(\gamma), \quad (6)$$

where θ^*, r^*, denote the coordinates of the point T, the functions f and g are the right parts of the system (4), (5) after the transition to the reverse time t,

$$\Omega = 2.338107, \quad \Gamma_0 = \sqrt{\frac{2}{|g_{\theta\theta}(\theta^*, r^*) g_r(\theta^*, r^*)|}} \, |f(\theta^*, r^*)|,$$

$$\Gamma_1 = \frac{6 g_{\theta\theta}(\theta^*, r^*) f_\theta(\theta^*, r^*) - 2 g_{\theta\theta\theta}(\theta^*, r^*) f(\theta^*, r^*)}{3 g_{\theta\theta}^2(\theta^*, r^*)}.$$

The coordinates θ^*, r^* can be found from the system $g(\theta^*, r^*) = g_\theta(\theta^*, r^*) = 0$. Substituting all the found values into (6) and setting $r = 1$ since the point C has the coordinate $r = 1$, we obtain the equation for calculation the critical parameter value $\varepsilon_1 = \varepsilon_1^*$ in the form

$$\varepsilon_1^* = \varepsilon_{10} + \gamma^{\frac{2}{3}} \varepsilon_{11} + \gamma \ln \frac{1}{\gamma} \varepsilon_{12} + O(\gamma).$$

A direct calculation gives

$$\varepsilon_{10} = \frac{pe(\psi - 1)}{\varepsilon_2}, \quad \varepsilon_{11} = -54\Omega(\psi - 1)^3 \sqrt[3]{\frac{2}{9}} \mu_1^{\frac{4}{3}} \mu_{23}^{-\frac{2}{3}} \varepsilon_{10}^{-\frac{4}{3}} \varepsilon_2^{-\frac{4}{3}}$$

$$\times \left\{(1-6\beta)\left[\frac{\psi-1}{\varepsilon_2}(\mu_{23}^3-1)-1\right]e+2\varepsilon_{10}\mu_{23}\beta\right\}^{-\frac{1}{3}}\left[3\mu_{23}^2 e\frac{\psi-1}{\varepsilon_2}+\varepsilon_{10}(1+\beta)\right]^{-\frac{1}{3}},$$

$$\varepsilon_{12}=\frac{\varepsilon_{10}\varepsilon_2}{3\mu_{23}}\left\{(1-14\beta)\left[\frac{\psi-1}{\varepsilon_2}(\mu_{23}^3-1)-1\right]e+4\varepsilon_{10}\mu_{23}\beta\right\}$$

$$\times \left\{(1-6\beta)\left[\frac{\psi-1}{\varepsilon_2}(\mu_{23}^3-1)-1\right]e+2\varepsilon_{10}\mu_{23}\beta\right\}^{-2},$$

where

$$q=\frac{\varepsilon_2+\psi-1}{\psi-1},\ \mu_1=\sqrt{\frac{q^2}{4}+\frac{\varepsilon_{10}^3\varepsilon_2^3}{27(\psi-1)^3}},\ \mu_2=\sqrt[3]{\frac{q}{2}+\mu_1},\ \mu_3=\sqrt[3]{\frac{q}{2}-\mu_1},\ \mu_{23}=\mu_2+\mu_3,$$

and p is a root of the equation

$$1=\sqrt[3]{\frac{q}{2}+\sqrt{\frac{q^2}{4}+\frac{p^3}{27}}}+\sqrt[3]{\frac{q}{2}-\sqrt{\frac{q^2}{4}+\frac{p^3}{27}}}.$$

It should underline that the main feature of the critical regime is that during it the temperature of the combustible mixture can reach a high value within the framework of a safe process, see Fig. 1b.

References

1. I. Goldfarb, V. Gol'dshtein, I. Shreiber, and A. Zinoviev, Liquid drop effects on self-ignition of combustible gas, *Proceedings of the 26th Symposium (International) on Combustion* (Pittsburgh: The Combustion Institute, 1996), pp. 1557–1563
2. E.F. Mishechenko, NKh Rozov, *Differential equations with small parameters and relaxation oscillations* (Plenum Press, New York, 1980)
3. E. Shchepakina and V. Sobolev, "Black swans and canards in laser and combustion models", in *Singular Perturbations and Hysteresis*, ed by M. Mortell, R. O'Malley, A. Pokrovskii, V. Sobolev. *SIAM* 207–256 (2005)
4. E. Shchepakina, V. Sobolev, M.P. Mortell, Singular perturbations. Introduction to system order reduction methods with applications. *Lect. Notes in Math.* **2114**, Cham: Springer (2014)
5. V.A. Sobolev, E.A. Shchepakina, Self-ignition of dusty media. Combust. Explos. Shock. Waves **29**(3), 378–381 (1993)

A New Approach to Canards Chase in $3D$

Elena Shchepakina

Abstract A new approach to the canards chase in 3D for some class of singularly perturbed systems is suggested. The proposed approach is discussed by the use of a competitive model of population dynamics. The presence of an exact black swan (a stable/unstable slow invariant manifold) makes it possible to find a new kind of trajectories with multiple stability changes.

1 Introduction

In this paper, a new approach to the canards chase for a class of singularly perturbed systems with two slow and one fast variables is proposed. This approach is based on the geometric theory of invariant manifolds of singularly perturbed systems [4–6]. Recall, that canards are trajectories of a singularly perturbed system which at first move along a stable slow invariant manifold and then continue for a while along an unstable slow invariant manifold. A slow invariant manifold is defined as an invariant surface of slow motions.

It should be noted that a 3D canard is a result of gluing the stable and unstable slow invariant manifolds at one point of the breakdown surface [7]. For a fixed gluing point, this is possible due to a proper choice of an additional scalar parameter of the differential system. In the proposed approach, two additional parameters are used to construct a 3D canard. Both these parameters correspond to the canards for 2D-projections of the original system. The proposed approach is illustrated via a model of competing populations.

This work was funded by RFBR and Samara Region (project 16-41-630529-p) and the Ministry of Education and Science of the Russian Federation under the Competitiveness Enhancement Program of Samara University (2013–2020).

E. Shchepakina (✉)
Department of Differential Equations and Control Theory, Samara National Research University, Moskovskoye shosse 34, Samara 443086, Russian Federation
e-mail: shchepakina@ssau.ru

A. Korobeinikov et al. (eds.), *Extended Abstracts Spring 2018*,
Trends in Mathematics 11, https://doi.org/10.1007/978-3-030-25261-8_18

2 A Competing Predators Model

Consider two predator species competing for a single prey in a constant and uniform environment. The singular perturbed model of the processes is the following (see [1]):

$$\dot{x} = x\left(\frac{m_1 z}{\beta_1 + z} - d_1\right), \tag{1}$$

$$\dot{y} = y\left(\frac{m_2 z}{\beta_2 + z} - d_2\right), \tag{2}$$

$$\varepsilon \dot{z} = z\left(1 - z - \frac{m_1 x}{\beta_1 + z} - \frac{m_2 y}{\beta_2 + z}\right). \tag{3}$$

Here, x and y are the dimensionless population densities of the predators; z is the dimensionless population density of the prey; $\varepsilon = 1/\gamma$, where γ is the intrinsic rate of growth of the prey; for $i = 1, 2$, $m_i > 0$ is the maximal growth or birth rate of the ith predator; $\beta_i = a_i/K$, where a_i is the half-saturation constant for the i-th predator, K is the carrying capacity of the prey; $d_i > 0$ is the death rate of the i-th predator.

3 2*D* Canards

Consider the case of the absence of one of the predators, i.e., when, for example, $y \equiv 0$. In this case the system (1)–(3) takes the form

$$\dot{x} = x\left(\frac{m_1 z}{\beta_1 + z} - d_1\right) := f(x, z), \tag{4}$$

$$\varepsilon \dot{z} = z\left(1 - z - \frac{m_1 x}{\beta_1 + z}\right) := g(x, z). \tag{5}$$

If we put $\varepsilon = 0$ into the fast subsystem, we get the *degenerate equation*

$$z\left(1 - z - \frac{m_1 x}{\beta_1 + z}\right) = 0,$$

which describes the *slow curve* S of (4) and (5); see [5, 6]. The curve S consists of the straight line $z = 0$ and the parabola. Two breakdown points,

$$A_1\left(x = \frac{\beta_1}{m_1}, z = 0\right), \quad A_2\left(x = \frac{(1 + \beta_1)^2}{4m_1}, z = \frac{1 - \beta_1}{2}\right),$$

divide S into the stable subsets (S_1^s and S_2^s) and the unstable subsets (S_1^u and S_2^u), see Fig. 1.

In an ε–neighborhood of the stable (unstable) subset S_2^s (S_2^u) of the slow curve, there exists the stable (unstable) slow invariant manifold $S_{2,\varepsilon}^s$ ($S_{2,\varepsilon}^u$). We can glue together $S_{2,\varepsilon}^s$ and $S_{2,\varepsilon}^u$ at the point A_2, using the standard procedure [4–7], to get a canard. For this, we consider d_1 as a gluing parameter. The canard and the corresponding parameter value $d_1 = d_1^c(\varepsilon)$ allow asymptotic expansions in powers of the small parameter ε:

$$z = h(x, d_1^c(\varepsilon), \varepsilon) = h_0(x, d_{10}) + \varepsilon h_1(x, d_{10}, d_{11}) + O(\varepsilon^3), \tag{6}$$

$$d_1^c(\varepsilon) = d_{10} + \varepsilon d_{11} + O(\varepsilon^2). \tag{7}$$

We can calculate the functions h_0, h_1, etc., from the *invariance equation*

$$\varepsilon \frac{\partial h}{\partial x} f\left(x, h(x, d_1^c(\varepsilon), \varepsilon)\right) = g\left(x, h(x, d_1^c(\varepsilon), \varepsilon)\right),$$

which follows from the system (4) and (5) and the asymptotic expansions (6) and (7). However, all functions in (6) have a discontinuity at the breakdown point A_2. A proper choice of d_{10}, d_{11}, etc., enables us to avoid this discontinuity. The outcome of this procedure is a canard shown by the green curve in Fig. 1. The canard corresponds to the canard point $d_1 = d_1^c$, where

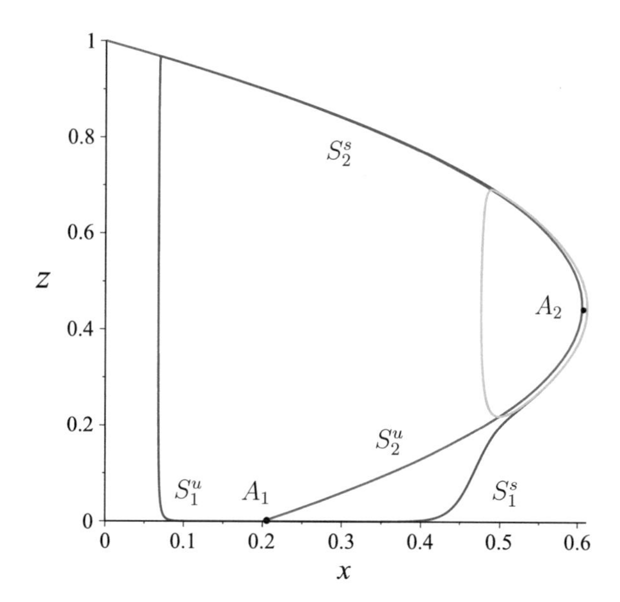

Fig. 1 The slow curve (red), the canard (green), and the canard doublet (blue) of (4) and (5)

$$d_1^c = \frac{m_1(1-\beta_1)}{1+\beta_1} - \varepsilon\frac{\beta_1^2(1+\beta_1)}{2(1-\beta_1)^2} + O(\varepsilon^2).$$

It should be noted that $z = 0$ is the exact canard of the system (4) and (5). In this special case, the trajectories of the system, starting in the basin of attraction of S_1^s, will continue their movement for a while along S_1^u. Therefore, we can transform the single canard, corresponding to the canard point $d_1 = d_1^c$, to a shape of a *canard doublet* (see the blue curve in Fig. 1) [2, 8]. Recall, that in the case of a planar system, the canards are exponentially close to each other near the slow curve and have the same asymptotic expansion (6) in powers of ε. An analogous assertion is true for corresponding parameter values (7). Namely, any two values of the parameter d_1, for which canards exist, have the same asymptotic expansions, and the difference between them is given by $\exp(-1/c\varepsilon)$, where c is some positive number. For example, for $\beta_1 = 0.1$, $m_1 = 0.5$, and $\varepsilon = 0.1$, the values of d_1 corresponding to the canard and the canard doublet in Fig. 1 are 0.408498400000 and 0.408498356366, respectively.

The results of this section can be extended to the case $x \equiv 0$ due to the competitive symmetry between the predators in (1)–(3). Similar reasoning gives the canard point d_2^c of the parameter d_2, where

$$d_2^c = \frac{m_2(1-\beta_2)}{1+\beta_2} - \varepsilon\frac{\beta_2^2(1+\beta_2)}{2(1-\beta_2)^2} + O(\varepsilon^2).$$

4 3*D* Canards

We now return to the 3D system (1)–(3). Substituting the canards points for the parameters d_1 and d_2 into the complete system (1)–(3), we get a canard in 3D. It should be noted that the discussed approach makes it possible to easily obtain various forms of 3D canards. It can be done by slightly changing the values d_1^c or/and d_2^c.

Note that $z \equiv 0$ is the exact slow invariant manifold, which is divided by the line $1 - m_1x/\beta_1 - m_2y/\beta_2 = 0$ into the stable and the unstable parts. Thus, $z \equiv 0$ is the *black swan* [3, 4]. The presence of the exact black swan allows us to obtain a new kind of trajectories with multiple changes of stability, a *cascade of 3D canard doublets*.

To obtain the trajectory shown in Fig. 2a, we transform the canard on yOz-plane to a shape of a canard doublet keeping the canard on xOz-plane. A shape of this trajectory can be modified, from the cascade of 3D canards without head to the cascade of 3D canard doublets that shown in Fig. 2b.

It should be noted that the considered situation, when a differential system possesses an exact black swan is typical for many biological models with two slow and one fast variables.

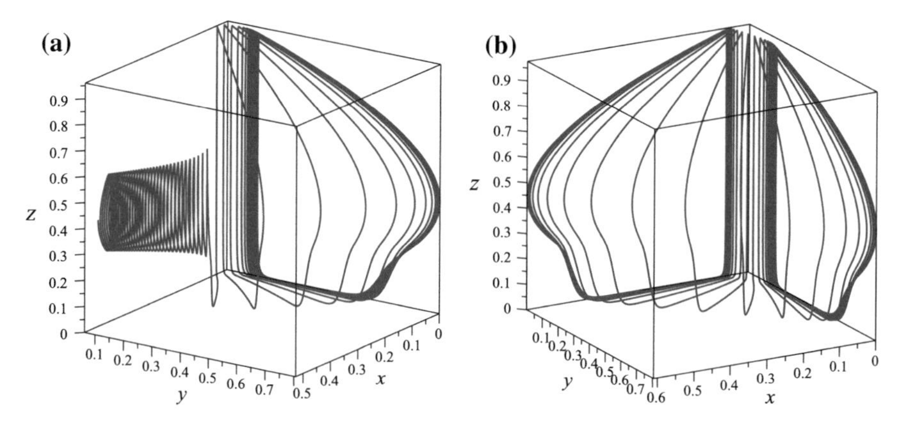

Fig. 2 The cascades of 3D canard doublets of the system (1)–(3). $\varepsilon = 0.1$, $\beta_1 = 0.1$, $\beta_2 = 0.13$, $m_1 = 0.5$, $m_2 = 0.4$, and **a** $d_1 = 0.408515869462$, $d_2 = 0.307288368584$; **b** $d_1 = 0.408498356366$, $d_2 = 0.307288368584$

References

1. W. Liu, D. Xiao, Y. Yi, Relaxation oscillations in a class of predatorprey systems. J. Differ. Equ. **188**, 306–331 (2003)
2. A. Pokrovskii, E. Shchepakina, V. Sobolev, Canard doublet in a Lotka–Volterra type model. J. Phys. Conf. Ser. **138**, 012019 (2008)
3. E. Shchepakina, Black swans and canards in self-ignition problem. Nonlinear Anal. Real World Appl. **4**, 45–50 (2003)
4. E. Shchepakina, V. Sobolev, Integral manifolds, canards and black swans. Nonlinear Anal. Theory Methods Appl. **44**, 897–908 (2001)
5. E. Shchepakina, V. Sobolev, Black swans and canards in laser and combustion models, in *Singular perturbations and hysteresis*, ed. by M.P. Mortell, R. O'Malley, A. Pokrovskii, V. Sobolev (SIAM, 2005), pp. 207–255
6. E. Shchepakina, V. Sobolev, M.P. Mortell, *Singular Perturbations: Introduction to System Order Reduction Methods with Applications*. Lecture Notes in Mathematics, vol. 2114 (Springer, Berlin, 2014)
7. V.A. Sobolev, E.A. Shchepakina, Duck trajectories in a problem of combustion theory. Differ. Equ. **32**, 1177–1186 (1996)
8. V. Sobolev, Canard cascades. Discret. Contin. Dyn. Syst. Ser. B **18**, 513–521 (2013)

Slow Invariant Manifolds in the Problem of Order Reduction of Singularly Perturbed Systems

Elena Tropkina and Vladimir Sobolev

Abstract The method of integral manifolds is used to study singularly perturbed systems of differential equations. The algorithms for the construction of the slow invariant manifolds in the case with different dimensions of the fast and slow variables was derived.

1 Introduction

Consider the system of differential equations

$$\dot{x} = f(x, y, \varepsilon), \tag{1}$$

$$\varepsilon \dot{y} = g(x, y, \varepsilon), \tag{2}$$

where $x \in \mathbb{R}^n$, $y \in \mathbb{R}^n$, ε is a small positive parameter, $0 < \varepsilon \ll 1$, functions f and g are continuous with respect to (x, y) for all $x \in \mathbb{R}^n$, $y \in D \subset \mathbb{R}^m$ $(D \subset \mathbb{R}^m)$. We will consider a situation where the system (1), (2) has an integral manifold, that is, when the following conditions are fulfilled (see [1, 4]):

(i) the equation $g(x, y, 0) = 0$ has an isolated solution $y = \psi_0(x)$ for $x \in \mathbb{R}^n$;
(ii) the functions f and g are uniformly continuous and bounded together with their partial derivatives with respect to all variables up to $(k + 2)$th order inclusively

This work was funded by RFBR and Samara Region (project 16-41-630529-p) and the Ministry of Education and Science of the Russian Federation under the Competitiveness Enhancement Program of Samara University (2013–2020).

E. Tropkina (✉) · V. Sobolev
Department of Differential Equations and Control Theory, Samara National Research University, Moskovskoye Shosse, 34, Samara 443086, Russian Federation
e-mail: elena_a.85@mail.ru

V. Sobolev
e-mail: v.sobolev@ssau.ru

A. Korobeinikov et al. (eds.), *Extended Abstracts Spring 2018*,
Trends in Mathematics 11, https://doi.org/10.1007/978-3-030-25261-8_19

$(k \geq 0)$ in some region $\Omega_0 = \{(x, y, \varepsilon) : x \in \mathbb{R}^n,\ \|y - \psi_0(x)\| < \rho,\ 0 \leq \varepsilon \leq \varepsilon_0\}$;

(iii) the eigenvalues of the matrix

$$B(x) = \frac{\partial g}{\partial y}(x, \psi_0(x), 0)$$

satisfy the inequality $Re\lambda_i(x) \leq -2\gamma < 0$.

Note that some interesting aspects of the theory of slow integral manifolds and the behavior of solutions in their neighborhood were presented in [2, 3].

The degenerated system regarding to (1), (2) has a form

$$\begin{aligned} \dot{x} &= f(x, y, 0), \\ 0 &= g(x, y, 0). \end{aligned} \tag{3}$$

It should be noted that the equations of system (3) can often be either transcendental or polynomials of a high degree with respect to y. In these cases, a solution of the system cannot be found in explicit form as $y = \psi_0(x)$. In these cases, for the system-order reduction, it is possible to use a parametric form for the representation of the slow invariant manifolds [4, 5]. Below we consider the three major cases, where either fast variables, or only a fraction of the fast variables, or fast variables supplemented by a certain number of slow variables, play a role of the parameters.

2 The Case $n = m$

Consider the case of the dimensions equality of the fast and slow variables. Suppose that the system (3) can be solved with respect to x in the form $x = \varphi_0(y)$. In this case, the fast vector-variable y can play a role of a parameter for the representation of the slow invariant manifolds in the parametric form

$$x = \varphi(y, \varepsilon) = \varphi_0(y) + \varepsilon\varphi_1(y) + \cdots + \varepsilon^k\varphi_k(y) + \cdots . \tag{4}$$

The corresponding invariance equation is obtained by substituting (4) in (1):

$$\frac{\partial \varphi}{\partial y} g(\varphi, y, \varepsilon) = \varepsilon f(\varphi, y, \varepsilon). \tag{5}$$

For all functions in (5), we write the formal asymptotic expansions in the powers of the small parameter ε:

$$f\Big(\sum_{k\geq 0} \varepsilon^k \varphi_k, y, \varepsilon\Big) = \sum_{k\geq 0} \varepsilon^k f^{(k)}(\varphi_0, \ldots, \varphi_k, y),$$

$$g\Big(\sum_{k\geq 0}\varepsilon^k\varphi_k, y, \varepsilon\Big) = g^{(0)}(\varphi_0, y) + B(y)\sum_{k\geq 1}\varepsilon^k\varphi_k + \sum_{k\geq 1}\varepsilon^k g^{(k)}(\varphi_0, \dots, \varphi_{k-1}, y),$$

where $g^{(0)}(\varphi_0, y) = g(\varphi_0, y, 0)$ and the nondegenerate matrix $B(y) = g_x(\varphi_0, y, 0)$; see [1, 4]. Taking these expansions into account, the invariance equation (5) takes the form:

$$\sum_{k\geq 0}\varepsilon^k\frac{\partial\varphi_k}{\partial y}\Big(g^{(0)} + B\sum_{k\geq 1}\varepsilon^k\varphi_k + \sum_{k\geq 1}\varepsilon^k g^{(k)}\Big) = \varepsilon\sum_{k\geq 0}\varepsilon^k f^{(k)}.$$

Equating the coefficients at the same of like powers of ε in the last equation, we get the expressions, which uniquely define the coefficients in (4) when $\det(\partial\varphi_0/\partial y) \neq 0$.

Indeed, for ε^0 we have $g(\varphi_0, y, 0) = 0$, which give the function $\varphi_0(y)$. For ε^1, we get

$$\varphi_1 = \Big(\frac{\partial\varphi_0}{\partial y}B\Big)^{-1}\Big(f^{(0)} - \frac{\partial\varphi_0}{\partial y}g^{(1)}\Big).$$

Likewise, for ε^k we obtain

$$\varphi_k = \Big(\frac{\partial\varphi_0}{\partial y}B\Big)^{-1}\Big[f^{(k-1)} - \frac{\partial\varphi_0}{\partial y}g^{(k)} - \sum_{i=1}^{k-1}\frac{\partial\varphi_i}{\partial y}\big(B\varphi_i + g^{(k-i)}\big)\Big].$$

Thus, the parametric representation of the slow invariant manifold of (1), (2) is found in the form (4).

3 The Case $n < m$

Consider the case where the number of fast variables in the system (1), (2) exceeds the number of slow variables. Then, the system (3) contains m equations for n unknowns and $n < m$. We take all components of vector x ($\dim(x) = n$) complemented by $m - n$ components of vector y, as the unknowns. Thereby, the number of equations and unknowns in the system (3) will coincide.

Suppose that the solution of (3) can be written in the form

$$x = \varphi_0(y_2), \quad y_1 = \psi_0(y_2),$$

with a parameter y_2, where $y = (y_1, y_2)^T$, $\dim y_1 = m - n$, $\dim y_2 = n$. The system (1), (2) start the sentence with this can be rewritten in a more convenient form:

$$\dot{x} = f(x, y_1, y_2, \varepsilon), \tag{6}$$

$$\varepsilon\dot{y}_1 = g_1(x, y_1, y_2, \varepsilon), \tag{7}$$

$$\varepsilon \dot{y}_2 = g_2(x, y_1, y_2, \varepsilon). \tag{8}$$

We will find the slow integral manifold in the form

$$x = \varphi(y_2, \varepsilon), \tag{9}$$

$$y_1 = \psi(y_2, \varepsilon). \tag{10}$$

Substituting (9), (10) into (6) and (7), and taking into account (8), we obtain the invariance equations

$$\frac{\partial \varphi}{\partial y_2} g_2(\varphi, \psi, y_2, \varepsilon) = \varepsilon f(\varphi, \psi, y_2, \varepsilon),$$

$$\frac{\partial \psi}{\partial y_2} g_2(\varphi, \psi, y_2, \varepsilon) = g_1(\varphi, \psi, y_2, \varepsilon).$$

For the functions $\varphi(y_2, \varepsilon)$, and $\psi(y_2, \varepsilon)$ we write the formal asymptotic expansions:

$$\varphi(y_2, \varepsilon) = \varphi_0(y_2) + \varepsilon\varphi_1(y_2) + \cdots + \varepsilon^k \varphi_k(y_2) + \cdots, \tag{11}$$

$$\psi(y_2, \varepsilon) = \psi_0(y_2) + \varepsilon\psi_1(y_2) + \cdots + \varepsilon^k \psi_k(y_2) + \cdots \tag{12}$$

Equating the coefficients of the same powers of ε in the invariance equations, we get the expressions, which uniquely define the coefficients in (11) and (12) when $\det \partial\varphi_0/\partial y_2 \neq 0$ and $\det \partial\psi_0/\partial y_2 \neq 0$.

4 The Case $n > m$

Consider the case when the dimension of slow variables is greater than the dimension of fast variables. We call attention to the degenerate subsystem (3). It contains m equations for n unknowns, where $n > m$. To find the parametric representation of the slow invariant manifold of (1), (2), we take all components of y complemented by $n - m$ components of the vector x, as the parameters. Then a solution of the system (3) can be written in the parametric form $x_1 = \varphi_0(x_2, y)$, where $x = (x_1, x_2)^T$, $\dim x_1 = m$, $\dim x_2 = n - m$.

The system (1), (2) in this case can be rewritten in more convenient form as

$$\begin{aligned} \dot{x}_1 &= f_1(x_1, x_2, y, \varepsilon), \\ \dot{x}_2 &= f_2(x_1, x_2, y, \varepsilon), \\ \varepsilon \dot{y} &= g_2(x_1, x_2, y, \varepsilon). \end{aligned} \tag{13}$$

We will find the slow integral manifold in the form

$$x_1 = \varphi(x_2, y, \varepsilon) = \varphi_0(x_2, y) + \varepsilon\varphi_1(x_2, y) + \cdots + \varepsilon^k\varphi_k(x_2, y) + \cdots. \quad (14)$$

The invariance equation

$$\varepsilon\frac{\partial\varphi}{\partial x_2} f_2(\varphi, x_2, y, \varepsilon) + \frac{\partial\varphi}{\partial y} g(\varphi, x_2, y, \varepsilon) = \varepsilon f_1(\varphi, x_2, y, \varepsilon) \quad (15)$$

is yielded from (13) and (14). Equating the coefficients at the same powers of ε in the last equation, we get the expressions, which uniquely define the coefficients in (14) for the case when $\det\left(\partial\varphi_0/\partial y \; G\right) \neq 0$.

Thus, formula (14) defines the slow integral manifold of the system in the parametric form.

References

1. M.P. Mortell, R.E. O'Malley, A. Pokrovskii, V.A. Sobolev (eds.), Singular perturbation and hysteresis, in *SIAM, 2005*
2. E. Shchepakina, O. Korotkova, Condition for canard explosion in a semiconductor optical amplifier. JOSA B Opt. Phys. **28**(8), 1988–1993 (2011)
3. E. Shchepakina, O. Korotkova, Canard explosion in chemical and optical systems. DCDS-B **18**(2), 495–512 (2013)
4. E.A. Shchepakina, V.A. Sobolev, M.P. Mortell, *Singular Perturbations. Introduction to System Order Reduction Methods with Applications*. Lecture Notes in Mathematics, vol. 2114 (Springer, Berlin, 2014)
5. V.A. Sobolev, E.A. Tropkina, Asymptotic expansions of slow invariant manifolds and reduction of chemical kinetics models. Comput. Math. Math. Phys. **52**(1), 75–89 (2012)

Breathing as a Periodic Gas Exchange in a Deformable Porous Medium

Michela Eleuteri, Erica Ipocoana, Jana Kopfová and Pavel Krejčí

Abstract We propose to model the mammalian lungs as a viscoelastic deformable porous medium with a hysteretic pressure–volume relationship described by the Preisach operator. Breathing is represented as an isothermal time-periodic process with the gas exchange between the interior and exterior of the body. The main result consists of proving the existence of a periodic solution under an arbitrary periodic forcing in suitable function spaces.

1 Introduction

As pointed out in [6], the first measurements which showed a hysteretic pressure–volume characteristic in mammalian lungs were obtained in [2] in 1913. A mechanical system combining linear viscoelasticity with the rate-independent Prandtl model of elastoplasticity was used by J. Hildebrandt in [5] to describe the breathing process of cats. Here we refer to the analysis carried out by D. Flynn in [4], where the Preisach operator is shown to be an appropriate model for the pressure–volume hysteresis relationship in lungs.

Supported by the Project of Excellence CZ.02.1.01/0.0/0.0/16_019/0000778 of the Ministry of Education, Youth and Sports of the Czech Republic.

M. Eleuteri (✉) · E. Ipocoana
Università Degli Studi di Modena e Reggio Emilia, Modena, Italy
e-mail: michela.eleuteri@unimore.it

E. Ipocoana
e-mail: 207531@studenti.unimore.it

J. Kopfová
Silesian University Opava, Opava, Czech Republic
e-mail: jana.kopfova@math.slu.cz

P. Krejčí
Faculty of Civil Engineering, Czech Technical University, Prague, Czech Republic
e-mail: krejci@math.cas.cz

A. Korobeinikov et al. (eds.), *Extended Abstracts Spring 2018*,
Trends in Mathematics 11, https://doi.org/10.1007/978-3-030-25261-8_20

Our work focuses on representing the breathing as an isothermal, time-periodic process described by a PDE system with hysteresis. It consists of the momentum balance equation and the mass balance, similarly as in a more general study of deformable porous media in [3], but with different boundary conditions. Instead of prescribing boundary displacement as in [3], we prescribe here mechanical reaction between lungs and their surroundings. Since viscosity is present in our model, we also do not have any restriction on the input amplitude. The mathematical problem thus consists of proving that our PDE system with a degenerating Preisach operator under the time derivative admits a periodic solution for every periodic boundary forcing with a given regularity.

2 The Model

Let u denote the displacement vector in the solid, σ the stress tensor, q the gas mass flux and s the gas mass content in the pores. Similarly as in [1], we assume that the system is governed by the momentum balance equation

$$\rho u_{tt} = \mathrm{div}\sigma, \tag{1}$$

where ρ is the solid mass density, and by the gas mass balance

$$s_t + \mathrm{div} q = 0, \tag{2}$$

where q is the mass flux. Then we introduce two constitutive relations. In the first one we have

$$\sigma = \mathbf{B}\nabla_s u_t + \mathbf{A}\nabla_s u - p\ \delta, \tag{3}$$

where $\mathbf{B}$, representing viscosity, and $\mathbf{A}$, representing elasticity, are symmetric positive-definite constant tensors of order 4, the symbol ∇_s denotes the symmetric gradient, p is the air pressure, and δ is the Kronecker tensor. The second constitutive relation links pressure and volume in the form

$$f(p) + G[p] = \frac{1}{\rho_a} s - \mathrm{div}\ u, \tag{4}$$

where $\rho_a > 0$ is the referential air mass density at standard pressure, $f : \mathbb{R} \to (0, \infty)$ is an increasing function, and G is a Preisach operator.

Under the small deformation hypothesis, the term div u represents the void volume difference with respect to the reference state, it means that, at constant pressure, if div u increases, then s/ρ_a increases at the same rate. Similarly, at constant void volume, the mass content s is an increasing function (with different inflation and deflation curves) of the pressure. Eventually, at constant gas mass content, the pressure increases if the void volume decreases.

For the mass flux, we assume the Darcy law

$$q = -\rho_a \mu(x) \nabla p, \tag{5}$$

where $\mu(x) > 0$ is a permeability coefficient depending on space.

According to the previous analysis, the model reads

$$\rho u_{tt} = \mathrm{div}(\mathbf{B}\nabla_s u_t + \mathbf{A}\nabla_s u) - \nabla p\,, \tag{6}$$

$$(f(p) + G[p])_t = -\mathrm{div} u_t + \mathrm{div}\mu(x)\nabla p\,, \tag{7}$$

for x in a bounded connected Lipschitzian domain $\Omega \subset \mathbb{R}^3$ and for $t \geq 0$.

On the boundary $\partial\Omega$, we prescribe the following boundary conditions:

$$-\sigma \cdot n\big|_{\partial\Omega} = \beta(x)(\mathbf{C}u + \mathbf{D}u_t - g) + pn\,, \tag{8}$$

$$\frac{1}{\rho_a} q \cdot n\big|_{\partial\Omega} = \alpha(x)(p - h) - u_t \cdot n\,, \tag{9}$$

where n is the unit outward normal vector, $\beta \geq 0$ is the relative elasticity modulus of the boundary at the point $x \in \partial\Omega$, $\mathbf{C}$ and $\mathbf{D}$ are symmetric positive- definite 3×3 matrices, $g = g(x, t)$ is a given external force acting on the body Ω, $h = h(x, t)$ is the given outer air pressure, and $\alpha(x) \geq 0$ is the boundary permeability at the point $x \in \partial\Omega$.

The physical meaning of the first boundary condition in (8) is that on the part of the boundary where β is positive, the body Ω interacts with the exterior, which is viscoelastic with stiffness $\mathbf{C}$, viscosity $\mathbf{D}$, and active component g. There is no mechanical interaction with the exterior on the part of boundary where β vanishes. Similarly, the second boundary condition in (8) reflects the assumption that gas exchange proportional to the inner and outer pressure difference takes place on the part of the boundary where α is positive.

We now write Problems (6)–(7) in variational form for all test functions $\phi \in W^{1,2}(\Omega; \mathbb{R}^3)$ and $\psi \in W^{1,2}(\Omega)$ as follows:

$$\int_\Omega (\rho u_{tt}\phi + (\mathbf{B}\nabla_s u_t + \mathbf{A}\nabla_s u) : \nabla_s \phi + \nabla p\, \phi)\mathrm{d}x + \int_{\partial\Omega} \beta(x)(\mathbf{C}u + \mathbf{D}u_t - g)\phi \mathrm{d}s(x) = 0\,, \tag{10}$$

$$\int_\Omega ((f(p) + G[p])_t \psi + (\mu(x)\nabla p - u_t)\nabla\psi)\mathrm{d}x + \int_{\partial\Omega} \alpha(x)(p - h)\psi \mathrm{d}s(x) = 0\,, \tag{11}$$

and the identities (10)–(11) are supposed to hold for a. e. $t > 0$.

2.1 Setting

Before presenting the main result of the present work, we must introduce the setting we need to study our problem. In particular, we have to make appropriate mathematical assumptions about different terms involved in the model.

- *Preisach operator*: Let $\gamma \in L^\infty((0,\infty) \times \mathbb{R})$ be a given function, $\gamma(r,v) \geq 0$ a. e., and there exists $B > 0$ such that $\gamma(r,v) = 0$ for $r + |v| \geq B$. We define

$$G[p] = \int_0^\infty \int_0^{\xi_r} \gamma(r,v)\mathrm{d}v\mathrm{d}r,$$

  where $\xi_r = \mathfrak{p}_r[p]$ is the output of the play operator applied to p. Note that for a fixed initial distribution of the play operators and input $p \in L^q(\Omega; C_T)$, the output $G[p]$ of the Preisach operator is T-periodic for $t \geq T$, so that we can consider G as a (Lipschitz continuous) mapping $L^q(\Omega; C_T) \to L^q(\Omega; C_T)$.
- *Periodic spaces*: We fix a period $T > 0$ and denote by L^q_T the L^q-space of T-periodic functions $v : \mathbb{R} \to \mathbb{R}$ for $q \geq 1$, and by C_T the space of continuous real T-periodic functions on $\mathbb{R}$. It is now quite natural to introduce $L^q_T(W^{1,2}(\Omega))$ and $L^q_T(W^{1,2}(\Omega, \mathbb{R}^3))$ of T-periodic L^q-functions $v : \mathbb{R} \to W^{1,2}(\Omega)$ and $v : \mathbb{R} \to W^{1,2}(\Omega, \mathbb{R}^3)$, respectively, as well as with the spaces $L^q(\Omega; C_T)$.
- *Nonlinearity*: $f : \mathbb{R} \to \mathbb{R}$ is a C^1-function such that there exist $0 < f_0 < f_1$ and $\omega \geq 0$ with the property

$$\frac{f_0}{1+p^2} \leq f'(p) \leq f_1(1+p^2)^\omega,$$

  for all $p \in \mathbb{R}$.

  Moreover, we assume that

  (i) the permeability coefficient μ belongs to $L^\infty(\Omega)$ and there exists a constant $\mu_0 > 0$ such that $\mu(x) \geq \mu_0$ a. e.;
  (ii) the nonnegative functions α and β belong to $L^\infty(\partial\Omega)$ and do not identically vanish, that is, $\int_{\partial\Omega} \beta(x)\mathrm{d}s(x) > 0$, $\int_{\partial\Omega} \alpha(x)\mathrm{d}s(x) > 0$;
  (iii) the functions g, g_t belong to $L^2_T(L^2(\partial\Omega; \mathbb{R}^3))$, h, h_t belong to $L^2_T(L^2(\partial\Omega))$, $h \in L^\infty(\partial\Omega \times (0,T))$;
  (iv) the symmetric positive-definite constant tensors $\mathbf{A}$, $\mathbf{B}$ and symmetric positive-definite constant matrices $\mathbf{C}$, $\mathbf{D}$ are given.

2.2 Main Result

Theorem 1 *Let the assumptions from Sect. 2.1 hold. Then system* (10)–(11) *has a solution* (u, p) *such that* $u,\, u_t,\, \nabla_s u,\, \nabla_s u_t \in L^2_T(L^2(\Omega;\mathbb{R}^3)) \cap L^\infty(T, 2T; L^2(\Omega;\mathbb{R}^3))$, $u_{tt} \in L^2_T(L^2(\Omega;\mathbb{R}^3))$, $p_t, \nabla p \in L^2_T(L^2(\Omega))$, $p \in L^\infty(\Omega \times (T, 2T))$.

Proof The main ideas of the proof are the following:

(i) We take a cut-off function for f;
(ii) We use the Galerkin method, both in space and time, so we take suitable orthonormal bases;
(iii) As a result, we get an algebraic system which has a solution by a homotopy argument;
(iv) We derive a priori estimates with the help of Preisach energy inequality and Korn and Poincaré inequalities;
(v) In order to get uniform estimates in time, we test the cut-off equations by u_t and u_{tt}, regularizing in time when needed;
(vi) Eventually, we are able to remove the cut-off parameter using the Moser method. □

References

1. B. Albers, P. Krejčí, Unsaturated porous media flow with thermomechanical interaction. Math. Methods Appl. Sci. **39**(9), 2220–2238 (2016)
2. M. Cloetta, Untersuchungen über die Elastizität der Lunge und deren Bedeutung für die Zirkulation. Pflüger's Archiv für die gesamte Physiologie des Menschen und der Tiere **152**, 339–364 (1913)
3. B. Detmann, P. Krejčí, E. Rocca, Periodic waves in unsaturated porous media with hysteresis, in *Proceedings of ECM 2016*, eds. by V. Mehrmann, M. Skutella (EMS Publishing House, Zürich, 2018), pp. 219–234
4. D. Flynn, A survey of hysteresis models of mammalian lungs. Rend. Sem. Mat. Politec. Torino **72**, 17–36 (2014)
5. J. Hildebrandt, Pressure-volume data of cat lung interpreted by a plastoelastic, linear viscoelastic model. J. Appl. Physiol. **28**, 365–372 (1970)
6. J. Mead, Mechanical properties of lungs. Physiol. Rev. **41**, 281–330 (1961)

Invariant Objects on Lattice Systems with Decaying Interactions

Rubén Berenguel and Ernest Fontich

Abstract In this note, we describe the setting and main results concerning the existence of invariant tori and a class of invariant manifolds for differentiable skew product systems in lattices having interactions with spatial decay among all nodes. We obtain decay of the parameterizations of the objects we find.

1 Introduction

We consider systems defined on lattices with each node having its own dynamics, and an interaction with the other nodes. The origin of the study of lattice systems can be found in the first models of the dynamics of chains of particles under the action of some potential, with an interaction to nearest neighbours, models which were first considered by Prandtl [14] and Dehlinger [5]. Later these models were also considered by Frenkel and Kontorova for specific cases [8, 9]. See also the book [2]. As a motivating example, we mention the Klein–Gordon lattice. It consists of a one-dimensional array of particles governed by a potential V and having interactions with the nearest neighbour through a spring type force.

The equations can be written as

$$\ddot{q}_n + V'(q_n) = \varepsilon(q_{n+1} + q_{n-1} - 2q_n),$$

$n \in \mathbb{Z}$, where $\varepsilon > 0$ is the so-called coupling constant. It admits a (formal) Hamiltonian formulation

This work has been supported by grants MTM2016-80117-P, MDM2014-0445 (Spain) and 2017-SGR-1374 (Catalonia).

R. Berenguel (✉) · E. Fontich
Departament de Matemàtiques i Informàtica, Barcelona Graduate School of Mathematics (BGSMath), Universitat de Barcelona (UB), Gran Via, 585, 08007 Barcelona, Spain
e-mail: ruben@mostlymaths.net

E. Fontich
e-mail: fontich@ub.edu

A. Korobeinikov et al. (eds.), *Extended Abstracts Spring 2018*,
Trends in Mathematics 11, https://doi.org/10.1007/978-3-030-25261-8_21

$$H((q_n, p_n)_{n\in\mathbb{Z}}) = \sum_{n\in\mathbb{Z}} \Big(\frac{1}{2}p_n^2 + V(q_n) + \frac{\varepsilon}{2}(q_n - q_{n+1})^2\Big),$$

where $(q_n, p_n) \in \mathbb{R}^2$ or $(q_n, p_n) \in \mathbb{T} \times \mathbb{R}$. This can be easily generalized to n-dimensional nodes, to N dimensional lattices and to long range interaction. The corresponding Hamiltonian would be

$$H((q_n, p_n)_{n\in\mathbb{Z}^N}) = \sum_{n\in\mathbb{Z}^N} \Big(\frac{1}{2}|p_n|^2 + V(q_n) + \varepsilon \sum_{j\in\mathbb{Z}^N} W_j(q_n - q_{n+j})\Big),$$

where W_j is a collection of functions such that $|W_j|$ decreases to 0 when $|j| \to \infty$.

Stroboscopic maps of these systems are (families) of diffeomorphisms $F_\varepsilon : \mathcal{M} \to \mathcal{M}$, where $\mathcal{M}$ is a suitable phase space described below, of the form

$$F_\varepsilon = F_0 + \varepsilon F_{1,\varepsilon},$$

where $F_{1,\varepsilon}$ describes the interactions between the nodes.

More generally, we consider skew product systems to be able to model quasi-periodic force interactions. Our goal is to obtain invariant tori, some of their invariant manifolds, and to study the decay of the parameterizations of these objects in terms of the decay of the maps.

We will be working with maps, although some results for differential equations follow as corollaries.

2 Setting and Notation

We will consider the space of sequences

$$\ell^\infty = \ell^\infty(\mathbb{R}^n) = (\mathbb{R}^n)^{\mathbb{Z}^N} = \{x = (x_i)_{i\in\mathbb{Z}^N} \in \prod_{i\in\mathbb{Z}^N} \mathbb{R}^n \mid \ \|x\| := \sup_{i\in\mathbb{Z}^N} |x_i| < \infty\},$$

or an open subset of it. Note that ℓ^∞ is a Banach space neither separable nor reflexive.

Moreover, it turns out that the matrix representation of a linear map of ℓ^∞

$$(Au)_i = \sum_{j\in\mathbb{Z}^N} A_{ij}u_j, \qquad A_{i,j} = \pi_i \circ A \circ \mathrm{emb}_j \in L(\mathbb{R}^n, \mathbb{R}^n),$$

$i, j \in \mathbb{Z}^N$, does not completely determine the map; see [6]. Here, $\pi_i : \ell^\infty \to \mathbb{R}^n$ and $\mathrm{emb}_j : \mathbb{R}^n \to \ell^\infty$ are the canonical projection and embedding, respectively.

We will use a useful class of decay functions introduced by Jiang–de-la-Llave [12].

Definition 1 A decay function is a map $\Gamma\colon \mathbb{Z}^N \to \mathbb{R}^+$ satisfying

(i) $\sum_{j\in\mathbb{Z}^N} \Gamma(j) \le 1$,
(ii) $\sum_{j\in\mathbb{Z}^N} \Gamma(i-j)\Gamma(j-k) \le \Gamma(i-k)$.

Example 2 Given $\alpha > 0$, $\beta \ge 0$, there exists $a_0 = a_0(\alpha, \beta, N) > 0$ such that if $0 < a < a_0$

$$\Gamma(i) = \begin{cases} a|i|^{-\alpha}e^{-\beta|i|} & \text{if } i \ne 0, \\ a & \text{if } i = 0, \end{cases}$$

is a decay function. Note that, $\Gamma(i) = a\, e^{-\beta|i|}$, with $a > 0$, $\beta > 0$, is not a decay function.

These functions are used as weights to control the strength of the interaction between nodes i, j, separated a distance $|i-j|$.

We start by introducing linear maps with decay, or more precisely Γ-linear maps. We define

$$L_\Gamma(\ell^\infty, \ell^\infty) = \{A \in L(\ell^\infty, \ell^\infty) \mid \|A\|_\Gamma < \infty\},$$

where

$$\|A\|_\Gamma = \max\{\|A\|, \gamma(A)\} \qquad \text{and} \qquad \gamma(A) = \sup_{i,j\in\mathbb{Z}^N} \Gamma(i-j)^{-1}|A_{ij}|.$$

In this way, if $A \in L_\Gamma$ then the matrix blocks of A satisfy $|A_{ij}| \le \gamma(A)\,\Gamma(i-j)$.

The properties of Γ imply that $\|AB\|_\Gamma \le \|A\|_\Gamma\|B\|_\Gamma$ which, in turn, implies that the space L_Γ is a Banach algebra. We have $\|\mathrm{Id}\|_\Gamma = \Gamma^{-1}(0) > 1$.

Obviously, we have that $L_\Gamma(\ell^\infty, \ell^\infty) \subset L(\ell^\infty, \ell^\infty)$ as sets. However, $L_\Gamma(\ell^\infty, \ell^\infty)$ is not a closed subalgebra of $L(\ell^\infty, \ell^\infty)$.

The space $L^k_\Gamma(\ell^\infty, \ell^\infty)$ is defined as the space of k-linear maps that are Γ-linear with respect to each variable. Compositions and contractions of k-linear maps with Γ decay are also multilinear maps with Γ decay; see details in [6].

Next we define some spaces of functions. Given $j \in \mathbb{Z}^m$ and $r \ge 0$, we define

$$\begin{aligned} S^r_{j,\Gamma} &= S^r_{j,\Gamma}(\mathbb{T}^d) \\ &= \Big\{\sigma \in C^r(\mathbb{T}^d, \ell^\infty) \mid \|\sigma\|_{S^r_{j,\Gamma}} := \max_{0\le|k|\le r}\, \sup_{i\in\mathbb{Z}^m}\, \sup_{\theta\in\mathbb{T}^d} \|\partial^k_\theta \sigma_i(\theta)\|\Gamma(i-j)^{-1} < \infty\Big\}. \end{aligned}$$

We need to deal with two different sets of variables, x and θ, which play a slightly different role. Therefore, we consider spaces of anisotropic regularity.

Given U an open set of $\ell^\infty(\mathbb{R}^n)$ and two non-negative integers t, r, we define

$$\begin{aligned} C^{t,r}(U \times \mathbb{T}^d, \ell^\infty) = \{F \in C^0(U \times \mathbb{T}^d, \ell^\infty) \mid{}& D^i_x \partial^k_\theta F \text{ exist and are} \\ & \text{continuous and bounded for } 1 \le i \le t,\ 0 \le |k| \le r\}, \end{aligned}$$

where $k \in (\mathbb{Z}^+)^d$ and ∂^k_θ stands for $\partial/\partial\theta_1^{k_1} \cdots \partial\theta_d^{k_d}$ and

$$C^{t,r}_\Gamma(U \times \mathbb{T}^d, \ell^\infty) = \{F \in C^{t,r}(U \times \mathbb{T}^d, \ell^\infty) |\ D^i_x \partial^k_\theta F \text{ exist and } D^i_x \partial^k_\theta F \in C^0(U \times \mathbb{T}^d, L^i_\Gamma(\ell^\infty, \ell^\infty)), 1 \le i \le t,\ 0 \le |k| \le r,\ \|F\|_{C^{t,r}_\Gamma} < \infty\},$$

with norm

$$\|F\|_{C^{t,r}_\Gamma} = \max\big(\max_{0\le|k|\le r} \|\partial^k_\theta F\|_{C^0},\ \max_{\substack{1\le i\le t \\ 0\le|k|\le r}} \sup_{\substack{x\in U \\ \theta\in\mathbb{T}^d}} \|D^i_x \partial^k_\theta F(x,\theta)\|_\Gamma \big).$$

For maps depending only on x, the corresponding analogous space is $C^t(U, \ell^\infty)$. In a similar way to [4], where the authors have to deal with variables x and parameters λ, we also introduce other classes of spaces. We define

$$C^{t,r}_{j,\Gamma}(U \times \mathbb{T}^d, \ell^\infty) = \{F \in C^{t,r}_\Gamma(U \times \mathbb{T}^d, \ell^\infty) \ |\ F(x,\cdot) \in S^r_{j,\Gamma},\ x \in U, \|F\|_{C^{t,r}_{j,\Gamma}} = \max\big(\|F\|_{C^{t,r}_\Gamma}, \sup_{x\in U} \|F(x,\cdot)\|_{S^r_{j,\Gamma}}\big) < \infty\}.$$

Let $\Sigma_{t,r} = \{(k,i) \in (\mathbb{Z}^+)^{d+1} \ |\ |k| \le r,\ i + |k| \le t + r\}$. We define

$$C^{\Sigma_{t,r}}(U \times \mathbb{T}^d, \ell^\infty) = \{F \in C^0(U \times \mathbb{T}^d, \ell^\infty) \,|\, D^i_x \partial^k_\theta F \text{ exist and are continuous and bounded for } (k,i) \in \Sigma_{t,r}, \|F\|_{C^{\Sigma_{t,r}}} = \max\big(\|F\|_{C^0}, \max_{(k,i)\in\Sigma_{t,r}} \sup_{\substack{x\in U \\ \theta\in\mathbb{T}^d}} \|D^i_x \partial^k_\theta F(x,\theta)\|\big) < \infty\},$$

and the decay version

$$C^{\Sigma_{t,r}}_\Gamma(U \times \mathbb{T}^d, \ell^\infty) = \{F \in C^{\Sigma_{t,r}}(U \times \mathbb{T}^d, \ell^\infty) \,| D^i_x \partial^k_\theta F \in C^0(U \times \mathbb{T}^d, L^i_\Gamma(\ell^\infty, \ell^\infty)), (k,i) \in \Sigma_{t,r},\ i \ge 1,\ \|F\|_{C^{t,r}_\Gamma} < \infty\},$$

with norm

$$\|F\|_{C^{\Sigma_{t,r}}_\Gamma} = \max\big(\max_{0\le|k|\le r} \|\partial^k_\theta F\|_{C^0},\ \max_{\substack{(k,i)\in\Sigma_{t,r}, \\ i\ge1}} \sup_{\substack{x\in U \\ \theta\in\mathbb{T}^d}} \|D^i_x \partial^k_\theta F(x,\theta)\|_\Gamma \big).$$

3 Invariant Tori and Invariant Manifolds

The skew product systems we deal with are defined on $\ell^\infty(\mathbb{R}^n) \times \mathbb{T}^d$ and have the form

$$(x, \theta) \mapsto (F(x,\theta), \theta + \omega),$$

where $\omega \in \mathbb{R}^d$ is a fixed vector we refer to as frequency vector, and

$$F(x, \theta) = F_0(x) + F_1(x, \theta),$$

$x \in \ell^\infty(\mathbb{R}^n)$, $\theta \in \mathbb{T}^d$. We will call F_0 the unperturbed system. It is uncoupled, i.e., the dynamics of any node under F_0 are given by some map $f\colon \mathbb{R}^n \to \mathbb{R}^n$ with no interaction between nodes. Actually, we will have $(F_0(x))_i = f(x_i)$, for all $i \in \mathbb{Z}^N$. We assume that f has a hyperbolic fixed point at $0 \in \mathbb{R}^n$. Our first goal is, assuming F_1 is small in some sense, to find invariant tori near $0 \in \ell^\infty(\mathbb{R}^n)$.

According to the parameterization method [7, 10, 11, 13] we look for the tori as graphs of functions $W_0\colon \mathbb{T}^d \to \ell^\infty(\mathbb{R}^n)$. Since the skew product sends $(W_0(\theta), \theta)$ to $(F(W_0(\theta), \theta), \theta + \omega)$ the graph of W_0 is invariant if and only if

$$F(W_0(\theta), \theta) = W_0(\theta + \omega). \tag{1}$$

The main result concerning the existence of tori is

Theorem 3 *Assume*

(i) $F_0 \in C^t(U, \ell^\infty)$,
(ii) $F_1 \in C^{t,r}(U \times \mathbb{T}^d, \ell^\infty)$ *with* $t \geq r + 1$, $r \geq 0$,
(iii) $\|F_1\|_{C^{t,r}}$ *small enough.*

Then, the functional equation (1) *has a solution* $W_0 \in C^r(\mathbb{T}^d, \ell^\infty)$ *close to 0.*

Moreover, if $F_1 \in C^{t,r}_{j,\Gamma}(U \times \mathbb{T}^d, \ell^\infty)$ *with* $t \geq r + 2$, $r \geq 0$ *and* $\|F_1\|_{C^{t,r}_{j,\Gamma}}$ *is small enough then* $W_0 \in S^r_{j,\Gamma}(\mathbb{T}^d, \ell^\infty)$ *which implies that the torus is localized close to the node* j.

We can also find invariant manifolds of these invariant tori. We will look for non-resonant manifolds. Let us first introduce them in the simpler setting of fixed points of maps f of $\mathbb{R}^n$. For example, the matrix

$$A = \begin{pmatrix} 1/2 & 1 & & & \\ 0 & 1/2 & & & \\ & & 1/3 & & \\ & & & 1/4 & 1 \\ & & & 0 & 1/4 \end{pmatrix}.$$

A has many invariant subspaces. Denoting by E_i the i-th coordinate axis, we have that $E_1, E_3, E_4, E_1 \oplus E_2, E_4 \oplus E_5$, or sums of these spaces, e.g., $E_1 \oplus E_4$, $E_1 \oplus E_2 \oplus E_4, \ldots$ are invariant. If $f\colon \mathbb{R}^n \to \mathbb{R}^n$ is at least C^1 and $Df(0) = A$ we could expect to obtain invariant manifolds tangent to these subspaces. However, for them to exist, we need some conditions on the eigenvalues and on the regularity of f. See [3]. Non-resonant manifold in lattices for analytic systems are studied in [1].

In our setting, first we translate the torus to the origin $x = 0$ and denote the translated system again by F. Since $W_0 = O(\|F_1\|)$ we can write $F(x, \theta) = F_0(x) + \tilde{F}_1(x, \theta)$ and $M(\theta) := D_x F(0, \theta) = DF_0(0) + D_x \tilde{F}_1(0, \theta)$ with $\tilde{F}_1$ small. Let $M_0 = D_x F_0(0)$ and $\tilde{M}(\theta) = D_x \tilde{F}_1(0, \theta)$ and consider the decomposition $\ell^\infty = \mathcal{E}^1 \oplus \mathcal{E}^2$ such that with respect to it $M_0 = \begin{pmatrix} A_{11} & 0 \\ 0 & A_{22} \end{pmatrix}$. Let $\mathbb{D}$ be the open unit disc in $\mathbb{C}$. Given a set S we denote the annulus generated by it as the set $\mathcal{A} S = \{e^{i\theta} s \mid \theta \in [0, 2\pi),\ s \in S\}$.

Now we will address the result concerning non-resonant manifolds. In general, $D_x F(0, \theta)$ is not in block triangular form as it is required in the theorem below. In such a case, one can prove that with the additional hypothesis $\mathcal{A}\,\mathrm{Spec}\,A_{1,1} \cap \mathcal{A}\,\mathrm{Spec}\,A_{2,2} = \emptyset$ there is a C^r_Γ linear transformation turning $M(\theta)$ into block triangular form. Of course, if the linear map is already in block triangular form, this condition can be skipped.

Theorem 4 *In the setting described above, assume that for some L the following hypotheses hold:*

(i) $F \in C^{\Sigma_{t,r}}_\Gamma\left(\ell^\infty(\mathbb{R}^n) \times \mathbb{T}^d, \ell^\infty(\mathbb{R}^n)\right)$ *with* $t \geq r+1$, M_0, $\widetilde{M}(\theta) \in L_\Gamma(\ell^\infty(\mathbb{R}^n), \ell^\infty(\mathbb{R}^n))$ *is block upper triangular,* $\sup_{\theta\in\mathbb{T}^d} \|\widetilde{M}(\theta)\|_\Gamma$ *small;*
(ii) $\mathcal{A}\,\mathrm{Spec}\,(A_{1,1}) \subset \mathbb{D}\backslash\{0\}$*;*
(iii) $0 \notin \mathrm{Spec}\,(A_{2,2})$*;*
(iv) $\mathcal{A}\,\mathrm{Spec}\,(A_{1,1})^{L+1} \cdot \mathcal{A}\,\mathrm{Spec}\,(M_0^{-1}) \subset \mathbb{D}$*;*
(v) $\mathcal{A}\,\mathrm{Spec}\,(A_{1,1})^i \cap \mathcal{A}\,\mathrm{Spec}\,(A_{2,2}) = \emptyset$ *for* $2 \leq i \leq L$*;*
(vi) $L+1 \leq t$.

Then, we can determine a polynomial bundle map $R\colon \mathcal{E}^1 \times \mathbb{T}^d \to \mathcal{E}^1$ *of degree not larger than L in* $C^{\infty,r}_\Gamma(\mathcal{E}^1 \times \mathbb{T}^d, \mathcal{E}^1)$ *such that* $R(0,\theta) = 0$, $\|D_s R(0,\theta) - A_{1,1}(\theta)\|$ *is small and a bundle map* $W\colon B(0,1) \times \mathbb{T}^d \subset \mathcal{E}^1 \times \mathbb{T}^d \to \ell^\infty(\mathbb{R}^n)$ *in* $C^{\Sigma_{t,r}}_\Gamma(B(0,1) \times \mathbb{T}^d \subset \mathcal{E}^1 \times \mathbb{T}^d, \ell^\infty)$ *such that*

$$F(W(s,\theta),\theta) = W(R(s,\theta), \theta+\omega), \tag{2}$$

where $W(0,\theta) = 0$, $\Pi_{\mathcal{E}^1} D_s W(0,\theta) = \mathrm{Id}_{\mathcal{E}^1}$ *and* $\Pi_{\mathcal{E}^2} D_s W(0,\theta) = 0$.

Equation (2) forces the image of W to be invariant by F, and F restricted to it to be conjugated to R.

To solve Eq. (2) in the appropriate space first we look for approximations $W^{\leq}$ and $R^{\leq}$ such that

$$W^{\leq}(s,\theta) = 0 + W_1(\theta)s + \cdots + W_L(\theta)s^{\otimes L},$$
$$R^{\leq}(s,\theta) = R_1(\theta)s + \cdots + R_L(\theta)s^{\otimes L},$$

satisfying

$$F(W^{\leq}(s,\theta),\theta) = W^{\leq}(R^{\leq}(s,\theta), \theta+\omega) + \mathcal{O}(\|s\|^{L+1}).$$

To determine W_k and R_k for $k \geq 2$ we have to solve the following equations, commonly known as cohomological equations:

$$M(\theta)W_k(\theta) = W_1(\theta+\omega)R_k(\theta) + W_k(\theta+\omega)R_1^{\otimes k}(\theta) + \hat{Q}_k(\theta), \tag{3}$$

$k \geq 2$, where $\hat{Q}_k(\theta)$ comes inductively and depends on $D_x^j F(0,\theta)$, $j \leq k$, and $W_j(\theta)$ and $R_j(\theta)$ for $j < k$. The solvability of (3) depends on the spectral properties of the so-called Sylvester operators in homogeneous polynomial bundle maps. The spectral properties in the statement of the theorem imply that we can solve (3) recursively from $k = 2$ to $k = L$.

When $W^{\leq}$ is determined, we look for $W^{>}$ such that

$$F((W^{\leq} + W^{>})(s, \theta), \theta) = (W^{\leq} + W^{>})(R^{\leq}(s, \theta), \theta + \omega)$$

in the space $C_{\Gamma}^{\Sigma_{t,r}}(B(0, 1) \times \mathbb{T}^d \subset \mathcal{E}^1 \times \mathbb{T}^d, \ell^{\infty}(\mathbb{R}^n))$ by dealing with a suitable fixed point equation.

References

1. D. Blazevski, R. de la Llave, Localized stable manifolds for whiskered tori in coupled map lattices with decaying interaction. Ann. Henri Poincaré **15**(1), 29–60 (2014)
2. O.M. Braun, Y.S. Kivshar, *The Frenkel–Kontorova model: concepts, methods, and applications*. Texts and Monographs in Physics (Springer, Berlin, 2004)
3. X. Cabré, E. Fontich, R. de la Llave, The parameterization method for invariant manifolds, I. Manifolds associated to non-resonant subspaces. Indiana Univ. Math. J. **52**(2), 283–328 (2003)
4. X. Cabré, E. Fontich, R. de la Llave, The parameterization method for invariant manifolds, II. Regularity with respect to parameters. Indiana Univ. Math. J. **52**(2), 329–360 (2003)
5. U. Dehlinger, Zur theorie der rekristallisation reiner metalle. Annalen der Physik **394**(7), 749–793 (1929)
6. E. Fontich, R. de la Llave, P. Martín, Dynamical systems on lattices with decaying interaction I: a functional analysis framework. J. Differ. Equ. **250**(6), 2838–2886 (2011)
7. E. Fontich, R. de la Llave, Y. Sire, Construction of invariant whiskered tori by a parameterization method, I. Maps and flows in finite dimensions. J. Differ. Equ. **246**(8), 3136–3213 (2009)
8. Y.I. Frenkel, T.A. Kontorova, The model of dislocation in solid body. Zh. Eksp. Teor. Fiz **8** (1938)
9. Y.I. Frenkel, T.A. Kontorova, On the theory of plastic deformation and twinning. Acad. Sci. U.S.S.R. J. Phys. **1**, 137–149 (1939)
10. A. Haro, M. Canadell, J.L. Figueras, A. Luque, J.M. Mondelo, The parameterization method for invariant manifolds. From rigorous results to effective computations. Appl. Math. Sci. **195** (2016). Springer
11. A. Haro, R. de la Llave, A parameterization method for the computation of invariant tori and their whiskers in quasi-periodic maps: rigorous results. J. Differ. Equ. **228**(2), 530–579 (2006)
12. M. Jiang, R. de la Llave, Smooth dependence of thermodynamic limits of SRB-measures. Comm. Math. Phys. **211**(2), 303–333 (2000)
13. R. de la Llave, A. González, À. Jorba, J. Villanueva, KAM theory without action-angle variables. Nonlinearity **18**(2), 855–895 (2005)
14. L. Prandtl, A conceptual model to the kinetic theory of solid bodies. Z. Angew. Math. Mech. **8**, 85–106 (1928)

Wave-Pinning by Global Feedback in the Bistable Schlögl Model

F. Font, E. Moreno and S. Alonso

Abstract In this work, we introduce a wave-pinning mechanism in the bistable Schlögl model. Wave-pinning is induced by dynamically varying the unstable fixed point with a spatial global feedback. We present numerical simulations of the model in one and two dimensions for typical parameter values. The wave-pinning mechanism presented here can be used to reproduce the limited presence of phosphatidylinositol (3,4,5)-trisphosphate (PIP3) in the membrane of *Dictyostelium discoideum* cells, which plays a crucial role in the polarization and motility of the cell.

1 Introduction

Pattern formation is an ubiquitous phenomena in nature. Examples range from stripe pattern on a zebra's coat [3], to dissolution or growth of crystals in solutions [4]. Typically, these systems are modeled by means of nonlinear reaction–diffusion equations. A minimal model for pattern formation was formulated by Friedrich Schlögl to describe nonequilibrium phase transitions [5]. Although the model is commonly known as the Schlögl model, the same model was previously formulated by Zel'dovich and Frank–Kamenetskii to describe flame propagation [6]. In the past 50 years, the Schlögl model has been adapted to describe many other systems in physics and biology, including gas discharge between two glass plates or cardiac dynamics.

The bistable Schlögl model describing the evolution of a concentration field $u(x, t)$ is given by the reaction–diffusion equation

F. Font (✉) · E. Moreno · S. Alonso
Department of Physics, Universitat Politècnica de Catalunya, Barcelona, Spain
e-mail: ffont@crm.cat

E. Moreno
e-mail: eduardo.moreno.ramos@upc.edu

S. Alonso
e-mail: s.alonso@upc.edu

A. Korobeinikov et al. (eds.), *Extended Abstracts Spring 2018*,
Trends in Mathematics 11, https://doi.org/10.1007/978-3-030-25261-8_22

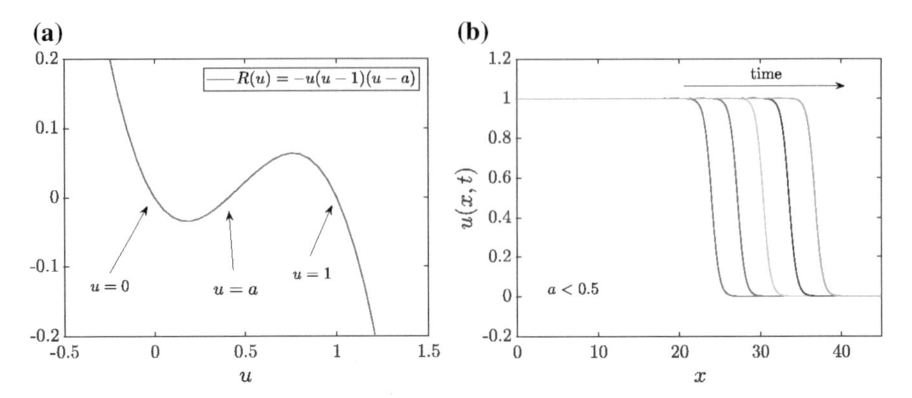

Fig. 1 **a** Reaction term for the Schlögl model as a function of u. The fixed points are indicated with arrows (in this case $a = 0.4$). **b** Concentration profiles for different times

$$\frac{\partial u}{\partial t} = D_u \nabla^2 u - k\, u(u-1)(u-a), \tag{1}$$

where D and k are the diffusion and reaction coefficients, respectively. In our case, $u(x, t)$ stands for the concentration of PIP3 in the cell membrane. The reaction term $R(u) = -k\, u(u-1)(u-a)$ can be interpreted as the derivative of a potential field, i.e., $R(u) = -\partial_u V(u)$. In the one-dimensional case, the model has the analytical (traveling wave) solution

$$u(x,t) = \frac{1}{2}\Big[1 - \tanh\Big(\frac{1}{2}\sqrt{\frac{k}{2D_u}}(x - ct)\Big)\Big], \quad c = \sqrt{\frac{2D_u}{k}}(1-2a). \tag{2}$$

From (2), one can see that the wave velocity c is positive, zero, or negative depending on the value of the unstable fixed point a. In Fig. 1, we show the reaction term of the Schlögl model and indicate the fixed points (panel (a)), and plot the traveling wave solution at different times for typical parameter values (panel (b)).

In the next section, we introduce a global feedback relation in model (1) to control the size of the wavefronts, in Sect. 3 we present and discuss numerical simulations of the model with global feedback (in 1D and 2D) and, finally, we draw our conclusions in Sect. 4.

2 Wave-Pinning by Global Feedback

A control of the size of the wave is important to model pattern formation in *Dictyostelium discoideum* cells, where the area covered by the pattern is limited and it never covers the entire cell membrane [1]. Thus, we introduce a global feedback control mechanism in the Schlögl model that stops the wavefront when a critical size is reached. The governing equation will now read

$$\frac{\partial u}{\partial t} = D_u \nabla^2 u - ku(u-1)(u-a(u)), \tag{3}$$

with the feedback-control mechanism given by

$$a = a_0 + \Delta a \Big(\int_A u\, dA - p\, A \Big), \tag{4}$$

where A is the area of the domain, Δa the strength of the global feedback input, and p the critical fraction of area covered by the wavefront. Note the way in which the global feedback is introduced in the model differs from that used in previous studies [2, 4], where the global feedback induces a vertical shift in the reaction term R and, therefore, the value of the stable fixed points (in our case $u = 0$, $u = 1$) also change.

3 Results and Discussion

In this section, we present numerical simulations of the model (3)–(4) in one and two dimensions with no flux boundary conditions using an explicit finite difference scheme. The parameter values used are $D_u = 0.1$, $k = 1$, $a_0 = 0.5$, $L = 45$. Although these values have physical meaning and corresponding units we omit their description for brevity.

3.1 *Simulations in* 1*D*

In Fig. 2, we present the results of the simulations in 1D (setting $\nabla = \partial_x$, $A = L$ and $dA = dx$ in (3)–(4)), using the initial condition: $u(x, 0) = 1$ for $x \in [0, L/2]$ and $u(x, 0) = 0$ for $x \in (L/2, L]$. In panel (a), we show the evolution of the PIP3 concentration wave for the case $p = 0.65$ and $\Delta a = 0.02$. We observe how the wave travels forward until $\int u\, dx = 0.65\, L$ and then stops. The global feedback is pushing the fixed point toward $a = 0.5$, which is precisely the value of the unstable fixed point in the Schlögl model leading to a velocity of the traveling wave equal to 0 (see Eq. (2)). To better visualize this phenomena, known as wave-pinning, we show in panel (b) the position and velocity of the point in the domain where $u = 0.5$ that we define as $x = s(t)$, i.e., $u(s(t), t) = 0.5$. We observe that the speed of the wave, represented by $\dot{s}(t)$, increases during a short transient period and then decreases toward zero.

In panels (c) and (d), we show the reaction term and the potential, respectively, at three different times during wave propagation. Initially, the fixed point $u = 1$ is more stable than $u = 0$. As time increases, $(\int u\, dx - p\, L) \to 0$ and $a \to 0.5$ making the potential symmetric with respect $u = 0.5$ and the fixed points $u = 0$ and $u = 1$ become equally stable.

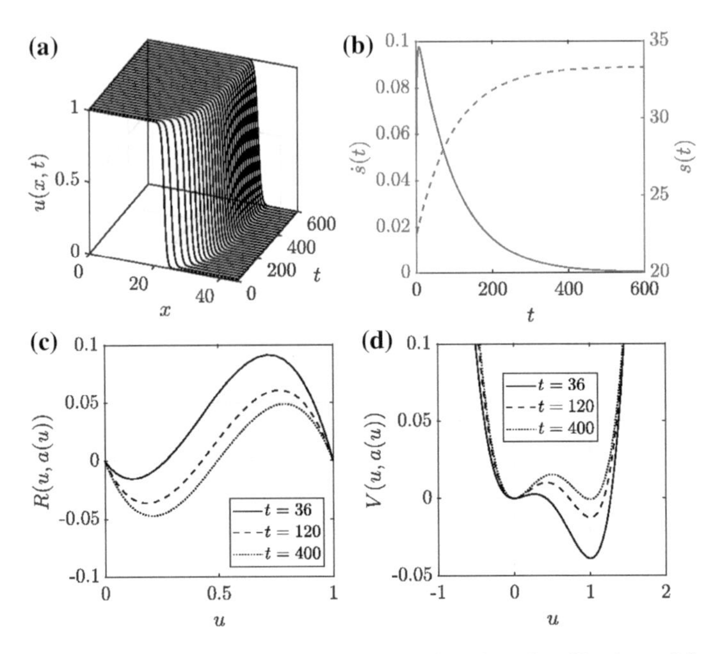

Fig. 2 **a** Space and time evolution of the concentration of PIP3 predicted by the model with global feedback. **b** Speed (left y-axes) and position (right y-axes) of the wavefront as a function of time. **c** Reaction term and **d** Potential at different times

3.2 *Simulations in 2D*

In Fig. 3, we compare 2D numerical simulations of the Schlögl model with and without global feedback at several times. In this case, the imposed initial condition is $u(x, y, 0) = 1$ for $r \leq L/2$ and $u(x, y, 0) = 0$ for $r > L/2$, where $r = \sqrt{x^2 + y^2}$. The simulations for the Schlögl model without global feedback (panels (a)–(d)) show how a PIP3 concentration wave travels unperturbed through the medium and at $t = 720$ has already covered most of the domain (the entirety of the domain is covered around $t \approx 1000$). In the case of the model with global feedback (panels (e)–(h)), the wave evolves initially fast (in agreement with the observations for the 1D case), then slows down, and eventually stops when the area covered equals the critical area $0.5\,L^2$ (note for these simulations we have used $p = 0.5$ and and $\Delta a = 5 \cdot 10^{-4}$). We propose this mechanism as a mass conservation constraint to model the limited availability of PIP3 on the cell membrane.

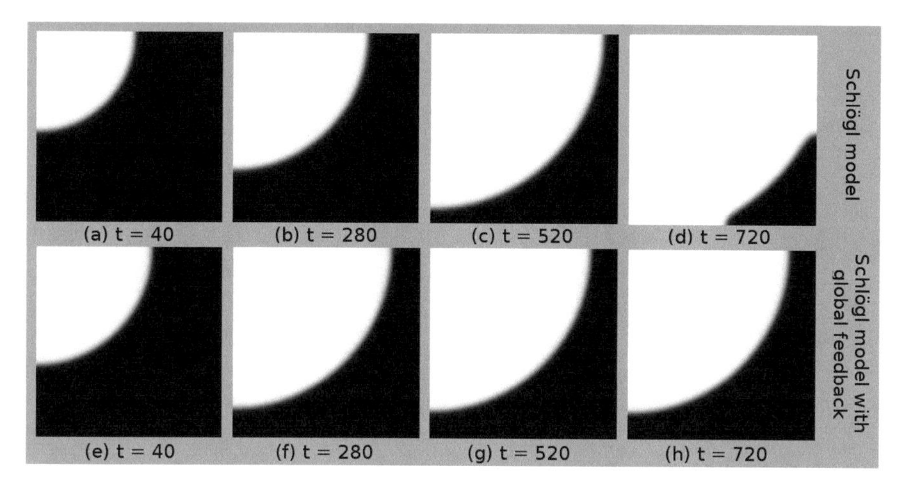

Fig. 3 Propagation of a wave in the Schlögl model (**a**–**d**) and in the Schlögl model with global feedback with $p = 0.5$ (**e**–**h**)

4 Conclusions

In this work, we have introduced a wave-pinning mechanism in the bistable Schlögl model. The mechanism consists of the control of the total size of the wavefront by means of a global feedback that varies dynamically the value of the unstable fixed point of the model. The wave-pinning mechanism presented provides a route to model the limited availability of PIP3 to form patterns that result in the polarization and motility of *Dictyostelium discoideum* cells.

References

1. M. Gerhardt, M. Ecke, M. Walz, A. Stengl, C. Beta, G. Gerisch, Actin and PIP3 waves in giant cells reveal the inherent length scale of an excited state. J. Cell Sci. **127**, 4507–4517 (2014)
2. A.W. Liehr, *Dissipative Solitons in Reaction Diffusion Systems*. Springer Series in Synergetics, vol. 70 (2013)
3. J.D. Murray, *Mathematical Biology II: Spatial Models and Biomedical Applications*, 3rd edn. (Springer, 2003)
4. L. Schimansky-Geier, C. Zülicke, E.Z. Schöll, Domain formation due to Ostwald ripening in bistable systems far from equilibrium. Phys. B Condens. Matter **84**, 433–441 (1991)
5. F. Schlögl, Chemical reaction models for non-equilibrium phase transition. Zeitschrift für Physik **253**, 147–161 (1972)
6. Y.B. Zel'dovich, D.A. Frank-Kamenetskii, On the theory of uniform flame propagation. Dokl. Akad. Nauk. SSSR **19**, 693–798 (1938)

Cooperativity in Neurons–Astrocytes Coupled Dynamics

J.-P. Françoise and Hongjun Ji

Abstract Our aim in this article is to study properties of a generalized dynamical system modeling brain lactate kinetics, with N neuron compartments and A astrocyte compartments. In particular, we prove the uniqueness of the stationary point and its asymptotic stability. Furthermore, we check that the system is positive and cooperative.

1 Introduction

The system of ODEs

$$\begin{aligned} \frac{dx}{dt} &= J - T\Big(\frac{x}{k+x} - \frac{y}{k'+y}\Big), \quad T,\ k,\ k',\ J > 0, \\ \epsilon\frac{dy}{dt} &= F(L-y) - T\Big(\frac{y}{k'+y} - \frac{x}{k+x}\Big), \quad \epsilon,\ F,\ L > 0, \end{aligned} \tag{1}$$

where ϵ is a small parameter was proposed and studied as a model for brain lactate kinetics (see [1, 2, 5–7]). In this context, $x = x(t)$ and $y = y(t)$ correspond to the lactate concentrations in an interstitial (i.e. extra-cellular) domain and in a capillary domain, respectively. Furthermore, the non-linear term $T\big(x/(k+x) - y/(k'+y)\big)$ stands for a co-transport through the brain–blood boundary (see [4]). Finally, J and F are forcing and input terms, respectively, assumed frozen. The model has a unique stationary point which is asymptotically stable. Recently, in [3, 8], a PDE's system obtained by adding diffusion of lactate was introduced. The authors proved existence

The first author is very grateful to the CRM, the Dynamical System group of UAB and the organizing committee of the Murphys conference for their invitation and financial support.

J.-P. Françoise (✉) · H. Ji
Sorbonne Université, Laboratoire Jacques-Louis Lions, UMR 7598 CNRS, 4 Pl. Jussieu, Tour 16–26, 75252 Paris, France
e-mail: Jean-Pierre.Francoise@upmc.fr

H. Ji
e-mail: Hongjun.Ji@upmc.fr

A. Korobeinikov et al. (eds.), *Extended Abstracts Spring 2018*,
Trends in Mathematics 11, https://doi.org/10.1007/978-3-030-25261-8_23

and uniqueness of nonnegative solutions and obtained linear stability results. A more general ODE's model for brain lactate kinetics, where the intracellular compartment splits into neuron and astrocyte, was considered in [5, 6]. It displays

$$\begin{aligned}
\frac{dx}{dt} &= I_0 + T_1\Big(-\frac{x}{k+x} + \frac{u}{k_n+u}\Big) + T_2\Big(-\frac{x}{k+x} + \frac{v}{k_a+v}\Big) - T\Big(\frac{x}{k+x} - \frac{y}{k'+y}\Big)\\
\frac{du}{dt} &= I_1 - T_1\Big(-\frac{x}{k+x} + \frac{u}{k_n+u}\Big)\\
\frac{dv}{dt} &= I_2 - T_2\Big(-\frac{x}{k+x} + \frac{v}{k_a+v}\Big) - T_a\Big(\frac{v}{k_a+v} - \frac{y}{k'+y}\Big)\\
\frac{dy}{dt} &= F(L-y) + T\Big(\frac{x}{k+x} - \frac{y}{k'+y}\Big) + T_a\Big(\frac{v}{k_a+v} - \frac{y}{k'+y}\Big),
\end{aligned} \tag{2}$$

where all the constants are nonnegative. It also includes transports through cell membranes and a direct transport from capillary to intracellular astrocyte. It was proved in [5, 6] that this 4-dimensional system displays a unique stationary point but its nature was left open. The stability of the unique stationary point is an important issue as it relates with therapeutic protocols developed in the Refs. [5, 6]. Another important issue is the boundedness of the lactate concentrations related with the viability domain (cf. [5, 6]). We can in fact consider a natural extension of this system into a more general $N + A + 2$ system. For this generalized system, we prove both unicity and asymptotic stability of the stationary point. In this article we do not consider fast-slow limits and we stick to the 4-dimensional case ($A = 1$, $N = 1$).

1.1 Conditions for the Positivity of the Stationary Point

The stationary point $s^* = (s'', y^*)$ does not belong necessarily to $\mathbb{R}^4_+$ as it was observed already for the 2-dimensional system in [2, 5]. Following the notations of Eq. (1), the stationary point belongs to $\mathbb{R}^2_+$ if and only if

$$T > J[1 + \frac{1}{k'}(L + \frac{J}{F})]. \tag{3}$$

Similar explicit conditions can be given for the 4-dimensional system as shown in [5, 6].

2 Cooperative Dynamics and Asymptotic Stability of the Stationary Point

Theorem 1 *Denote J_0 the Jacobian matrix J_F of the system (2) for the input $F = 0$. All off-diagonal elements of the matrix J_F (and of J_0) are nonnegative.*

Following [10], such matrices are called Metzler matrices.

Theorem 2 *The stationary point of system* (2) *is asymptotically stable.*

Proof As we can see, there are no zero elements at the first row and the first column in matrix J_F (and J_0). This means that in the graph associated with the matrix, there is a sequence of directed edges leading from N_i to N_j for all $i, j \in (1, \ldots, 4)$. Hence, $G(J_F)$ is strongly connected, so J_F (and J_0) is an irreducible matrix. Note that the strictly positive vector $w \in \mathbb{R}^4$

$$w = \left(\frac{(x+k)^2}{k}, \frac{(u+k_n)^2}{kn}, \frac{(v+k_a)^2}{ka}, \frac{(y+k')^2}{k'}\right)^T \tag{4}$$

solves $J_0 w = 0$. By [9, (ii)], the vector w is necessarily proportional to the positive eigenvector v which corresponds to the spectral abscissa. Hence, we obtain that $\mu(J_0) = 0$. By [9, (iii)], $\mu(J_F) < \mu(J_0) = 0$. This shows that all the real parts of eigenvalues of the Jacobian matrix J_F are negative, which means that the stationary point of system (2) is asymptotically stable.

3 Remarks and Perspectives

A natural question (for instance for the 4-dimensional system) is whether the conditions on the non-existence of stationary point inside the domain Ω implies that there is no bounded positive solutions.

There is a non-autonomous version of the Brain Lactate Dynamics for which the entries $J(t)$ and the forcing term $F(t)$ are time dependent. Further studies on the cooperative nature of these dynamics will be developed.

It should be interesting to analyse the reaction–diffusion PDE system obtained by adding diffusion to the 4-dimensional system (2) from the viewpoint of cooperative systems.

References

1. A. Aubert, R. Costalat, Interaction between astrocytes and neurons studied using a mathematical model of compartmentalized energy metabolism. J. Cereb. Blood Flow Metab. **25**, 1476–1490 (2005)
2. R. Costalat, J.P. Françoise, C. Menuel, M. Lahutte, J.N. Vallée, G. de Marco, J. Chiras, R. Guillevin, Mathematical modeling of metabolism and hemodynamics. Acta Biotheor. **60**, 99–107 (2012)
3. R. Guillevin, A. Miranville, A. Perrillat-Mercerot, On a reaction-diffusion system associated with brain lactate kinetics. Electron. J. Differ. Equ. **23**, 1–16 (2017)
4. J. Keener, J. Sneyd, *Mathematical Physiology*. Interdisciplinary Applied Mathematics, vol. 8, 2nd edn. (Springer, New York, 2009)

5. M. Lahutte-Auboin, Modélisation biomathématique du métabolisme énergétique cérébral: réduction de modèle et approche multi-échelle, application à l'aide à la décision pour la pathologie des gliomes, Ph.D. thesis, Université Pierre et Marie Curie (2015)
6. M. Lahutte-Auboin, R. Costalat, J.P. Françoise, R. Guillevin, Dip and buffering in a fast-slow system associated to brain lactate kinetics. arXiv: 1308.0486v1
7. M. Lahutte-Auboin, R. Guillevin, J.P. Françoise, J.N. Vallée, R. Costalat, On a minimal model for hemodynamics and metabolism of lactate: application to low grade glioma and therapeutic strategies. Acta Biotheor **61**, 79–89 (2013)
8. A. Miranville, A singular reaction-diffusion equation associated with brain lactate kinetics. Math. Methods Appl. Sci. **40**(7), 2452465 (2017)
9. H.L. Smith, On the asymptotic behavior of a class of deterministic models of cooperating species. SIAM J. Appl. Math. **46**, 368375 (1986)
10. H.L. Smith, Monotone dynamical systems: an introduction to the theory of competitive and cooperative systems, in *Mathematical Surveys and Monographs*. American Mathematical Society, Providence, Rhode Island, USA (1995)

Resonance-Based Mechanisms of Generation of Relaxation Oscillations in Networks of Non-oscillatory Neurons

Andrea Bel and Horacio G. Rotstein

Abstract We investigate a minimal network model consisting of a 2D linear (non-oscillatory) resonator and a 1D linear cell, mutually inhibited with piecewise-linear graded synapses. We demonstrate that this network can produce oscillations in certain parameter regimes and the corresponding limit gradually transition from regular oscillations (of non-relaxation type) to relaxation oscillations as the levels of mutual inhibition increase.

1 Introduction

Membrane potential (subthreshold) resonance (MPR) refers to the ability of a neuron to exhibit a peak in their voltage amplitude response to oscillatory input currents at a preferred (resonant) input frequency (f_{res}); see [4–7] and Fig. 1a. MPR results from the interplay of an autocatalytic process (positive feedback) and a slower negative

---

This work was partially supported by the National Science Foundation grant DMS-1608077 (HGR), the NJIT Faculty Seed Grant 211278 (HGR) and the Universidad Nacional del Sur Grant PGI 24/L096. HGR is grateful to the Courant Institute of Mathematical Sciences at New York University and the Centre de Recerca Matemàtica, Barcelona. The authors are grateful to Antoni Guillamon for useful comments on an earlier version of this manuscript.

A. Bel (✉)
Departamento de Matemática, Universidad Nacional del Sur, Bahía Blanca, Argentina
e-mail: andrea.bel@uns.edu.ar

INMABB, CONICET, Bahía Blanca, Argentina

H. G. Rotstein
Federated Department of Biological Sciences, Rutgers University and New Jersey Institute of Technology, Newark, NJ, USA
e-mail: horacio@njit.edu

Institute for Brain and Neuroscience Research, New Jersey Institute of Technology, Newark, NJ, USA

Graduate Faculty, Behavioral Neurosciences Program, Rutgers University, Newark, NJ, USA

Investigador Correspondiente,CONICET, Bahía Blanca, Argentina

A. Korobeinikov et al. (eds.), *Extended Abstracts Spring 2018*,
Trends in Mathematics 11, https://doi.org/10.1007/978-3-030-25261-8_24

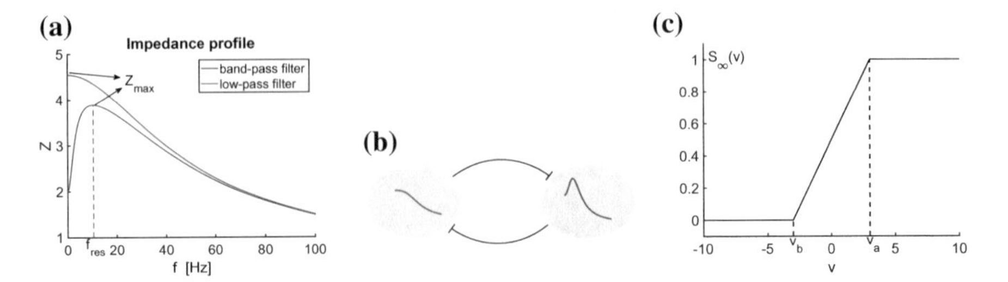

Fig. 1 **a** Representative impedance (Z) profiles for a band-pass (blue) and low-pass (red) filters. For linear systems receiving sinusoidal inputs with frequency f, the output is also a sinusoidal function with the same frequency and phase-shifted. **b** Network diagram of a mutually inhibited resonator (band-pass filter) and a non-resonator (low-pass filter). **c** Representative PWL connectivity function for the graded synapses

feedback effect. For neurons, these are provided by the participating currents. Neurons may also exhibit membrane potential (subthreshold) oscillations either damped or sustained in the absence of any time-dependent input. However, MPR and intrinsic oscillations are different phenomena governed by different mechanisms as demonstrated by the fact that 2D linear systems may exhibit MPR in the absence of damped oscillations [5–7]. We refer to the neurons that exhibit MPR as resonators. Here, we focus on resonators that are not damped oscillators.

MPR has been measured in a variety of neuron types and it has been investigated theoretically in [4–6, 9], and references therein. However, the role that MPR play in the generation of network oscillations is not well understood (but see [1, 3, 8]). In this paper, we demonstrate by means of a numerical simulation example that a minimal network model (Fig. 1b) consisting of a 2D linear resonator (e.g. Fig. 1a, blue) and 1D linear passive cell (e.g. Fig. 1a, red) mutually inhibited with piecewise-linear (PWL) graded synapses (Fig. 1c) can produce oscillations in certain parameter regimes. The corresponding limit cycles experience a transition from regular oscillations (of non-relaxation type) to relaxation oscillations as the levels of mutual inhibition increase.

2 Model: Networks of Linearized Cells with Piecewise-Linear Graded Synapses

We used linearized biophysical (conductance based) models for the individual cells and piecewise-linear (PWL) graded synaptic connections. The linearization process for conductance-based models (around the resting potential for the voltage variable) for single cells has been previously described in [5, 7]. We refer the reader to these references for details.

The dynamics of a network of two mutually inhibitory cells are described by

$$C_k \frac{dv_k}{dt} = -g_{L,k}\, v_k - g_k\, w_k - G_{in,jk}\, S_\infty(v_j)(v_k - E_{in}), \tag{1}$$

$$\tau_k \frac{dw_k}{dt} = v_k - w_k, \tag{2}$$

for $k = 1, 2,\ j \neq k$. In Eqs. (1)–(2), t is time, v_k represents the voltage (mV), w_k represents the normalized gating variable for the resonant ionic current, $C_k = 1$ is the capacitance, $g_{L,k}$ is the linearized leak maximal conductance, g_k is the ionic current linearized conductance, τ_k is the linearized time constant and the last term in Eq. (1) is the graded synaptic current modulated by the activity of the other cell where $G_{in,jk}$ is the maximal synaptic conductance, $E_{in} = -20$ is the synaptic reversal potential (referred to the resting potential) and $S_\infty(v)$ is a PWL function of sigmoid type (Fig. 1c) of the form

$$S_\infty(v) = \begin{cases} 0 & \text{if } v < v_b \\ (v_a - v_b)^{-1}\,(v - v_b) & \text{if } v_b < v < v_a \\ 1 & \text{if } v > v_a, \end{cases} \tag{3}$$

where v_a and v_b are constants. In this paper we use $g_2 = 0$ (cell 2 is 1D), $v_b = -v_a$ and $G_{in} = G_{in,12} = G_{in,21}$.

We use the following units: mV for v_k and w_k, ms for t, μF/cm^2 for capacitance, μA/cm^2 for current and mS/cm^2 for the maximal conductances.

The numerical solutions were computed by using the modified Euler method (Runge–Kutta, order 2) [2] with a time step $\Delta t = 0.1$ ms in MATLAB (The Mathworks, Natick, MA). Smaller values of Δt have been used to check the accuracy of the results.

3 Results

Figure 2 shows the results of our numerical simulations for representative values of G_{in}. Because the network is mutually inhibitory the two cells oscillate in antiphase. The network oscillations emerge for $G_{in} \sim 0.1296$. As G_{in} increases the oscillation amplitude increases, first abruptly and then gradually (Fig. 3a). As this happens, the network oscillation frequency decreases (Fig. 3b). The network oscillations are terminated for $G_{in} \sim 0.176$ (not shown).

These oscillations (Fig. 2) are a network phenomena since for the parameter values we used, the resonator is not a damped oscillator and the passive cell is 1D. Sustained (limit cycle) oscillations require the interplay of a resonant (negative feedback) and amplifying (positive feedback) processes. For the network oscillations in Fig. 2, the resonant process is provided by the resonator and the amplifying process is provided by the network connectivity mediated by the passive cell [1].

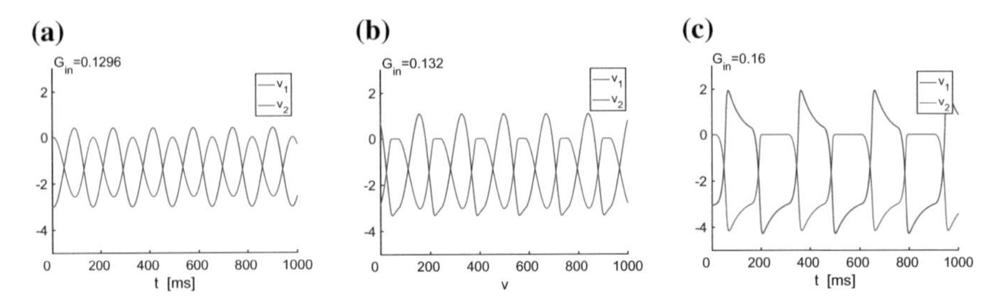

Fig. 2 Representative voltage traces for the resonator/passive cell mutually inhibitory network (Fig. 1b). **a** $G_{in} = 0.1296$. **b** $G_{in} = 0.132$. **c** $G_{in} = 0.16$. The resonator has $f_{res} \sim 10.4$. We used the following parameter values: $C_1 = C_2 = 1, g_{L,1} = 0.25, g_1 = 0.25, \tau_1 = 100, g_{L,2} = 0.5, v_a = 3, v_b = -3, E_{in} = -20$, and $G_{in} = G_{in,12} = G_{in,21}$

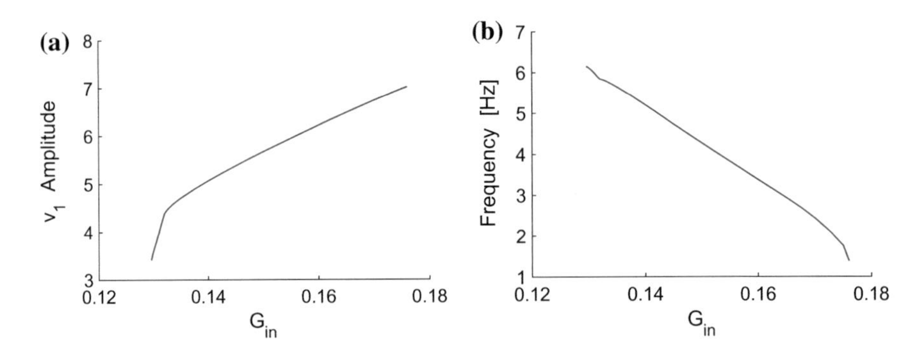

Fig. 3 Dependence of the oscillations amplitude and frequency on the levels of mutual inhibition for the resonator/passive cell mutually inhibitory network (Fig. 1b). The parameter values are as in Fig. 2. **a** Amplitude versus G_{in} curve. We plotted the amplitude of v_1. **b** Network oscillations frequency versus G_{in}

The transition from regular oscillations (non-relaxation type) to relaxation oscillations as G_{in} increases in Fig. 2 is a network phenomenon. There is a time scale separation between the activator (v_1) and the inhibitor (w_1) in the resonator ($\tau_1 = 100$). However, for this time scale separation at the individual cell level to be communicated to the network level to produce network relaxation oscillations the levels of mutual inhibition have to be relatively large.

4 Discussion

We have demonstrated that a minimal network model (2D resonator, 1D linear cell and mutual inhibition) can produce sustained network oscillations in certain parameter regimes. These oscillations crucially depend on the negative feedback provided by the resonator. Mutual inhibition mediated by the passive cell is responsible for the amplification necessary to support the existence of a limit cycle. Our results provide an example of an oscillatory network of non-oscillatory cells, where resonance and amplification at different levels of organization interact to produce network oscillations. For high enough levels of mutual inhibition, the time scale between the two variables in the resonator is communicated to the network level to produce relaxation oscillations. If the levels of mutual inhibition are not high enough, this time scale separation remains occluded.

Our results highlight the role of MPR in isolated neurons for the generation of network oscillations, and have implications for neuronal network dynamics described either by conductance-based models or firing rate models with adaptation.

References

1. A. Bel, H.G. Rotstein, Membrane potential resonance in non-oscillatory neurons interacts with synaptic connectivity to produce network oscillations. *In preparation* (2018)
2. R.L. Burden, J.D. Faires, *Numerical Analysis* (PWS Publishing Company, Boston, 1980)
3. Y. Chen, X. Li, H.G. Rotstein, F. Nadim, Membrane potential resonance frequency directly influences network frequency through electrical coupling. J. Neurophysiol. **116**(4), 1554–1563 (2016)
4. B. Hutcheon, Y. Yarom, Resonance, oscillations and the intrinsic frequency preferences of neurons. Trends Neurosci. **23**(5), 216–222 (2000)
5. M.J.E. Richardson, N. Brunel, V. Hakim, From subthreshold to firing-rate resonance. J. Neurophysiol. **89**, 2538–2554 (2003)
6. H.G. Rotstein, Frequency preference response to oscillatory inputs in two-dimensional neural models: a geometric approach to subthreshold amplitude and phase resonance. J. Math. Neurosci. **4**, 11 (2014)
7. H.G. Rotstein, F. Nadim, Frequency preference in two-dimensional neural models: a linear analysis of the interaction between resonant and amplifying currents. J. Comput. Neurosci. **37**, 9–28 (2014)
8. R.A. Tikidji-Hamburyan, J.J. Martínez, J.A. White, C. Canavier, Resonant interneurons can increase robustness of gamma oscillations. J. Neurosci. **35**, 15682–15695 (2015)
9. R. Zemankovics, S. Káli, O. Paulsen, T.F. Freund, N. Hájos, Differences in subthreshold resonance of hippocampal pyramidal cells and interneurons: the role of h-current and passive membrane characteristics. J. Physiol. **588**, 2109–2132 (2010)

New Advances on the Lyapunov Constants of Some Families of Planar Differential Systems

Iván Sánchez-Sánchez and Joan Torregrosa

Abstract This note presents some advances regarding the Lyapunov constants of some families of planar polynomial differential systems, as a first step toward the resolution of the center and cyclicity problems. First, a parallelization approach is computationally implemented to achieve the 14th Lyapunov constant of the complete cubic family. Second, a technique based on interpolating some specific quantities so as to reconstruct the structure of the Lyapunov constants is used to study a Kukles system, some fifth-degree homogeneous systems, and a quartic system with two invariant lines.

1 Introduction

Let us consider a real polynomial differential system in the plane with some parameters, $\lambda \in \mathbb{R}^d$, written in complex coordinates as

$$\begin{cases} \dot{z} = iz + Z(z, w, \lambda), \\ \dot{w} = -iw + W(z, w, \lambda), \end{cases} \tag{1}$$

where $w = \bar{z}$ and $Z(z, w, \lambda)$, $W(z, w, \lambda) = \bar{Z}(z, w, \lambda)$ are polynomial perturbations having neither linear nor constant terms in z, w. The center problem consists in identifying whether the origin of (1) is a center or a focus, when the origin is a

This work has been realized thanks to the Spanish Ministerio de Ciencia, Innovación y Universidades and the Agencia Estatal de Investigación grant numbers MTM2016-77278-P (FEDER) and FPU16/04317; the Catalan AGAUR grant number 2017 SGR 1617; and the Marie Skłodowska-Curie European grant agreement H2020-MSCA-RISE-2017-777911.

I. Sánchez-Sánchez (✉) · J. Torregrosa
Departament de Matemàtiques, Universitat Autònoma de Barcelona, 08193 Bellaterra, Barcelona, Catalonia, Spain
e-mail: isanchez@mat.uab.cat

J. Torregrosa
e-mail: torre@mat.uab.cat

A. Korobeinikov et al. (eds.), *Extended Abstracts Spring 2018*,
Trends in Mathematics 11, https://doi.org/10.1007/978-3-030-25261-8_25

monodromic nondegenerate equilibrium point. This problem is related to the local cyclicity problem, which aims to determine the maximum number of small limit cycles which bifurcate from the origin when perturbing the system in a polynomial class of fixed degree. All of them are relevant studies in the 16th Hilbert problem.

To deal with this problem let us consider the Poincaré map Π, which maps a given point ρ in a section Σ transversal to the orbit γ to the first intersection $\Pi(\rho)$ of Σ and γ in positive time. The Poincaré map can be analytically extended to $\rho = 0$, so we can consider its Taylor expansion and define the displacement map

$$d(\rho) := \Pi(\rho) - \rho = V_3\,\rho^3 + V_4\,\rho^4 + V_5\,\rho^5 + V_6\,\rho^6 + \cdots = \sum_{n=3}^{\infty} V_n\,\rho^n, \quad (2)$$

for certain values V_n. Observe that the center problem is equivalent to determine whether all V_n are zero or not, since periodic orbits are fixed points of the Poincaré map. If not all V_n vanish, the first nonzero V_n must have odd subindex. These V_n with odd $n \geq 3$ are known as Lyapunov constants, and they will be denoted by $L_{(n-1)/2} := V_n$. According to [6], the Lyapunov constants are polynomials, whose variables are the parameters of system (1). In this case for which not all V_n vanish, the origin of the system is a focus and its stability is determined by the first nonzero Lyapunov constant. As a consequence, the center problem reduces to the problem of finding and vanishing all the Lyapunov constants L_k, that is solving the nonlinear system $\{L_1 = L_2 = \cdots = 0\}$. They also are an essential tool in the study of the cyclicity problem, since the limit cycles correspond to the isolated zeroes of (2). In this short paper, we will briefly present two methods to compute these Lyapunov constants or study some of their useful properties to deal with the center and cyclicity problems.

2 A Parallelization of Lyapunov Method

An algorithm to find Lyapunov constants is the so-called Lyapunov method, which is based on the utilization of a first integral of system (1). The computations could be made using real values (see [9]), but we consider complex coordinates because the obtained expressions are shorter. The objective is then to find a formal first integral $F = F_2 + F_3 + F_4 + \cdots$ of system (1), with $F_k(z, w) := \sum_{j=0}^{k} h_{k-j,j}\, z^{k-j} w^j$ homogeneous degree k polynomials. We aim to study whether $\dot{F}$ vanishes or not. We compute

$$\dot{F} = F_z\,\dot{z} + F_w\,\dot{w} = F_z\,(iz + Z(z, w, \lambda)) + F_w\,(-iw + W(z, w, \lambda)) = \sum_{k \geq 1} L_k(\lambda)\,(zw)^{k+1}. \quad (3)$$

The last equality is a consequence of a result which states that if there exists such F, then in suitable coordinates it is analytic on zw; see [3]. Therefore, $\dot{F}$ is also analytic on zw. Observe that if all $L_k(\lambda)$ vanish then $\dot{F} = 0$, and therefore F is a first integral, so the origin is a center. These $L_k(\lambda)$ are actually the Lyapunov constants, which are polynomials in the parameters λ; see [4]. For the sake of simplicity, we will denote them simply as L_k. F can be found recursively, starting by imposing equation (3) and performing formal operations as follows:

$$\begin{aligned}
&(F_{2z} + F_{3z} + F_{4z} + \cdots)\,(iz + Z_2 + Z_3 + Z_4 + \cdots) + \\
&+ (F_{2w} + F_{3w} + F_{4w} + \cdots)\,(-iw + W_2 + W_3 + W_4 + \cdots) = \\
&= L_1\,(zw)^2 + L_2\,(zw)^3 + L_3\,(zw)^4 + \cdots .
\end{aligned}$$

Here we solve the linear system obtained by equating the same degree coefficients in z, w. This way one can find the coefficients $h_{k-j,j}$ of each term F_k of the first integral and use it to find the corresponding Lyapunov constants.

The above technique has been implemented in PARI/GP (or simply PARI) programming language; see [5]. As the computation of Lyapunov constants is a highly computationally expensive procedure, this algorithm has been optimized and improved by means of parallelization, which allows to significantly increase computation velocity. The idea is to find each of the Lyapunov constants and the coefficients $h_{k-j,j}$ of F_k of degree k in terms of the coefficients of lower degree. This part is relatively fast computationally speaking since the manipulated expressions are not too large. Then we parallelize the substitution of those coefficients with their actual value, and here parallelization is essential because this process deals with very large expressions.

The results of this parallelization technique are amazing, and its efficiency has allowed our method to find Lyapunov constants in a relatively short time for cases which had not been solved before due to the huge amount of time and computational complexity required. In particular, we have applied this method to the complete cubic system

$$\begin{cases}
\dot{z} = iz + \hat{r}_{20}z^2 + \hat{r}_{11}zw + \hat{r}_{02}w^2 + \hat{r}_{30}z^3 + \hat{r}_{21}z^2w + \hat{r}_{12}zw^2 + \hat{r}_{03}w^3, \\
\dot{w} = -iw + \hat{s}_{20}w^2 + \hat{s}_{11}wz + \hat{s}_{02}z^2 + \hat{s}_{30}w^3 + \hat{s}_{21}w^2z + \hat{s}_{12}wz^2 + \hat{s}_{03}z^3.
\end{cases} \tag{4}$$

We have observed that if time is rescaled by dividing by the imaginary unit i, computations are much more efficient and the calculation time decreases. Actually, the computations we describe here cannot be performed if this time rescaling is not done. If we denote $r_{jk} = \hat{r}_{jk}/i$ and $s_{jk} = \hat{s}_{jk}/i$, the system in the new time variable can be written as

$$\begin{cases}
z' = z + r_{20}z^2 + r_{11}zw + r_{02}w^2 + r_{30}z^3 + r_{21}z^2w + r_{12}zw^2 + r_{03}w^3, \\
w' = -w + s_{20}w^2 + s_{11}wz + s_{02}z^2 + s_{30}w^3 + s_{21}w^2z + s_{12}wz^2 + s_{03}z^3.
\end{cases} \tag{5}$$

Up to our knowledge, the highest known Lyapunov constant for the above system is the 10th; see [7]. But with our parallelization technique, we have been able to reach the 14th. To perform this computation we have used the computer network of

Table 1 Size of the computed Lyapunov constants

Lyapunov constant	Size (MB)
11	111
12	261
13	588
14	1282

our department. The parallelization has been done with the software PBala; see [8]. This server has eight nodes with Intel Xeon 2.60 GHz processors. The total memory is 640 GB and 96 threads can be run at the same time. The size of the new Lyapunov constants is shown in Table 1. The total computing time was around 22 h.

It is well-known, see for example [2], that the solution of the center problem for the general cubic differential system needs at least 11 Lyapunov constants. Now, the obstacle to obtain a complete characterization of the cubic centers is how can we solve the nonlinear system $\{L_1 = L_2 = \cdots = 0\}$, and not how to construct it because we think that we have computed enough Lyapunov constants.

3 Interpolation and Reconstruction Technique

Let us consider the ideal in $\mathbb{C}[\lambda]$ generated by all Lyapunov constants $\langle L_1, L_2, L_3, \ldots\rangle$ associated to the differential system (1) with some fixed degree. Due to the Hilbert Basis Theorem, this ideal is finitely generated, so there must exist $m \in \mathbb{N}$ such that

$$\langle L_1, L_2, L_3, \ldots\rangle = \langle L_1, L_2, L_3, \ldots, L_m\rangle. \tag{6}$$

To know this m would significantly simplify the problem, because by computing the first m Lyapunov constants we would obtain the center conditions. Nevertheless, as [1] states, there are no general methods to find this m and this is the reason why the center problem has been solved only for certain polynomial families.

Let $\mathcal{B}_k := \langle L_1, \ldots, L_k\rangle$ be the Bautin ideal generated by the first k Lyapunov constants. The method we suggest aims to check whether a certain Lyapunov constant L_n belongs to $\mathcal{B}_{n-1}$, and therefore it vanishes when the previous are equal to zero. It is important to remark that to apply this technique at this step we assume that we have been able to compute the first n Lyapunov constants.

Let us start by writing

$$L_n = \sum_{j=1}^{n-1} A_j L_j, \tag{7}$$

where A_j are polynomials whose variables are the parameters of (1). Our method consists in trying to see whether we can determine these polynomials A_j, since this will tell if expression (7) is possible or not.

Using the notation of (5) for the parameters, let us consider a monomial $M = \prod_{k,\ell} r_{k\ell}^{p_{k\ell}} s_{k\ell}^{q_{k\ell}}$, where $r_{k\ell}$ and $s_{k\ell}$ correspond to the coefficients of $z^k w^\ell$ of $Z(z, w, \lambda)$ and $W(z, w, \lambda)$, respectively. We define the quasi-degree of M as $\sum_{k,\ell}(k + \ell - 1)(p_{k\ell} + q_{k\ell})$ and its weight as $\sum_{k,\ell}(1 - k - \ell)(p_{k\ell} - q_{k\ell})$. Then, by [4], the monomials of a Lyapunov constant L_j satisfy that they have quasi-degree $2j$ and weight 0. Now using these properties together with the degree of L_j, we can select which monomials are candidates to be part of each A_j, but with undetermined coefficients. Thus, we have that A_j are polynomials whose monomials have been selected and have undetermined coefficients, and these coefficients of A_j are what we try to compute.

Now knowing the structure of A_j, we would substitute it in (7), expand the products and the sum and finally equate the coefficients of monomials with the same literal part. This gives a set of linear equations consisting of the coefficients of equality (7). If this system of linear equations is compatible, then the polynomials A_j do exist and L_n vanishes when $L_1, \ldots, L_{n-1}$ are zero; otherwise, if the system is incompatible then the polynomials A_j do not exist and L_n does not belong to $\mathcal{B}_{n-1}$. Instead of explicitly solving the system of equations, we have compared ranks of the system matrices to see if they are equal or not.

With this method we have studied three different families and we have obtained the following results.

Proposition 1 *Consider the Kukles differential system*

$$\begin{cases} \dot{x} = -y, \\ \dot{y} = x + b_{20}x^2 + b_{11}xy + b_{02}y^2 + b_{30}x^3 + b_{21}x^2y + b_{12}xy^2 + b_{03}y^3, \end{cases}$$

which in complex coordinates is written as

$$\begin{cases} \dot{z} = iz + r_{20}z^2 + r_{11}zw + r_{02}w^2 + r_{30}z^3 + r_{21}z^2w + r_{12}zw^2 + r_{03}w^3, \\ \dot{w} = -iw - r_{20}z^2 - r_{11}zw - r_{02}w^2 - r_{30}z^3 - r_{21}z^2w - r_{12}zw^2 - r_{03}w^3. \end{cases}$$

Then L_9 does not belong to $\mathcal{B}_8$. In the case $r_{12} = 0$, L_9 again does not belong to $\mathcal{B}_8$, but L_{10} does belong to $\mathcal{B}_9$ since there exist A_j such that $L_{10} = \sum_{j=1}^{9} A_j L_j$.

From the above result, we can guess that only the first 9 Lyapunov constants are enough to solve the center problem for this family when $r_{12} = 0$. This interpolation method works better than the standard approach using a Groebner basis for the simplifications.

The center problem for degree 5 homogeneous perturbations of the linear oscillator is an open problem, and even the value m in (6) is unknown. The mechanism proposed here fails for the general family due to the size of the computations. The next result presents some particular cases.

Proposition 2 *Consider the linear plus homogeneous degree 5 polynomial differential system*

$$\begin{cases} \dot{z} = iz + r_{41}z^4w + r_{32}z^3w^2 + r_{23}z^2w^3 + r_{14}zw^4 + r_{05}w^5, \\ \dot{w} = -iw + s_{41}w^4z + s_{32}w^3z^2 + s_{23}w^2z^3 + s_{14}wz^4 + s_{05}z^5. \end{cases}$$

Then the next properties hold:

(i) if $r_{41} = s_{41} = 0$, then L_{10} does not belong to $\mathcal{B}_9$ but L_{11} does belong to $\mathcal{B}_{10}$;
(ii) if $r_{32} = s_{32} = 0$, then L_{11} and L_{12} do not belong to $\mathcal{B}_{10}$ and $\mathcal{B}_{11}$, respectively.

The last considered family is a special quartic differential system with four invariant straight lines.

Proposition 3 *Consider the following system with two parallel invariant lines*

$$\begin{cases} \dot{x} = (1 - x^2)(-y + a_{20}x^2 + a_{11}xy + a_{02}y^2), \\ \dot{y} = (1 - y^2)(x + b_{20}x^2 + b_{11}xy + b_{02}y^2), \end{cases}$$

which can be written in complex coordinates as

$$\begin{cases} \dot{z} = iz + r_{40}z^4 + r_{31}z^3w + r_{22}z^2w^2 + r_{14}zw^3 + r_{04}w^4 + r_{21}z^2w + r_{03}w^3 + \\ \quad r_{20}z^2 + r_{11}zw + r_{02}w^2, \\ \dot{w} = -iw + s_{40}w^4 + s_{31}w^3z + s_{22}w^2z^2 + s_{14}wz^3 + s_{04}z^4 + s_{21}w^2z + s_{03}z^3 + \\ \quad s_{20}w^2 + s_{11}wz + s_{02}z^2. \end{cases}$$

For this system both L_7, L_8, and L_9 do not belong to $\mathcal{B}_6$, $\mathcal{B}_7$, and $\mathcal{B}_8$, respectively. However, when $r_{11} = s_{11} = 0$, L_7 does belong to $\mathcal{B}_6$.

References

1. L.A. Cherkas, V.G. Romanovski, The center conditions for a Liénard system. Comput. Math. Appl. **52**, 363–374 (2006)
2. C. Christopher, Estimating limit cycle bifurcations from centers, in *Differential Equations with Symbolic Computation* (Trends Math., Birkhäuser, Basel, 2005), pp. 23–35
3. F. Dumortier, J. Llibre, J.C. Artés, *Qualitative Theory of Planar Differential Systems* (Universitext, Springer, Berlin, 2006)
4. A. Gasull, A. Guillamon, V. Mañosa, An analytic-numerical method for computation of the Liapunov and period constants derived from their algebraic structure. SIAM J. Numerical Anal. **36**, 1030–1043 (1999)
5. The PARI Group, PARI/GP version 2.9.1, University of Bordeaux (2018). http://pari.math.u-bordeaux.fr/
6. V.G. Romanovski, D.S. Shafer, *The Center and Cyclicity Problems* (Birkhäuser Boston Inc., Boston, MA, 2009)

7. V.G. Romanovski, Private Communication (2017)
8. O. Saleta, PBALA version `6.0.2`, Universitat Autònoma de Barcelona, Bellaterra (2017). http://github.com/oscarsaleta/
9. S. Songling, A method of constructing cycles without contact around a weak focus. J. Differ. Equ. **41**, 301–312 (1981)

Canards Existence in the Hindmarsh–Rose Model

Jean-Marc Ginoux, Jaume Llibre and Kiyoyuki Tchizawa

Abstract In two previous papers, we have proposed a new method for proving the existence of "canard solutions" on one hand for three- and four-dimensional singularly perturbed systems with only one *fast* variable and, on the other hand, for four-dimensional singularly perturbed systems with two *fast* variables; see [4, 5]. The aim of this work is to extend this method, which improves the classical ones used till now to the case of three-dimensional singularly perturbed systems with two *fast* variables. This method enables to state a unique generic condition for the existence of "canard solutions" for such three-dimensional singularly perturbed systems which is based on the stability of *folded singularities* (*pseudo singular points* in this case) of the *normalized slow dynamics* deduced from a well-known property of linear algebra. Applications of this method to a famous neuronal bursting model enables to show the existence of "canard solutions" in the Hindmarsh–Rose model.

This work is supported by the Ministerio de Economía, Industria y Competitividad, Agencia Estatal de Investigación grants MTM2016-77278-P (FEDER) and MDM-2014-0445, the Agència de Gestió d'Ajuts Universitaris i de Recerca grant 2017SGR1617, and the H2020 European Research Council grant MSCA-RISE-2017-777911. The first author would like to thank Prof. J. Llibre for his kind invitation to visit Universitat Autònoma de Barcelona.

J.-M. Ginoux (✉)
Laboratoire LSIS, CNRS, UMR 7296,
Université de Toulon, BP 20132, 83957 La Garde cedex, France
e-mail: ginoux@univ-tln.fr

J. Llibre
Facultat de Ciències, Departament de Matemàtiques, Edifici C,
Universitat Autònoma de Barcelona, 08193 Bellaterra, Barcelona, Spain
e-mail: jllibre@mat.uab.cat

K. Tchizawa
Institute of Administration Engineering Ltd., Tokyo 101-0021, Japan
e-mail: tchizawakiyoyuki@aim.com

A. Korobeinikov et al. (eds.), *Extended Abstracts Spring 2018*,
Trends in Mathematics 11, https://doi.org/10.1007/978-3-030-25261-8_26

1 Introduction

The concept of "canard solutions" for three-dimensional singularly perturbed systems with two *slow* variables and one *fast* has been introduced at the beginning of the 80s by Benoît [2] and Benoît–Lobry [3]. Their existence has been proved by Benoît [2, p. 170] in the framework of "Non-Standard Analysis" according to a theorem which states that canard solutions exist in such systems provided that the *pseudo singular point* of the *slow dynamics*, i.e., of the *reduced vector field*, is of the *saddle* type. Nearly twenty years later, Szmolyan–Wechselberger [12] provided a "standard version" of Benoît's theorem [2]. Recently, Wechselberger [15] generalized this theorem for n-dimensional singularly perturbed systems with k *slow* variables and m *fast* (where $n = k + m$). The method they used require to implement a "desingularization procedure" which can be summarized as follows: first, they compute the *normal form* of such a singularly perturbed system, which is expressed according to some coefficients (a and b for dimension three and $\tilde{a}$, $\tilde{b}$ and $\tilde{c}_1$ for dimension four) depending on the functions defining the original vector field and their partial derivatives with respect to the variables. Secondly, they project the "desingularized vector field" (originally called "normalized slow dynamics" by Eric Benoît [2, p. 166]) of such a *normal form* on the tangent bundle of the critical manifold. Finally, they evaluate the Jacobian of the projection of this "desingularized vector field" at the *folded singularity* (originally called *pseudo singular points* by José Argémi [1, p. 336]). This lead Szmolyan–Wechselberger [12, p. 427] and Wechselberger [15, p. 3298] to a "classification of *folded singularities* (*pseudo singular points*)". Thus, they showed that for three-dimensional (resp. four-dimensional) singularly perturbed systems such *folded singularity* is of the *saddle type*, if the following condition is satisfied: $a < 0$ (resp. $\tilde{a} < 0$).

In a first paper entitled "Canards Existence in Memristor's Circuits" (see Ginoux–Llibre [4]), we presented a method enabling to state a unique "generic" condition for the existence of "canard solutions" for three- and four-dimensional singularly perturbed systems with only one fast variable, which is based on the stability of *folded singularities* of the *normalized slow dynamics* deduced from a well-known property of linear algebra. We proved that this unique condition is completely identical to that provided by Benoît [2], Szmolyan–Wechselberger [12] and Wechselberger [15].

In a second paper entitled: "Canards Existence in FitzHugh–Nagumo and Hodgkin–Huxley Neuronal Models" (see Ginoux–Llibre [5]) we extended this method to the case of four-dimensional singularly perturbed systems with $k = 2$ *slow* and $m = 2$ *fast* variables. Then, we stated that the provided condition for the existence of canards is "generic", since it is exactly the same for singularly perturbed systems of dimension three and four with one or two *fast* variables. The method we used led us to the following proposition: *If the normalized slow dynamics has a pseudo singular point of saddle type, i.e., if the sum σ_2 of all second-order diagonal minors of the Jacobian matrix of the normalized slow dynamics evaluated at the pseudo singular point is negative, i.e., if $\sigma_2 < 0$, then the three-dimensional (resp. four-dimensional) singularly perturbed system exhibits a canard solution, which evolves* *from the attractive part of the slow manifold towards its repelling part*. Then, on one hand, for three-dimensional singularly perturbed systems with only one fast variable, we proved that the condition for which the *pseudo singular point* is of the saddle type, i.e., $\sigma_2 < 0$, is identical to that proposed by Benoît [2, p. 171] in his theorem, i.e., $D < 0$, and also to that provided by Szmolyan–Wechselberger [12], i.e., $a < 0$. On the other hand, for four-dimensional singularly perturbed systems with one or two fast variables, we proved that the condition for which the *folded singularity* (resp. the *pseudo singular point*) is of the saddle type, i.e., $\sigma_2 < 0$, is identical to that proposed by Wechselberger [15, p. 3298] in his theorem, i.e., $\tilde{a} < 0$.

Note that there is no proof of the approximation: it is not established whether the time-scaled reduced system holds on the approximation for the original system in the case of k slow variables ($k \geq 3$) and m fast variables ($m \geq 2$). It was proved in the case $k = 2$ and $m = 1$ by Benoît via constructing a local model and obtaining its solution, and in the case $k = 2$ and $m = 2$ was also extensively proved by Tchizawa [13, 14]. For the case $k = 1$ and $m = 2$ (the Hindmarsh–Rose model), we shall construct a local model again and we shall obtain the solutions, thus providing a constructive proof for the approximation. The fact that the pseudo singular point a saddle or a node does not ensure the existence of canards, because it may not satisfy the approximation.

The aim of this paper is to extend this method to the case of three-dimensional singularly perturbed systems with one *slow* and two *fast* variables and to show that the provided condition for the existence of canards, i.e., $\sigma_2 < 0$, still holds and is consequently "generic".

The Hindmarsh–Rose model [8] describes the basic properties of individual neurons and appears as a reduction of the conductance based in the Hodgkin–Huxley model for neural spiking, see [9] for more details. Thus, the three-dimensional Hindmarsh–Rose polynomial ordinary differential system was originally written as

$$\begin{aligned}
\frac{dx}{dt} &= y - ax^3 + bx^2 - z + I, \\
\frac{dy}{dt} &= c - dx^2 - y, \\
\frac{dz}{dt} &= r\left[s\left(x - \alpha\right) - z\right],
\end{aligned} \tag{1}$$

where x is a transmembrane neuron potential, y and z are the characteristics of ionic currents dynamic and I is ambient current. The other parameters (a, b, c, d, I, s, α and r) reflect the physical features of the neurons and the dot indicates derivative with respect to the time t. We note that the parameter $r \ll 1$. Existence of canard solutions in system (1) has been originally suspected by Shilnikov et al. [11, p. 2149] and highlighted by Shchepakina [10]. Thus, according to the previous definitions, the Hindmarsh–Rose model may be written as a three-dimensional singularly perturbed system with $k = 1$ *slow* variable and $m = 2$ *fast* variables. By posing $x \to y_2$, $y \to y_1$, $z \to x_1$ and $t' \to \varepsilon t$ with $\varepsilon = r$, we obtain:

$$
\begin{aligned}
\dot{x}_1 &= f_1(x_1, y_1, y_2) = s(y_2 - \alpha) - x_1, \\
\varepsilon \dot{y}_1 &= g_1(x_1, y_1, y_2) = c - dy_2^2 - y_1, \\
\varepsilon \dot{y}_2 &= g_2(x_1, y_1, y_2) = y_1 - ay_2^3 + by_2^2 - x_1 + I,
\end{aligned}
\tag{2}
$$

where $x_1 \in \mathbb{R}$, $\vec{y} = (y_1, y_2)^t \in \mathbb{R}^2$, $0 < \varepsilon \ll 1$ and the functions f_i and g_i are assumed to be C^2 functions of (x_1, y_1, y_2) and the dot now indicates derivative with respect to the time t'.

We have proved the existence of different kind of canard solutions for system (2) see Figs. 1, 2, and 3.

In fact, Shchepakina [10] already found the canard shown in Fig. 1. We proved the existence of this canard showing the existence of a *pseudo singular point* of the saddle-type when the parameters satisfy $s < (c + I)/\alpha$. With $c = 1$, $I = 2.7$ and $\alpha = -1.2$, we find that $s < 3.0833$. Thus, Shchepakina highlighted a canard without head in the Hindmarsh–Rose model (see Fig. 1) for the "duck parameter" value $s = 3.08104454785581 41214 < 3.0833$.

In the inset of Fig. 1, the zoom in highlights a large distance between the canard solution and that of the *critical manifold*. This is due to the fact that this latter corre-

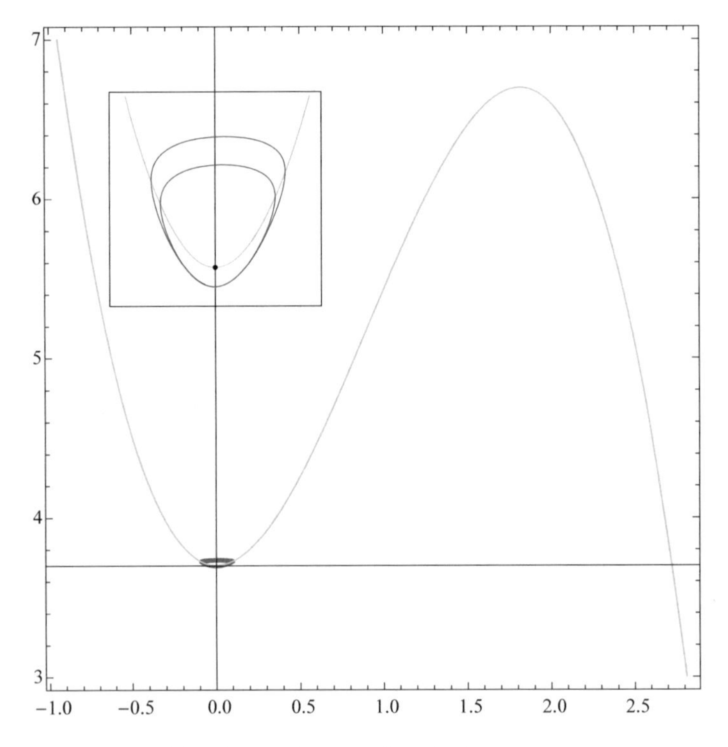

Fig. 1 Canard solution of the Hindmarsh–Rose (1) model in the (x, z) plane phase with the following parameter set: $a = 1$, $b = 3$, $c = 1$, $d = 0.275255$, $I = 2.7$, $\alpha = -1.2$ and for the "duck parameter" value $s = 3.08104454785581 41214$

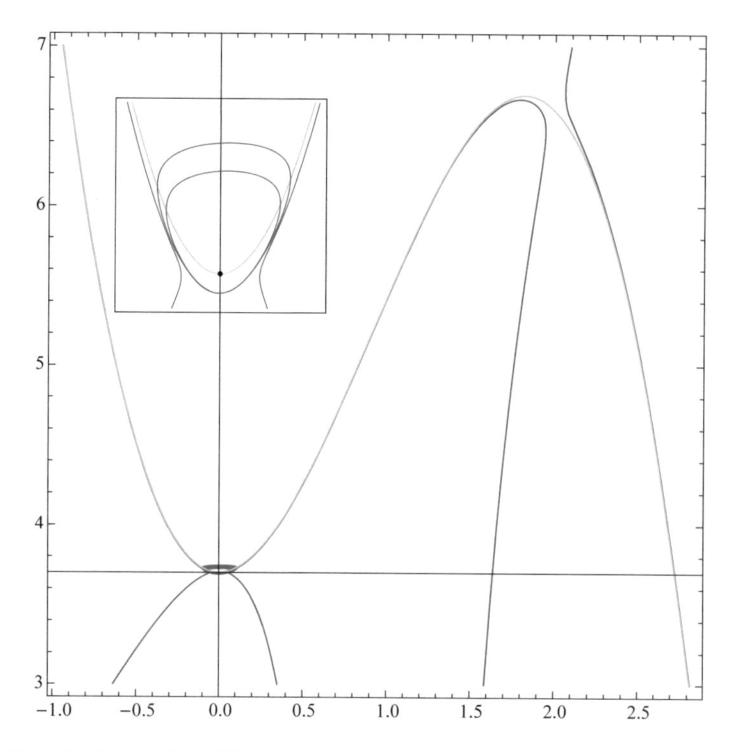

Fig. 2 Canard solution of the Hindmarsh–Rose (1) model in the (x, z) plane phase, its *critical manifold* (in green) and the second-order approximation in ε of the *slow invariant manifold* (in blue) with the following parameter set: $a = 1$, $b = 3$, $c = 1$, $d = 0.275255$, $I = 2.7$, $\alpha = -1.2$ and for the "duck parameter" value $s = 3.0810445478558141214$

sponds to zero-order approximation in ε of the *slow invariant manifold*. Nevertheless, while using the so-called *Flow Curvature Method* Ginoux–Rossetto [7] have already provided a second-order approximation in ε of the *slow invariant manifold* of the Hindmarsh–Rose model (1). The result is presented in Fig. 2.

With $c = 1$, $I = 2.7$ and $\alpha = -1.2$, we find that $s < 2.2200954$. Thus, we have highlighted a canard with head in the Hindmarsh–Rose model (see Fig. 3) for the "duck parameter" value $s = 2.220095 < 2.2200954$. For this parameters set the second-order approximation in ε of the *slow invariant manifold* of the Hindmarsh–Rose model (1) can be provided while using the *Flow Curvature Method* introduced by Ginoux–Rossetto [7]. The result is presented in Fig. 3.

All the details of the existence of these three different canards in the Hindmarsh–Rose model [8] can be found in [6].

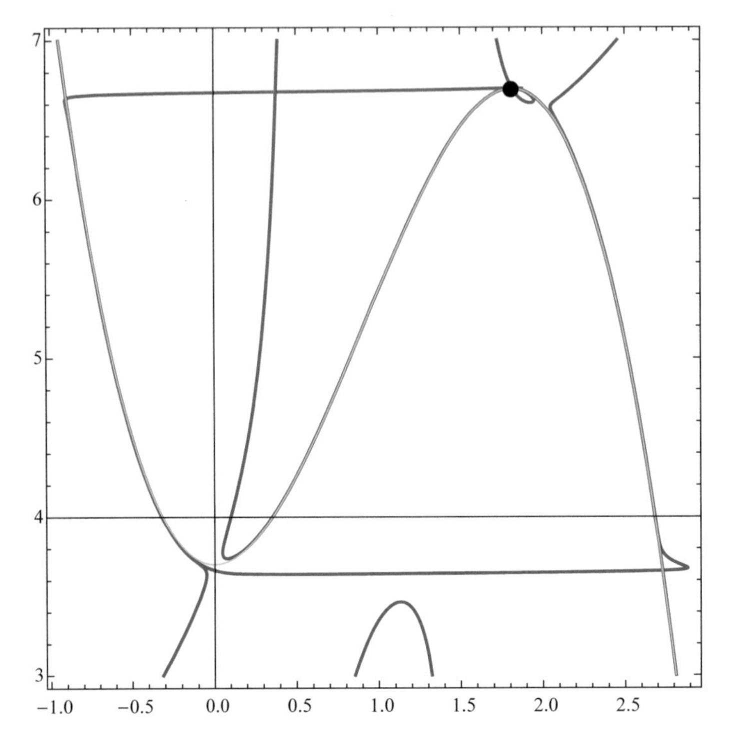

Fig. 3 Canard solution of the Hindmarsh–Rose (1) model in the (x, z) plane phase, its *critical manifold* (in green) and the second-order approximation in ε of the *slow invariant manifold* (in blue) with the following parameter set: $a = 1$, $b = 3$, $c = 1$, $d = 0.275255$, $I = 2.7$, $\alpha = -1.2$ and for the "duck parameter" value $s = 2.220095$

References

1. J. Argémi, Approche qualitative d'un problème de perturbations singulières dans $\mathbb{R}^4$, Equadiff, ed. by R. Conti, G. Sestini, G. Villari, vol. 1978, (1978) pp. 330–340
2. E. Benoît, Systèmes lents-rapides dans $\mathbb{R}^3$ et leurs canards. Astérisque **109–110**, 159–191 (1983)
3. E. Benoît, C. Lobry, Les canards de $\mathbb{R}^3$. Cr. Acad. Sc. Paris Série I **294**, 483–488 (1982)
4. J.M. Ginoux, J. Llibre, Canards in memristor's circuits, in *Qualitative Theory of Dynamical Systems* (2015), pp. 1–49
5. J.M. Ginoux, J. Llibre, Canards existence in FitzHugh–Nagumo and Hodgkin–Huxley neuronal models. Math. Prob. Eng. **15**, Article ID 342010, 17 (2015)
6. J.M. Ginoux, J. Llibre, K. Tchizawa, Canards existence in the Hindmarsh–Rose model, in *Proceedings of the workshop MURPHYS-HSFS-2018*
7. J.M. Ginoux, B. Rossetto, Slow manifold of a neuronal bursting model, in *Emergent Properties in Natural and Artificial Dynamical Systems, Understanding Complex Systems*, ed. by M.A. Aziz-Alaoui, C. Bertelle (Springer, Berlin, 2006), pp. 119–128
8. J.L. Hindmarsh, R.M. Rose, A model of neuronal bursting using three coupled first order differential equations. Proc. R. Soc. Lond. Ser. B Biol. Sci. **221**, 87–102 (1984)
9. A.L. Hodgkin, A.F. Huxley, A quantitative description of membrane current and its application to conduction and excitation in nerve. J. Physiol. Lond. **117**, 500–544 (1952)

10. E.A. Shchepakina, Three scenarios for changing of stability in the dynamic model of nerve conduction, in *Mathematical Modelling, Information Technology and Nanotechnology (ITNT-2016)*, vol. 1638 (2016), pp. 664–673
11. A. Shilnikov, M. Kolomiets, Methods of the qualitative theory for the Hindmarsh-Rose model: a case study, a tutorial. Int. J. Bifurc. Chaos **18**(8), 2141–2168 (2008)
12. P. Szmolyan, M. Wechselberger, Canards in $\mathbb{R}^3$. J. Diff. Equ. **177**, 419–453 (2001)
13. K. Tchizawa, On relative stability in 4-dimensional duck solutions. J. Math. Syst. Sci. **2**(9), 558–563 (2012)
14. K. Tchizawa, On the two methods for finding 4-dimensional duck solutions. Appl. Math. Sci. Res. Publ. **5**(1), 16–24 (2014)
15. M. Wechselberger, À propos de canards. Trans. Amer. Math. Soc. **364**, 3289–3309 (2012)

Effect of Delayed Harvesting on the Stability of Single-Species Populations

Daniel Franco, Juan Perán, Hartmut Logemann and Juan Segura

Abstract New results on the impact of harvesting times and intensities on the stability properties of Seno population models are presented. Special attention is given to the global stability of the positive equilibrium in terms of the harvest timing.

1 Introduction

The moment of intervention is a key question in harvest programmes and is currently generating increasing interest. However, little is known about its effect on the population stability.

We used a discrete-time equation introduced by Hiromi Seno in [4] to model the dynamics of populations harvested at any time during the reproductive season. For a wide family of population models described by unimodal maps, we showed that for high harvesting efforts—below the threshold above which all populations go eventually extinct—the moment of the intervention does not affect the stability of the positive equilibrium, which acts as a global attractor.

For many population models involving the Ricker map, which has been shown to be a good descriptor of the dynamics of many populations, local stability implies global stability. We showed that this is also the case for the Ricker–Seno model.

D. Franco · J. Perán
Departamento de Matemática Aplicada, E.T.S.I. Industriales, Universidad Nacional de Educación a Distancia (UNED), c/ Juan del Rosal 12, 28040 Madrid, Spain
e-mail: dfranco@ind.uned.es

J. Perán
e-mail: jperan@ind.uned.es

H. Logemann
Department of Mathematical Sciences, University of Bath, Bath BA2 7AY, UK
e-mail: h.logemann@bath.ac.uk

J. Segura (✉)
Departament d'Economia i Empresa, Universitat Pompeu Fabra, c/ Ramon Trias Fargas 25–27, 08005 Barcelona, Spain
e-mail: joan.segura@upf.edu

A. Korobeinikov et al. (eds.), *Extended Abstracts Spring 2018*,
Trends in Mathematics 11, https://doi.org/10.1007/978-3-030-25261-8_27

Additionally, we used this model to prove that timing can be stabilizing by itself. In other words, we showed that in some cases choosing an appropriate moment for removing individuals can induce an asymptotically stable positive fixed point in populations for which the same equilibrium would be unstable in case of triggering the intervention at the beginning or at the end on the reproductive season.

Our last result consists of pointing out that timing can be destabilizing for certain maps. We obtained specific mathematical counterexamples proving that Conjecture 3.5 in [1] is false.

2 Harvesting Model with Timing

Consider the discrete-time single-species population model

$$x_{t+1} = g(x_t)x_t, \tag{1}$$

where $x_t \in [0, \infty)$ is the population size at the beginning of the reproductive season t and $g\colon [0, \infty) \to \mathbb{R}$ is the per-capita production function. We are interested in populations satisfying the following conditions on g:

(i) $g'(x) < 0$ for all $x > 0$;
(ii) $g(0) > 1$;
(iii) there exists some $d > 0$ such that $xg(x)$ is strictly increasing on $(0, d)$ and strictly decreasing on (d, ∞).

Under these conditions, the dynamics are over-compensatory. On the other hand, harvesting a constant fraction $\gamma \in (0, 1)$ of the population at the end of every reproductive season corresponds to multiplication of the right-hand side of (1) by the survival fraction $(1 - \gamma)$,

$$x_{t+1} = (1 - \gamma)g(x_t)x_t. \tag{2}$$

Similarly, harvesting the same fraction at the beginning of the season leads to

$$x_{t+1} = g((1 - \gamma)x_t)(1 - \gamma)x_t. \tag{3}$$

In [4], Seno puts forward the following harvesting model, which encompasses the *limit* situations (2) and (3) by allowing the population to be harvested at any fixed point in time within the season. It reads

$$x_{t+1} = [\theta g(x_t) + (1 - \theta)g((1 - \gamma)x_t)](1 - \gamma)x_t, \tag{4}$$

where $\theta \in [0, 1]$ corresponds to the fixed harvesting moment. See [4] for a more detailed explanation and a graphical scheme of the population dynamics of this model.

Following the notation of [1], we rewrite the right-hand side of (4) as

$$\theta F_1(x_t) + (1-\theta)F_0(x_t) := F_\theta(x_t),$$

where $F_1(x) := (1-\gamma)g(x)x$ and $F_0(x) := g((1-\gamma)x)(1-\gamma)x$. Model (4) includes models (2) and (3) as special cases. Taking $\theta = 1$ corresponds to harvesting when the season ends, and $\theta = 0$ when it begins.

Over-compensatory models can exhibit positive unstable equilibria, which leads to fluctuating dynamics. We start by recalling a sufficient and necessary condition for the existence of such an equilibrium regardless of the intervention moment θ.

Proposition 1 (from Proposition 3.1 in [1]) *Assume that conditions (i)–(iii) hold. System (4) has a unique positive equilibrium (denoted by $K_\gamma(\theta)$) if and only if*

$$\gamma < \gamma^* := 1 - \frac{1}{g(0)}.$$

3 Results

3.1 Timing Does Not Affect Stability for High Harvesting Efforts

We showed that the asymptotic stability of $K_\gamma(0)$ implies the asymptotic stability of $K_\gamma(\theta)$ for $\theta \in [0, 1]$ if γ is chosen close enough to γ^* and g satisfies conditions (i)–(iii). Moreover, we obtained that $K_\gamma(\theta)$ is not only asymptotically stable, but attracts all solutions of (4) starting with a positive initial condition.

Proposition 2 *Assume that conditions (i)–(iii) hold. Then, there exists $\gamma_0 < \gamma^*$ such that for $\gamma \in [\gamma_0, \gamma^*)$ the fixed point $K_\gamma(\theta)$ of (4) is asymptotically stable for all $\theta \in [0, 1]$ and all positive solutions of (4) converge to $K_\gamma(\theta)$.*

3.2 Global Stability for Any Harvesting Time in the Ricker Case

Proposition 2 gives a sufficient condition for the global stability of the positive equilibrium of (4) in the Ricker case, for which $g(x) = \exp(r(1-x))$. But as the growth parameter r increases, the harvesting intensity has to be chosen higher and very close to the threshold $\gamma^* = 1 - e^{-r}$ above which all populations go extinct. This has two important drawbacks: (1) selecting harvesting efforts near such a threshold could be considered dangerous, and (2) attaining high harvesting intensities may be difficult in case of constraints of harvesting/thinning management. The following result proves that for the Ricker model the asymptotic stability of $K_\gamma(0)$ implies global stability of $K_\gamma(\theta)$ for all $\theta \in [0, 1]$.

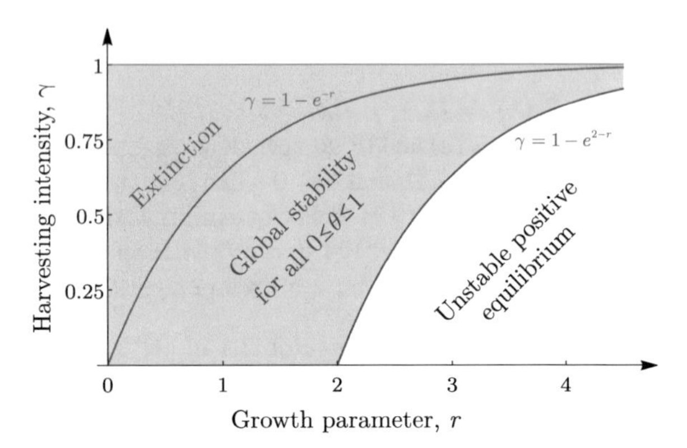

Fig. 1 In the blue area changing timing does not affect the global attraction of the positive equilibrium of model (4) for the Ricker map

Theorem 3 *Assume* $g(x) = e^{r(1-x)}$, $r > 0$, *and* $\gamma \in (0, 1)$ *such that* $1 - e^{2-r} \leq \gamma < 1 - e^{-r}$. *Then, for any* $\theta \in [0, 1]$, *the positive equilibrium of Eq. (4) is G.A.S.*

Figure 1 illustrates the region of parameters (r, γ) for which changing timing does not affect the global attraction of the positive equilibrium according to Theorem 3.

3.3 Timing Can Be Stabilizing by Itself

We proved that in the Ricker case it is possible to find $\theta \in (0, 1)$ such that $K_\gamma(\theta)$ for (4) is stable when $K_\gamma(0)$ is unstable.

Proposition 4 *Assume* $g(x) = e^{r(1-x)}$ *and* $r > 0$. *Then, there exists* $\gamma_c < \gamma_* := 1 - e^{2-r}$ *such that for any* $\gamma \in (\gamma_c, \gamma_*)$ *it is possible to find a timing interval* (θ_0, θ_1) *with the property that for each* $\theta \in (\theta_0, \theta_1)$ *the fixed point* $K_\gamma(\theta)$ *is asymptotically stable for (4).*

3.4 Timing Can Be Destabilizing

Proposition 4 shows that timing can be stabilizing by itself. In view of this, it is logical to ask the opposite question: *can timing be destabilizing?* Cid et al. conjectured in [1] that harvesting times θ in the interior of [0, 1] cannot be destabilizing if conditions (i)–(iii) are satisfied.

Conjecture 5 ([1, Conj. 3.5]) *Assume that conditions (i)–(iii) hold. If the positive equilibrium* $K_\gamma(0)$ *of (4) with* $\theta = 0$ *is asymptotically stable, then the fixed point* $K_\gamma(\theta)$ *is asymptotically stable for (4) for all* $\theta \in [0, 1]$.

A counterexample of this conjecture corresponds to the analytic function

$$g(x) = e^{6-15x+15x^2-\frac{11}{2}x^3}, \tag{5}$$

for which $\frac{dF_{0.6}}{dx}(K_{0.5}(0.6)) = F'_{0.6}(K_{0.5}(0.6)) \approx -1.278$ while $F'_0(K_{0.5}(0)) = F'_1(K_{0.5}(1)) \approx -0.207$.

4 Discussion and Conclusions

We studied the combined effect of harvesting intensity and harvesting time on the stability of a discrete population model proposed by Seno [4]. Under general conditions, we showed that timing has no negative effect on the stability of the positive equilibrium if the harvesting intensity is close enough to γ^*. Moreover, we proved that the latter stability is global. To the best of our knowledge, this is the first global stability result for (4) valid for general over-compensatory population models, since global stability results in [1] only cover under-compensatory models (such as the Beverton–Holt model) and the quadratic model.

For the Ricker–Seno model, we proved that there is global stability of the positive equilibrium regardless of the time of the intervention. Additionally, we showed that for this model timing can be stabilizing, that is, a harvesting intensity applied at an appropriate time of the season can asymptotically stabilize the positive equilibrium even when it cannot be stabilized at the beginning or at the end of the reproductive season with the same harvesting intensity.

Finally, we showed that timing can be destabilizing under natural conditions assumed on population production maps. This provides counterexamples for a conjecture recently published in [1]. However, these counterexamples are the result of mathematical constructions. Most of the population maps considered in the ecological literature satisfy additional conditions, as for example to have negative Schwarzian derivative, which may prevent any destabilizing effects of timing.

Our study leaves several open questions for future research. First, to find what extra conditions are necessary for Conjecture 5 to hold. Second, to provide general conditions for which timing is stabilizing by itself for population models different from the Ricker model.

Further details and proofs of the results provided here can be found in [2, 3].

References

1. B. Cid, F.M. Hilker, E. Liz, Harvest timing and its population dynamic consequences in a discrete single-species model. Math. Biosci. **248**, 78–87 (2014)
2. D. Franco, H. Logemann, J. Perán, J. Segura, Dynamics of the discrete Seno population model: combined effects of harvest timing and intensity on population stability. Appl. Math. Modell. **48**, 885–898 (2017)

3. D. Franco, J. Perán, J. Segura, Effect of harvest timing on the dynamics of the Ricker–Seno model, Math. Biosci. **306**, 180–185 (2018)
4. H. Seno, A paradox in discrete single species population dynamics with harvesting/thinning. Math. Biosci. **214**(1), 63–69 (2008)

Gevrey Asymptotics of Slow Manifolds in Singularly Perturbed Delay Equations

Karel Kenens and Peter De Maesschalck

Abstract We study a system of singularly perturbed delay differential equations and derive an equation for a slow manifold. To this equation, there exists a formal series solution which is Gevrey-1. By a Borel summation procedure, quasi-solutions are obtained from the formal solution, which determine the slow manifolds up to an exponentially small error.

1 Introduction

The setting of this paper is singularly perturbed delay differential equations, a setting that is encountered in applications on high-speed machinery [2] or mathematical neurology [8]. Compared to ordinary differential equations, the phase space of delay equations is infinite dimensional, making the study harder. Fortunately, center manifolds of delay equations are finitely dimensional, as well as slow manifolds of singularly perturbed ones and hence a system reduction to such a manifold is in many applications a key element in the global study. Here, we are concerned with the characterization of slow manifolds that arise in these systems. Our goal is to characterize such a manifold by formal series expansions, following [4]. The novelty here is that those expansions will be shown to be of Gevrey type, which allows us to construct functions, called quasi-solutions, which determine slow manifolds up to an exponentially small error. It is in many cases precise enough to obtain rigorous results. Moreover, quasi-solutions exhibit excellent smoothness and asymptotic properties (the exact meaning of this will become clear throughout the text).

The paper that we present here combines ideas from [4], on delay differential equations, with ideas from [3], on Gevrey asymptotics on ordinary differential equations. In the latter paper, the object of study is similar: characterization of the slow manifolds by means of Gevrey analysis of the formal power series expansion. We

K. Kenens (✉) · P. De Maesschalck
Department of Mathematics, Hasselt University, 3500 Hasselt, Belgium
e-mail: karel.kenens@uhasselt.be

P. De Maesschalck
e-mail: peter.demaesschalck@uhasselt.be

A. Korobeinikov et al. (eds.), *Extended Abstracts Spring 2018*,
Trends in Mathematics 11, https://doi.org/10.1007/978-3-030-25261-8_28

are able to pursue the analysis in the DDE setting up to the construction of quasi-invariant manifolds. The step from quasi-invariant manifold to invariant manifold is also done in the ODE setting in [3], but we leave this as a future topic of research since at the moment, as we have seen delicate smoothness issues in the DDE case, much like is seen in [7].

In this text, we consider the following system of singularly perturbed delay differential equations:

$$\begin{cases} \dot{x}(t) = \varepsilon\,(a - \gamma x(t)) \\ \dot{y}(t) = (1+J)\,y(t) - Jy(t-\tau) + x(t) - \frac{y^3(t)}{3}, \end{cases} \tag{1}$$

with $a \in \mathbb{R}$, $\gamma \in \mathbb{R}_0$, $J, \tau \in \mathbb{R}_0^+$. This model can be encountered in mathematical neurosciences, see [8]. This is very much a toy model, allowing us to exhibit the Gevrey expansion technique. Analog results can be obtained in a much broader class of equations, including, for example, systems of the form

$$\begin{cases} \dot{x}(t) = \varepsilon \\ \dot{y}(t) = ay(t) + by(t-\tau) + \varepsilon G\,(x(t), y(t), y(t-\tau), \varepsilon)\,, \end{cases}$$

where G is a holomorphic function.

A further application of the techniques could possibly be found in dynamical models of chatter in drilling processes. These typically exhibit delays due to the physical nature of the cutting tool. In [2], the authors study the model of Stone and Askari by referring to the theory of inertial and slow manifolds developed by Chicone, [4]. This work by Chicone relates invariant manifolds of the delay differential equation to so-called inertial manifolds, which are dealt with using asymptotic expansions. A combination of the techniques of Chicone with the knowledge from Gevrey asymptotic expansions could better characterize center manifolds that appear in similar models.

2 Statement of the Result

We are thus interested in slow manifolds of system (1), or equivalently center manifolds of the extended system

$$\begin{cases} \dot{x}(t) = \varepsilon(t)\,(a - \gamma x(t)) \\ \dot{y}(t) = (1+J)\,y(t) - Jy(t-\tau) + x(t) - \frac{y^3(t)}{3} \\ \dot{\varepsilon}(t) = 0. \end{cases} \tag{2}$$

There is clearly a curve of equilibria $\{p_b : b \in \mathbb{R}\}$, with

$$p_b := \left(-b + \frac{b^3}{3},\ b,\ 0\right);$$

see Fig. 1. Let us start the study of the center manifold with a determination of the linearization. The linearized system around such an equilibrium is given by

$$\begin{cases} \dot{x}(t) = \varepsilon(t)\left(a + \gamma\left(b - \frac{b^3}{3}\right)\right) \\ \dot{y}(t) = \left(1 - b^2\right) y(t) + J\left(y(t) - y(t-\tau)\right) + x(t) \\ \dot{\varepsilon}(t) = 0. \end{cases} \tag{3}$$

The characteristic equation associated to (3) at p_b is given by

$$\lambda^2\left(\lambda - \left(1 - b^2\right) - J\left(1 - e^{-\lambda\tau}\right)\right).$$

In the ODE setting ($\tau = 0$), the curve of singular points is normally hyperbolic almost at all points p_b (except for $b = \pm 1$) meaning that almost everywhere $\lambda = 0$ is a root of order 2 and there is one nonzero root. Also in the DDE setting, $p_{\pm 1}$ splits the curve of equilibria into three parts, each of which is a graph where 0 is a root of order 2. Let us denote these graphs by f_-, f_0, f_+ where $f_-(x) < -1 < f_0(x) < 1 < f_+(x)$; see Fig. 1. For all points on $f_0(x)$, $\lambda = 0$ is the only characteristic root on the imaginary axis. On $f_\pm$, there is a possibility for an extra pair of complex conjugated characteristic roots of an equilibrium to lie on the imaginary axis. This,

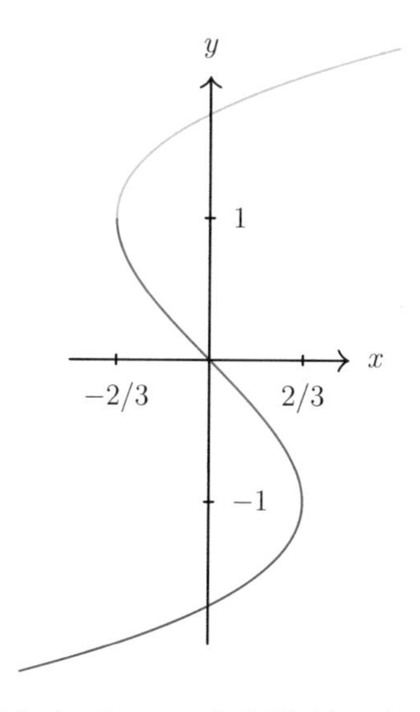

Fig. 1 Equilibria of system (2) in the plane $\varepsilon = 0$, divided into the curves f_-, f_0, f_+

however, can only happen in a finite number of points and it is not necessary for such points to even exist. If, for example, $J\tau \leq 1$, extra characteristic roots on the imaginary axis do not appear.

From here on out, we focus on one of the three graphs and denote it for simplicity by $f(x)$, we give the important remark that f is a holomorphic function and thus has an extension to a subset of the complex plane. Choose any x_0 in the domain of f for which $\lambda = 0$ is the only root on the imaginary axis. Translating the graph to the x axis and $(x_0, f(x_0))$ to the origin brings system (2) in the form

$$\begin{cases} \dot{x}(t) = \varepsilon(t)\left(a - \gamma x_0 - \gamma x(t)\right) \\ \dot{y}(t) = J\left(f\left(x(t) + x_0\right) - f\left(x(t-\tau) + x_0\right)\right) \\ \qquad + \left(1 + J - f^2\left(x(t) + x_0\right)\right) y(t) - J y(t-\tau) \\ \qquad - f\left(x(t) + x_0\right) y^2(t) - \frac{y^3(t)}{3} - \varepsilon(t) f'\left(x(t) + x_0\right)\left(a - \gamma x_0 - \gamma x(t)\right) \\ \dot{\varepsilon}(t) = 0. \end{cases} \tag{4}$$

One can calculate directly that the solution to the first equation satisfies

$$x(t-\tau) = x(t) + \left(x(t) - \frac{a - \gamma x_0}{\gamma}\right)\left(e^{\varepsilon\gamma\tau} - 1\right).$$

Thinking naively, one could then assume that a solution to the following equation,

$$\begin{aligned} &\varepsilon(a - \gamma x_0 - \gamma x)\frac{\partial Y}{\partial x}(x, \varepsilon) \\ &= J\left(f\left(x + x_0\right) - f\left(x + \left(x - \frac{a - \gamma x_0}{\gamma}\right)\left(e^{\varepsilon\gamma\tau} - 1\right) + x_0\right)\right) \\ &\quad + \left(1 + J - f^2\left(x + x_0\right)\right) Y(x, \varepsilon) - JY\left(x + \left(x - \frac{a - \gamma x_0}{\gamma}\right)\left(e^{\varepsilon\gamma\tau} - 1\right), \varepsilon\right) \\ &\quad - f\left(x + x_0\right) Y^2(x, \varepsilon) - \frac{Y^3(x, \varepsilon)}{3} - \varepsilon f'\left(x + x_0\right)\left(a - \gamma x_0 - \gamma x\right), \end{aligned} \tag{5}$$

satisfying $Y(x, 0) = 0$, would induce a center manifold of system (4).

In the following section, we show that this equation does indeed characterize a center manifold.

Our goal in this paper is to show the following, for the relevant terminology regarding Gevrey series, see Sect. 4.

Theorem 1 *There exists a unique formal series of the form* $\widehat{Y}(x, \varepsilon) = \sum_{n=1}^{\infty} y_n(x)\varepsilon^n$, *where all coefficients* y_n *are holomorphic on a neighborhood of* 0, *which formally solves Eq.* (5).

Moreover, for any open sector $S \subset \mathbb{C}$ *of opening less than* π, *there exists a function* $\widetilde{Y}(x, \varepsilon)$, *Gevrey-*1 *asymptotic to* $\widehat{Y}$ *w.r.t.* ε, *uniformly for* x *in a neighborhood of* 0, *which satisfies Eq.* (5) *up to an exponentially small error, i.e., there exists* $K, L > 0$ *such that*

$$\sup_x \Bigg| \varepsilon(a - \gamma x_0 - \gamma x)\frac{\partial \widetilde{Y}}{\partial x}(x, \varepsilon)$$
$$- J\left(f\left(x + x_0\right) - f\left(x + \left(x - \frac{a - \gamma x_0}{\gamma}\right)\left(e^{\varepsilon\gamma\tau} - 1\right) + x_0\right)\right)$$
$$- \left(1 + J - f^2\left(x + x_0\right)\right)\widetilde{Y}(x, \varepsilon) + J\widetilde{Y}\left(x + \left(x - \frac{a - \gamma x_0}{\gamma}\right)\left(e^{\varepsilon\gamma\tau} - 1\right), \varepsilon\right)$$
$$+ f\left(x + x_0\right)\widetilde{Y}^2(x, \varepsilon) + \frac{\widetilde{Y}^3(x, \varepsilon)}{3} + \varepsilon f'\left(x + x_0\right)\left(a - \gamma x_0 - \gamma x\right)\Bigg| \leq K e^{-\frac{L}{|\varepsilon|}}.$$

Remark 2 While our results will be local in nature, they can be easily applied to any compact subset of a normally hyperbolic part of the curve of equilibria.

3 Characterizing Center Manifolds

In this section, we follow [6] for the relevant definitions. Hale's characterization of center manifolds of delay differential equations is applied to our main equation; we prove that in this characterization, such center manifolds relate to solutions of (5).

The generalized eigenspace, when $a - \gamma x_0 \neq 0$, of the zero characteristic root of the linearisation of system (4) at (0, 0, 0) is given by

$$\left\{\left((a - \gamma x_0)(A + B\theta),\ -B(a - \gamma x_0)(1 - J\tau),\ B\right)^T \mid A, B \in \mathbb{R}\right\}.$$

Remark 3 For $a - \gamma x_0 = 0$, the generalized eigenspace is given by $\{(A,\ 0,\ B)^T \mid A, B \in \mathbb{R}\}$. This case does not essentially differ from when $a - \gamma x_0 \neq 0$ and we will thus not detail it any further.

Define $h_1 \colon \mathbb{R}^2 \to \mathcal{C}\left([-\tau, 0], \mathbb{R}\right)$, where $h_1\left(A, B\right)\left(\theta\right)$ is given by

$$\left(a - \gamma x_0\right) A\left(e^{-\gamma B\theta} - 1\right) - \frac{a - \gamma x_0}{\gamma}\left(e^{-\gamma B\theta} - 1 + \gamma B\theta\right)$$

and $\widetilde{h}_1 \colon \mathbb{R}^2 \to \mathcal{C}\left([-\tau, 0], \mathbb{R}\right)$ where $\widetilde{h}_1\left(A, B\right)\left(\theta\right)$ is given by

$$\left(a - \gamma x_0\right) A e^{-\gamma B\theta} - \frac{a - \gamma x_0}{\gamma}\left(e^{-\gamma B\theta} - 1\right).$$

The function is nothing more than a shorthand notation and is given by

$$\widetilde{h}_1\left(A, B\right)\left(\theta\right) = \left(a - \gamma x_0\right)\left(A + B\theta\right) + h_1\left(A, B\right)\left(\theta\right).$$

Furthermore, define $\widetilde{h}_2 \colon G \subset \mathbb{R}^2 \to \mathcal{C}\left([-\tau, 0], \mathbb{R}\right)$ given by

$$\widetilde{h}_2\left(A, B\right)\left(\theta\right) = Y\left(\widetilde{h}_1\left(A, B\right)\left(\theta\right), B\right),$$

where Y is a solution to (5) and G is a sufficiently small neighborhood of $(0, 0)$.
One calculates that

$$\begin{aligned}
\widetilde{h}_1\left(A, B\right)'\left(t\right) &= B\left(a - \gamma x_0 - \gamma \widetilde{h}_1\left(A, B\right)\left(t\right)\right), \\
\widetilde{h}_1\left(A, B\right)\left(t + \theta\right) &= \widetilde{h}_1\left(\frac{1}{\gamma} + \left(A - \frac{1}{\gamma}\right)e^{-\gamma B t}, B\right)(\theta), \\
\widetilde{h}_1\left(A, B\right)\left(\theta\right) + \left(\widetilde{h}\left(A, B\right)\left(\theta\right) - \frac{a - \gamma x_0}{\gamma}\right)&\left(e^{\gamma B \tau} - 1\right) = \widetilde{h}\left(A, B\right)\left(\theta - \tau\right).
\end{aligned}$$

This implies that supplementing system (4) with initial conditions,

$$\begin{aligned}
x_0\left(\theta\right) &= \widetilde{h}_1\left(A, B\right)\left(\theta\right), \\
y_0\left(\theta\right) &= \widetilde{h}_2\left(A, B\right)\left(\theta\right), \\
\varepsilon_0\left(\theta\right) &= B,
\end{aligned}$$

has solution given by

$$\begin{aligned}
x_t\left(\theta\right) &= \widetilde{h}_1\left(\frac{1}{\gamma} + \left(A - \frac{1}{\gamma}\right)e^{-\gamma B t}, B\right)\left(\theta\right), \\
y_t\left(\theta\right) &= \widetilde{h}_2\left(\frac{1}{\gamma} + \left(A - \frac{1}{\gamma}\right)e^{-\gamma B t}, B\right)\left(\theta\right), \\
\varepsilon_t\left(\theta\right) &= B.
\end{aligned}$$

Moreover, one can see rather easily that, if we define

$$h_2\left(A, B\right) = \widetilde{h}_2\left(A, B\right) + B\left(a - \gamma x_0\right)\left(1 - J\tau\right),$$

we have

$$h_1(0, 0) = 0,\ h_2(0, 0) = 0,\ Dh_1(0, 0) = 0,\ Dh_2(0, 0) = 0.$$

Consequently, the map

$$h \colon G \subset \mathbb{R}^2 \to \mathcal{C}\left([-\tau, 0], \mathbb{R}^3\right) : (A, B) \mapsto (h_1(A, B), h_2(A, B), 0)$$

is a center manifold. Moreover, it inherits the smoothness of Y.

4 A Primer on Gevrey Series and Asymptotics

Definition 4 Let $V \subset \mathbb{C}$ and $B > 0$. A formal series

$$\hat{f}(x, \varepsilon) = \sum_{n=0}^{\infty} f_n(x)\varepsilon^n$$

is Gevrey-1 in ε, uniformly for x in V, if $f_n \in \mathcal{O}(V)$ for all $n \in \mathbb{N}$ and there exists $A > 0$ such that

$$\sup_{x \in V} |f_n(x)| \leq AB^n n!\,.$$

For $\theta \in \left[0, 2\pi\right[$, $\delta \in \left]0, \pi\right[$ and $r > 0$, we denote the (open) sector in the direction θ with opening 2δ and radius r by

$$S(\theta, 2\delta, r) = \left\{\varepsilon \in \mathbb{C} \;\middle|\; 0 < |\varepsilon| < r,\ \mathrm{Arg}(\varepsilon e^{-i\theta}) \in \left]-\delta, \delta\right[\right\}.$$

Definition 5 Consider some sector S and a subset $V \subset \mathbb{C}^2$. Let $\hat{f}(x, \varepsilon) = \sum_{n=0}^{\infty} f_n(x)\varepsilon^n$ be a formal series in ε with coefficients in $\mathcal{O}(V)$. We say that a function $f(x, \varepsilon)$, holomorphic on $V \times S$, is Gevrey-1 asymptotic to the formal series $\hat{f}(x, \varepsilon)$, with respect to ε, uniformly for $(x, y) \in V$, if for every $\varepsilon \in S$ and every $N \in \mathbb{N}_0$ we have

$$\sup_{x \in V} \left| f(x, \varepsilon) - \sum_{n=0}^{N-1} f_n(x)\varepsilon^n \right| \leq CD^N N!\, |\varepsilon|^N$$

for certain $C, D > 0$. We denote this by

$$f(x, \varepsilon) \sim_1 \sum_{n=0}^{\infty} f_n(x)\varepsilon^n.$$

Remark 6 The characterization of Gevrey functions implies that the function can be extended to the vertex $\varepsilon = 0$ in a C^∞ smooth manner.

5 Quasi-Solutions—Proof of Theorem 1

5.1 Formal Solution

Since f is a holomorphic function at x_0, there exists an $R > 0$ such that $f(x + x_0)$ is holomorphic on $B(0, R)$.

Proposition 7 *There exists a unique formal series solution to* (5) *of the form* $\widehat{Y}(x, \varepsilon) = \sum_{n=1}^{\infty} y_n(x)\varepsilon^n$ *with* $y_n \in \mathcal{O}(B(0, R))$.

Proof Plugging the formal series $\widehat{Y}(x, \varepsilon) = \sum_{n=1}^{\infty} y_n(x)\varepsilon^n$ into Eq. (5), expanding $f(x + (x - (a - \gamma x_0)/\gamma)(e^{\varepsilon\gamma\tau} - 1) + x_0)$ in its Taylor series around $x + x_0$ and similarly expanding $y_n(x + (x - (a - \gamma x_0)/\gamma)(e^{\varepsilon\gamma\tau} - 1))$ around x, we can arrive at

$$\begin{aligned}
&\sum_{n=1}^{\infty} \left(1 - f^2(x + x_0)\right) y_n(x)\varepsilon^n \\
&= \varepsilon f'(x + x_0)(a - \gamma x_0 - \gamma x) + \sum_{k=1}^{\infty} \frac{J\left(x - \frac{a-\gamma x_0}{\gamma}\right)^k}{k!} f^{(k)}(x + x_0)\left(e^{\varepsilon\gamma\tau} - 1\right)^k \\
&+ \sum_{n=1}^{\infty}(a - \gamma x_0 - \gamma x)y_n'(x)\varepsilon^{n+1} + \sum_{n=1}^{\infty}\sum_{k=1}^{\infty} \frac{J\left(x - \frac{a\gamma x_0}{\gamma}\right)^k}{k!} y_n^{(k)}(x)\left(e^{\varepsilon\gamma\tau} - 1\right)^k \varepsilon^n \\
&+ f(x + x_0)\left(\sum_{n=1}^{\infty} y_n(x)\varepsilon^n\right)^2 + \frac{1}{3}\left(\sum_{n=1}^{\infty} y_n(x)\varepsilon^n\right)^3 .
\end{aligned} \tag{6}$$

Since the expansion in powers of ε of $e^{\varepsilon\gamma\tau} - 1$ has no constant term, for $n \geq 1$ the coefficient of ε^{n+1} on the RHS (right hand side) of (6) only depends on the functions $y_1, \ldots, y_n, f$ and their derivatives. Together with $1 - f^2(x + x_0) \neq 0$, indeed this follows immediately from $f(x) - f^3(x)/3 + x = 0$, we can thus recursively determine the coefficients of our formal solution.

Notice that since $f(x + x_0) \in \mathcal{O}(B(0, R))$, the same holds for the coefficients y_n.

5.2 Gevrey Property of Formal Solution

We will only state the result regarding the Gevrey property of the formal solution. For the proof, one can follow the general outlines of a proof for the ODE case, see, for example, [3]. The presence of a delay in Eq. (5) does present a small extra challenge in comparison with the ODE case; we are able to overcome this but do not delve deeper into the details for the sake of brevity.

Lemma 8 *Given the unique formal solution of the form*

$$\widehat{Y}(x, \varepsilon) = \sum_{n=1}^{\infty} y_n(x)\varepsilon$$

to Eq. (5), with $y_n \in \mathcal{O}(B(0,R))$. *For* $0 < T < R$, $\widehat{Y}(x,\varepsilon)$ *is a Gevrey-1 series in* ε, *uniformly for* $x \in B(0,T)$. *More specifically, there exist* $C, D > 0$ *such that*

$$\sup_{|x|<T} |y_n(x)| \leq C\left(\frac{D}{R-T}\right)^n n!\,.$$

5.3 Borel Summation

Starting from our formal solution, we construct a quasi-solution "nearly" solving (5). The exact properties such a function exhibits will become clear below.

The following theorem allows us to construct the desired function, for a proof [1] can be consulted, among others.

Theorem 9 (Borel–Ritt–Gevrey Theorem) *Given a formal series* $\widehat{g}(x,\varepsilon)$ *which is Gevrey-1 w.r.t.* ε, *uniformly for* $x \in B(0,T)$ *and a sector* $S(\theta,\delta,r)$ *with opening* $\delta < \pi$. *Then there exists a function* $g(x,\varepsilon)$, *analytic in* $B(0,T) \times S(\theta,\delta,r)$, *so that* $g(x,\varepsilon) \sim_1 \widehat{g}(x,\varepsilon)$.

By this result, there exists a function $\widetilde{Y}$, analytic on $B(0,T) \times S(\theta,\delta,r)$ satisfying $\widetilde{Y}(x,\varepsilon) \sim_1 \widehat{Y}(x,\varepsilon)$ with $\widehat{Y}$ the formal solution found in Proposition 7.

Since the function $\widetilde{Y}$ arises from a Borel summation procedure (and is certainly not unique), there is no guarantee that $\widetilde{Y}$ is a solution to (5). However, we do have the following.

Proposition 10 *There exist* $s > 0$, $0 < T' < T$ *such that the remainder term*

$$\begin{aligned}
\mathcal{R}(x,\varepsilon) :=\ & \varepsilon(a-\gamma x_0-\gamma x)\frac{\partial \widetilde{Y}}{\partial x}(x,\varepsilon) \\
& - J\left(f(x+x_0) - f\left(x+\left(x-\frac{a-\gamma x_0}{\gamma}\right)(e^{\varepsilon\gamma\tau}-1)+x_0\right)\right) \\
& - \left(1+J-f^2(x+x_0)\right)\widetilde{Y}(x,\varepsilon) + J\widetilde{Y}\left(x+\left(x-\frac{a-\gamma x_0}{\gamma}\right)(e^{\varepsilon\gamma\tau}-1),\varepsilon\right) \\
& + f(x+x_0)\widetilde{Y}^2(x,\varepsilon) + \frac{\widetilde{Y}^3(x,\varepsilon)}{3} + \varepsilon f'(x+x_0)(a-\gamma x_0-\gamma x)
\end{aligned} \tag{7}$$

is Gevrey-1 asymptotic to zero series, w.r.t. $\varepsilon \in S(\theta,\delta,s)$ *uniformly for* $x \in B(0,T')$, *i.e.,* $\mathcal{R}(x,\varepsilon) \sim_1 0$.

Proof One can rather easily prove this by applying the Ramis–Sibuya theorem, see, for example, [5]. We will, however, not give a proof since the exposition needed to introduce the Ramis–Sibuya theorem would take us to far. □

By Definition 5 of Gevrey asymptotics, $\mathcal{R}(x, \varepsilon) \sim_1 0$ implies that

$$\sup_{|x|<T'} |\mathcal{R}(x, \varepsilon)| \leq AB^N N! \, |\varepsilon|^N$$

for certain $A, B > 0$ and all $N \in \mathbb{N}_0$. It then follows readily that

$$\sup_{|x|<T'} |\mathcal{R}(x, \varepsilon)| \leq Ke^{-\frac{L}{|\varepsilon|}}$$

for certain $K, L > 0$. This proves Theorem 1.

References

1. W. Balser, *Formal Power Series and Linear Systems of Meromorphic Ordinary Differential Equations, Universitext* (Springer, New York, 2000), pp. xviii–299
2. S.A. Campbell, E. Stone, T. Erneux, Delay induced canards in a model of high speed machining. Dyn. Syst. **24**(3), 373–392 (2009)
3. M. Canalis-Durand, J.P. Ramis, R. Schäfke, Y. Sibuya, Gevrey solutions of singularly perturbed differential equations. J. Reine Angew. Math. **518**, 95–129 (2000)
4. C. Chicone, Inertial and slow manifolds for delay equations with small delays. J. Differ. Equ. **190**(2), 364–406 (2003)
5. A. Fruchard, R. Schäfke, *Composite Asymptotic Expansions*, Lecture Notes in Mathematics 2066 (Springer, Heidelberg, 2013), pp. x–161
6. J.K. Hale, L. Verduyn, M. Sjoerd, *Introduction to Functional Differential Equations*, Applied Mathematical Sciences, vol. 99 (Springer, New York, 1993)
7. F. Hartung, J. Turi, On differentiability of solutions with respect to parameters in state-dependent delay equations. J. Differ. Equ. **135**, 192–237 (1997)
8. M. Krupa, J. Touboul, Complex oscillations in the delayed FitzHugh-Nagumo equation. J. Nonlinear Sci. (2016)

Four-Dimensional Canards and Their Center Manifold

Kiyoyuki Tchizawa

Abstract We consider four-dimensional slow–fast systems, which can be represented either by one-dimensional slow vector field or three-dimensional fast vector field and denoted as $\mathbb{R}^{1+3}$, or $\mathbb{R}^{2+2}$, or $\mathbb{R}^{3+1}$. In each of these cases, the corresponding system can be well analyzed using blowing up the system and a time-scale reduction technique. Moreover, for each of these cases, by constructing a local model, the existence of a singular-limit solution (that are usually called Canards) is established. Some sufficient conditions for the existence of the canards are provided in this notice. What kind of four-dimensional canards are there?

1 Slow–Fast System in $\mathbb{R}^{2+2}$

We consider the following slow–fast system in $\mathbb{R}^{2+2}$:

$$\begin{aligned} \varepsilon \frac{dx}{dt} &= h(x, y, \varepsilon), \\ \frac{dy}{dt} &= f(x, y, \varepsilon), \end{aligned} \tag{1}$$

where $x \in \mathbb{R}^2$, $y \in \mathbb{R}^2$, $\varepsilon > 0$ is infinitesimal, $h \colon \mathbb{R}^{4+1} \to \mathbb{R}^2$, and $f \colon \mathbb{R}^{4+1} \to \mathbb{R}^2$. Note that ε is a non-standard number in a sense of Nelson, that is, using "idealization", "standardization", and "transfer" principles. For this system, we make the following five assumptions:

The author thanks Professors V. Sobolev and A. Korobeinikov for the invitation to the conference and Professor J. Llibre for his generous support of the author's visit. This work is partly supported by the Ministerio de Economía, Industria y Competitividad, Agencia Estatal de Investigación grant MTM2016-77278-P (FEDER), the Agència de Gestió d'Ajuts Universitaris i de Recerca grant 2017SGR1617, and the H2020 European Research Council grant MSCA-RISE-2017-777911.

K. Tchizawa (✉)
Institute of Administration Engineering Ltd, 101-0021 2-2-2 Sotokanda Chiyoda-Ku, Tokyo, Japan
e-mail: tchizawa@kthree.co.jp

A. Korobeinikov et al. (eds.), *Extended Abstracts Spring 2018*,
Trends in Mathematics 11, https://doi.org/10.1007/978-3-030-25261-8_29

(A1) $S = \{(x, y) \in \mathbb{R}^4 \mid h(x, y, 0) = 0\}$ is a two-dimensional differentiable manifold, and S intersects $T = \{(x, y) \in \mathbb{R}^4 \mid \det[\partial h(x, y, 0)/\partial x] = 0\}$ transversely, so that $PL = \{(x, y) \in S \cap T\}$ is a one-dimensional differentiable manifold.

(A2) $f(x, y, 0) \neq 0$ at $(x, y) \in PL$.

(A3) For $(x, y) \in S \setminus PL$, $\mathrm{rank}[\partial h(x, y)/\partial x] = 2$; for $(x, y) \in S$, $\mathrm{rank}[\partial h(x, y)/\partial y] = 2$. On set S, if $\det[\partial h(x, y)/\partial x] \neq 0$, the differentiation by t yields the following equality:

$$dx/dt = -[\partial h(x, y)/\partial x]^{-1}[\partial h(x, y)/\partial y] f(x, y, 0), \tag{2}$$

where a smooth function $y = g(x)$ exists. To avoid the degeneracy, we consider the time-scaled-reduced system

$$dx/dt = -\det[\partial h(x, y)/\partial x][\partial h(x, y)/\partial x]^{-1}[\partial h(x, y)/\partial y] f(x, y, 0). \tag{3}$$

Then, all the singular points of the time-scaled-reduced system are contained in the set $PS = \{(x, y) \in PL \mid -\det[\partial h(x, y)/\partial x][\partial h(x, y)/\partial x]^{-1}[\partial h(x, y)/\partial y] \cdot f(x, y, 0) = 0\}$, where $y = g(x)$. Points in the set PS are called pseudo-singular points in $\mathbb{R}^{2+2}$.

(A4) All the singular points of the reduced system are nondegenerate, i.e., the corresponding eigenvalues are nonzero.

(A5) The invariant manifold $INV(x, y)$ intersects the set PL transversely.

Note that in the case of k slow variables and m fast variables with $k \geq 3$ or $m \geq 2$, it is not established yet whether the time-scaled-reduced system holds an approximation for the original system. In the case $k = 2$ and $m = 1$, this was proved by Benoit [1] by constructing a local model and obtaining its solution, and in the case $k = 2$ and $m = 2$ this was proved extensively by Tchizawa [2].

2 Slow–Fast System in $\mathbb{R}^{1+3}$

Let us consider the approximation regarding the reduced system constructing a local model in the case $k = 1$ and $m = 3$.

Let the origin O be a "saddle" or a "node" point. By blowing up the coordinates $x = (x_1, x_2, x_3) \in \mathbb{R}^3$ and $y \in \mathbb{R}$ as follows: $x_1 = \alpha^2 u_1, x_2 = \alpha^2 u_2, x_3 = \alpha u_3$ and $y = \alpha^2 v$ with $\alpha \approx 0$, the system is reduced to the following:

$$\begin{aligned}
\varepsilon \frac{du_1}{dt} &= h_1(x, y, \varepsilon)/\alpha^2, \\
\varepsilon \frac{du_2}{dt} &= h_2(x, y, \varepsilon)/\alpha^2, \\
\varepsilon \frac{du_3}{dt} &= h_3(x, y, \varepsilon)/\alpha, \\
\frac{dv}{dt} &= f(x, y, \varepsilon)/\alpha^2.
\end{aligned} \tag{4}$$

Let us assume that rank$[\partial h(x, y)/\partial x] = 3$ and rank$[\partial h(x, y)/\partial y] = 1$, that is, there exists a vector function $x = \phi(y)$, $\phi = (\phi_1, \phi_2, \phi_3)$, which is invertible. Note that assumption (A3) is invalid in this case, and assumption (A1) holds, as like the set S is one-dimensional, and the set PL is zero-dimensional.

Assuming $\varepsilon = 0$, one get the following time-scaled-reduced system:

$$\begin{aligned}
\frac{dx}{dt} &= \det\left[\frac{\partial h(x, y)}{\partial x}\right]\left[\frac{\partial h(x, y)}{\partial x}\right]^{-1}\left[\frac{\partial h(x, y)}{\partial y}\right]\frac{dy}{dt}, \\
\frac{dy}{dt} &= f(x, y, 0).
\end{aligned} \tag{5}$$

Scaling $t = \alpha^2 \tau$ and $\varepsilon/\alpha^2 \approx 0$ yields the local model

$$\begin{aligned}
\delta \frac{du_1}{d\tau} &= (1/\alpha^2) h_1(x, y, \varepsilon) \\
&= h_1(0)/\alpha^2 + (\partial h_1(0)/\partial x_1)u_1 + (\partial h_1(0)/\partial x_2)u_2 \\
&\quad + (\partial h_1(0)/\partial x_3)u_3/\alpha + (\partial h_1(0)/\partial y)v + (\partial^2 h_1(0)/\partial x_3^2)u_3^2,
\end{aligned} \tag{6}$$

$$\begin{aligned}
\delta \frac{du_2}{d\tau} &= (1/\alpha^2) h_2(x, y, \varepsilon) \\
&= h_2(0)/\alpha^2 + (\partial h_2(0)/\partial x_1)u_1 + (\partial h_2(0)/\partial x_2)u_2 \\
&\quad + (\partial h_2(0)/\partial x_3)u_3/\alpha + (\partial h_2(0)/\partial y)v + (\partial^2 h_2(0)/\partial x_3^2)u_3^2,
\end{aligned} \tag{7}$$

$$\begin{aligned}
\delta \frac{du_3}{d\tau} &= (1/\alpha) h_3(x, y, \varepsilon) \\
&= h_3(0)/\alpha + (\alpha \partial h_3(0)/\partial x_1)u_1 \\
&\quad + (\alpha \partial h_3(0)/\partial x_2)u_2 + (\partial h_3(0)/\partial x_3)u_3 + (\alpha \partial h_3(0)/\partial y)v,
\end{aligned} \tag{8}$$

$$\begin{aligned}
\frac{dv}{d\tau} &= f(x, y, \varepsilon) \\
&= f(0) + \alpha^2(\partial f(0)/\partial x_1)u_1 + \alpha^2(\partial f(0)/\partial x_2)u_2 \\
&\quad + \alpha(\partial f(0)/\partial x_3)u_3 + \alpha^2(\partial f(0)/\partial y)v,
\end{aligned} \tag{9}$$

where $\delta = \varepsilon/\alpha^2 \approx 0$.

Let us denote $a_{ij} = \partial h_i(0)/\partial x_j$ and $b_{ij} = \partial^2 h_i(0)/\partial x_j^2$. Then the above system is under assumptions $f(0) \neq 0$, $h(0) = 0$, $a_{13} = a_{23} = 0$ and $a_{33} < 0$, which satisfies the trace condition. Taking $\delta = 0$, we get the system of algebraic equations

$$\begin{aligned} &a_{11}u_1 + a_{12}u_2 + (a_{13}/\alpha)u_3 + a_{14}v + (1/2)b_{13}u_3^2 = 0, \\ &a_{21}u_1 + a_{22}u_2 + (a_{23}/\alpha)u_3 + a_{24}v + (1/2)b_{23}u_3^2 = 0, \\ &a_{33}u_3 = 0, \\ &v = f(0)t. \end{aligned} \tag{10}$$

Solving these algebraic equations, we get

$$\begin{aligned} u_1 &= u_1^0 = (a_{12}a_{24} - a_{14}a_{22})f(0)t/(a_{11}a_{22} - a_{12}a_{21}), \\ u_2 &= u_2^0 = (a_{14}a_{21} - a_{11}a_{24})f(0)t/(a_{11}a_{22} - a_{12}a_{21}), \\ u_3 &= u_3^0 = 0. \end{aligned} \tag{11}$$

Now for $\delta \approx 0$ the solution is of the form:

$$u_1 = u_1^0 + L(\delta), \qquad u_2 = u_2^0 + L(\delta), \qquad u_3 = u_3^0 + L(\delta). \tag{12}$$

Theorem 1 *The time-scaled-reduced system in $\mathbb{R}^{1+3}$ has an approximation for the original system. If the pseudo-singular point is stable, and $a_{33} < 0$, tr$[\partial h(0,0)/\partial x] < 0$ hold, there exists a canard in the system $\mathbb{R}^{1+3}$.*

3 Slow–Fast System in $\mathbb{R}^{3+1}$

In the case of $k = 3$, $m = 1$, the analysis of the slow–fast system can be done by blowing up and constructing a local model. Consider the following slow–fast system:

$$\begin{aligned} \varepsilon\frac{dx}{dt} &= h(x, y, \varepsilon), \\ \frac{dy}{dt} &= f(x, y, \varepsilon), \end{aligned} \tag{13}$$

where $x \in \mathbb{R}$, $y \in \mathbb{R}^3$, $\varepsilon > 0$ is infinitesimal and $h\colon \mathbb{R}^{4+1} \to \mathbb{R}$, $f\colon \mathbb{R}^{4+1} \to \mathbb{R}^3$. For $\varepsilon = 0$, on the set S satisfying $h(x, y) = 0$ the differentiation by t yields

$$\left[\frac{\partial h(x, y)}{\partial y}\right]\frac{dy}{dt} + \frac{\partial h(x, y)}{\partial x}\frac{dx}{dt} = 0. \tag{14}$$

We assume that $\partial h(x, y)/\partial x \neq 0$. Then

$$\frac{dx}{dt} = -\left[\frac{\partial h(x, y)}{\partial y}\right] \frac{dy}{dt} \Big/ \frac{\partial h(x, y)}{\partial x}, \tag{15}$$

where there exists a smooth function $x = \phi(y)$. The reduced system is

$$\frac{dx}{dt} = \left[\frac{\partial h(x, y)}{\partial y}\right] f(x, y). \tag{16}$$

Let the origin O be a saddle or a node, again. Blowing up the coordinates $x \in \mathbb{R}$ and $y \in \mathbb{R}^3$ as follows: $x = \alpha^2 u$, $y = \alpha^2 v$ with $\alpha \approx 0$, where $y = (y_1, y_2, y_3)$ and $v = (v_1, v_2, v_3)$. Then the system is reduced to

$$\begin{aligned} \varepsilon \frac{du}{dt} &= \frac{1}{\alpha^2} h(x, y, \varepsilon), \\ \frac{dv}{dt} &= \frac{1}{\alpha^2} f(x, y, \varepsilon). \end{aligned} \tag{17}$$

Assume that $\mathrm{rank}[\partial h(x, y)/\partial y] = 3$, i.e., there exists a vector function $y = \psi(x)$, $\psi = (\psi_1, \psi_2, \psi_3)$, which is invertible. Note that assumption (A3) is also invalid, and assumption (A1) holds, as like the set S is three-dimensional, and the set PL is two-dimensional differentiable manifold. Scaling $t = \alpha^2 \tau$ and $\varepsilon/\alpha^2 \approx 0$ reduces the system to the system:

$$\begin{aligned} \delta \frac{du}{d\tau} &= \frac{1}{\alpha^2} h(x, y, \varepsilon), \\ \frac{dv}{d\tau} &= f(x, y, \varepsilon). \end{aligned} \tag{18}$$

Then, we obtain the local model as follows:

$$\begin{aligned} \delta \frac{du}{d\tau} &= a_{11} u + a_{12} v_1 + a_{13} v_2 + a_{14} v_3, \\ \frac{dv}{d\tau} &= f(0). \end{aligned} \tag{19}$$

We assume that $f(0) \neq 0$, $h(0) = 0$, $a_{11} \neq 0$, and $a_{14} \neq 0$. For $\delta = 0$, we get

$$\begin{aligned} &u = u^0 = -(a_{12} v_1 + a_{13} v_2 + a_{14} v_3)/a_{11}, \\ &v_1 = v_1^0 = f_1(0)t,\ v_2 = v_2^0 = f_2(0)t,\ v_3 = v_3^0 = f_3(0)t. \end{aligned} \tag{20}$$

Then, for $\delta \neq 0$, we have

$$u = u^0 + L(\delta), \qquad v = v^0 + L(\delta), \tag{21}$$

where $v^0 = (v_1^0, v_2^0, v_3^0)$.

Theorem 2 *The time-scaled-reduced system in $\mathbb{R}^{3+1}$ has an approximation for the original system. If the pseudo-singular point is a saddle (the eigenvalues are positive and negative in the slow vector field) or a node (all the eigenvalues are of the same sign), and $a_{11} < 0$ (or $a_{11} > 0$), $f(0) \neq 0$, under assumptions (A3), (A4), and (A5), then there exists a canard in the system $\mathbb{R}^{3+1}$.*

Remark 3 The trace condition, such that $\mathrm{tr}[\partial h(0,0)/\partial x] < 0$, ensures that one of the eigenvalues in the fast vector field changes the sign at the pseudo-singular point. In the case of $k = 1$, $m = 2$, put $y = 0$ and $h_2 = 0$. Then one can obtain the sufficient conditions for the existence of canards in the system $\mathbb{R}^{1+2}$. Furthermore, at the pseudo-singular point, if one of the eigenvalues on the slow manifold becomes infinitesimal, we have a very sensitive problem regarding the center manifold. (A5) provides the sufficient condition for the existence of the center manifold.

4 Center Manifold

The following system in $\mathbb{R}^{2+2}$ exhibits a sufficient condition for the existence of the center manifold. Let us return to system (1). Remember that for system (1), $x = (x_1, x_2)$, $y = (y_1, y_2)$, $h = (h_1, h_2)$ and $f = (f_1, f_2)$. Let us take

$$\begin{aligned} h_1(x, y) &= y_1 + x_2 - (1/3)x_1^3, \\ h_2(x, y) &= y_2 + x_1 - (1/3)x_2^3, \\ f_1(x, y) &= -x_1 - by_1, \\ f_2(x, y) &= -x_2 - by_2. \end{aligned} \tag{22}$$

This is based on the coupled FitzHugh–Nagumo equations. The time-scaled-reduced system is

$$\begin{aligned} \frac{dx_1}{dt} &= x_2^2(x_1 + b(-x_2 + (1/3)x_1^3)) - (x_2 + b(-x_1 + (1/3)x_2^3)), \\ \frac{dx_2}{dt} &= -(x_1 + b(-x_2 + (1/3)x_1^3)) - x_1^2(x_2 + b(-x_1 + (1/3)x_2^3)). \end{aligned} \tag{23}$$

For $b \approx 3/2$, the pseudo-singular points $PS \approx (1, 1), (-1, -1)$ are saddles. The invariant manifold $INV(x, y)$ is $INV(x, y) = \{(x, y) \in R^4 | x_1 - x_2 = 0, y_1 - y_2 = 0\}$. Furthermore, $\mathrm{tr}[\partial h(1, 1)/\partial x] = -2$ and $\det[\partial h(1, 1)/\partial x] = 0$. Note that the slow manifold is in the set $INV(x, y)$, satisfying $y_1 = -x_1 + (1/3)x_1^3$, and then it intersects with set $dy_1/dt = 0$. That is, $by_1 = x_1$ at $PS = (1, 1)$ for $b = 3/2$.

All these conditions are satisfied with the above assumptions. In this state, we obtain the center manifold, which does not include the pseudo-singular point.

References

1. E. Benoit, Systemes lents-rapides dans R^3 et leurs canards. Astérisque **190–110**, 159–191 (1983)
2. K. Tchizawa, On the two methods for finding 4-dim duck solutions. Appl. Math. Sci. Res. Publ. **5**(1), 16–24 (2014)

Modified Model for Proportional Loading and Unloading of Hypoplastic Materials

Erich Bauer, Victor A. Kovtunenko, Pavel Krejčí, Nepomuk Krenn, Lenka Siváková and Anna Zubkova

Abstract Classification of inner processes during loading and unloading tests in models of hypoplasticity developed by D. Kolymbas for the constitutive behavior of granular materials is the main aim of this work. We focus on a modified model proposed by Bauer. By introducing a dimensionless time parameter s, we transform the constitutive equation into a rate-independent form, and study the stress paths in different proportional loading regimes.

This work was supported by the Project No. 7AMB16AT035 within the MSMT Mobility Programme (Austria), by the Project SGS18/006/OHK1/1T/11 of the Czech Technical University, and by the Project of Excelence CZ.02.1.01/0.0/0.0/16_019/0000778 of the Ministry of Education, Youth and Sports of the Czech Republic.

E. Bauer (✉) · N. Krenn
Institute of Applied Mechanics, Graz University of Technology,
Technikerstrasse 4 / II, 8010 Graz, Austria
e-mail: erich.bauer@TUGraz.at

N. Krenn
e-mail: krenn.nepomuk@gmail.com

V. A. Kovtunenko · A. Zubkova
Institute for Mathematics and Scientific Computing, University of Graz,
Heinrichstrasse 36, 8010 Graz, Austria
e-mail: victor.kovtunenko@uni-graz.at

A. Zubkova
e-mail: anna.zubkova@uni-graz.at

P. Krejčí · L. Siváková
Faculty of Civil Engineering, Czech Technical University, Thákurova 7,
16629 Praha 6, Czech Republic
e-mail: krejci@math.cas.cz

L. Siváková
e-mail: lenka.sivakova@fbi.uniza.sk

A. Korobeinikov et al. (eds.), *Extended Abstracts Spring 2018*,
Trends in Mathematics 11, https://doi.org/10.1007/978-3-030-25261-8_30

1 Introduction

We pursue here the study started in [4] of the asymptotic behavior of stress trajectories under proportional loading and unloading in granular materials under the hypoplasticity hypothesis. Note that the idea of rate-independent hypoplasticity goes back to the works by Kolymbas, see, e.g., [10]. In engineering literature, this concept receives a lot of attention and it is appropriate to cite at least [5, 7, 8, 12, 14–16]. Its main purpose was to explain the phenomenon of *ratchetting* which is very strong in many existing rate-type constitutive models do not manifest a satisfactory agreement with real experiments. Rachetting is the process of accumulation of permanent deformation during cyclic loading and unloading. This behavior is characterized by progressively shifted loops in the strain–stress diagram.

Indeed, ratchetting is present in nonlinear kinematic hardening models of elastoplasticity of Armstrong–Frederick type, see [2], and the mathematical techniques developed in [6] for proving the well-posedness of these models have motivated the present study. Another mathematical approach to granular and multiphase media within the variational theory was proposed in [1, 9, 11].

An analytic identification of the asymptotic states in hypoplasticity whose existence was established in, e.g., in [13, Chap. 3.4], has been carried out in [4] for a simple one-parameter model suggested in [3]. Localization of the parameter domain which ensures Lyapunov stability of proportional strain paths was the main result there. Here, we obtain similar results for a modified model involving an additional physical parameter and discussed also in [3].

2 Description of the Model

2.1 Original Model

Our starting point is the model from [4] for inner processes in granular body under the strain–stress law

$$\dot{\boldsymbol{\sigma}}(t) = c_1 \left(\dot{\boldsymbol{\varepsilon}}(t) a^2 \operatorname{tr} \boldsymbol{\sigma} + \frac{\boldsymbol{\sigma}}{\operatorname{tr} \boldsymbol{\sigma}} (\boldsymbol{\sigma} : \dot{\boldsymbol{\varepsilon}}) + a(2\boldsymbol{\sigma} - \frac{1}{3} (\operatorname{tr} \boldsymbol{\sigma}) \mathbf{I}) \|\dot{\boldsymbol{\varepsilon}}\| \right), \tag{1}$$

where $\boldsymbol{\varepsilon}$ is the strain tensor and $\boldsymbol{\sigma}$ is the Cauchy stress tensor, $\mathbf{I}$ is the Kronecker tensor $\operatorname{tr} \boldsymbol{\sigma} = \boldsymbol{\sigma} : \mathbf{I}$ is the trace of $\boldsymbol{\sigma}$, $a > 0$ is a model parameter, and $c_1 < 0$ is a scaling parameter which, as we shall see, has no influence on the asymptotic behavior of the model. We consider proportional strain paths of the form

$$\boldsymbol{\varepsilon}(t) = \varepsilon(t)\mathbf{U}, \;\; \dot{\boldsymbol{\varepsilon}}(t) = \dot{\varepsilon}(t)\mathbf{U}, \tag{2}$$

where $\varepsilon(t)\colon [0,\infty) \to \mathbb{R}$ is a given monotone function, and $\mathbf{U}$ is a fixed symmetric tensor

$$\mathbf{U} = \begin{pmatrix} u_{11} & u_{12} & u_{13} \\ u_{21} & u_{22} & u_{23} \\ u_{31} & u_{32} & u_{33} \end{pmatrix}. \tag{3}$$

This is what we call a *proportional loading*.

2.2 Modified Model

The modification of Eq. (1) consists in including an additional physical parameter $b > 0$ for a refined modelling the volume strain behaviour. The enhanced equation reads:

$$\dot{\boldsymbol{\sigma}} = c_1\left(\dot{\boldsymbol{\varepsilon}} a^2 \operatorname{tr}\boldsymbol{\sigma} + \frac{\boldsymbol{\sigma}}{\operatorname{tr}\boldsymbol{\sigma}}(\boldsymbol{\sigma} : \dot{\boldsymbol{\varepsilon}}) + b \operatorname{tr}\boldsymbol{\sigma} \operatorname{tr}\dot{\boldsymbol{\varepsilon}}\mathbf{I} + a(2\boldsymbol{\sigma} - \frac{1}{3}\operatorname{tr}\boldsymbol{\sigma}\mathbf{I})\|\dot{\boldsymbol{\varepsilon}}\|\right). \tag{4}$$

Let us denote by $\langle\cdot,\cdot\rangle$ the canonical scalar product in the space of tensors $\langle\boldsymbol{\varepsilon},\boldsymbol{\sigma}\rangle = \boldsymbol{\varepsilon} : \boldsymbol{\sigma}$. With this notation and Hypothesis (2), Eq. (4) is of the form

$$\dot{\boldsymbol{\sigma}}(t) = c_1\dot{\varepsilon}(t)\big((a^2\mathbf{U} + \frac{\langle\boldsymbol{\sigma},\mathbf{U}\rangle}{\langle\boldsymbol{\sigma},\mathbf{I}\rangle}\boldsymbol{\sigma} + b\,\langle\mathbf{U},\mathbf{I}\rangle\,\mathbf{I})\,\langle\boldsymbol{\sigma},\mathbf{I}\rangle + a\|\mathbf{U}\|\operatorname{sign}\dot{\varepsilon}(t)\big(2\boldsymbol{\sigma} - \frac{1}{3}\langle\boldsymbol{\sigma},\mathbf{I}\rangle\,\mathbf{I}\big)\big). \tag{5}$$

Our analysis of Eq. (5) will be carried out under the following hypotheses:

(i) The material is initially compressed, that is, $\boldsymbol{\sigma}(0)$ is a given stress state such that $\langle\sigma(0),\mathbf{I}\rangle < 0$;
(ii) We investigate below the different dynamics of the model under increasing compression (or loading) corresponding to $\langle\mathbf{U},\mathbf{I}\rangle > 0$, decreasing compression (or unloading) corresponding to $\langle\mathbf{U},\mathbf{I}\rangle < 0$, and volume-preserving compression corresponding to $\langle\mathbf{U},\mathbf{I}\rangle = 0$;
(iii) $\varepsilon\colon [0,\infty) \to \mathbb{R}$ is absolutely continuous, $\dot{\varepsilon}(t) < 0$ for a.e. $t > 0$, and $\lim_{t\to\infty}\varepsilon(t) = -\infty$;

By introducing a time transformation $s(t)$ through the formula $\dot{s}(t) = c_1\dot{\varepsilon}(t)$, $s(0) = 0$, and $\boldsymbol{\sigma}'(s) = \mathrm{d}\boldsymbol{\sigma}/\mathrm{d}s$, we are able with the above assumptions to transform Eq. (5) into a rate-independent form:

$$\begin{aligned}\boldsymbol{\sigma}' &= a^2\,\langle\boldsymbol{\sigma},\mathbf{I}\rangle\,\mathbf{U} + \frac{\langle\boldsymbol{\sigma},\mathbf{U}\rangle}{\langle\boldsymbol{\sigma},\mathbf{I}\rangle}\boldsymbol{\sigma} - a\|\mathbf{U}\|\left(2\boldsymbol{\sigma} - \frac{1}{3}\langle\boldsymbol{\sigma},\mathbf{I}\rangle\,\mathbf{I}\right) + b\,\langle\boldsymbol{\sigma},\mathbf{I}\rangle\,\langle\mathbf{U},\mathbf{I}\rangle\,\mathbf{I} \\ &= \langle\boldsymbol{\sigma},\mathbf{I}\rangle\left(a^2\mathbf{U} + \frac{a}{3}\|\mathbf{U}\|\mathbf{I} + b\,\langle\mathbf{U},\mathbf{I}\rangle\,\mathbf{I}\right) + \boldsymbol{\sigma}\left(\frac{\langle\boldsymbol{\sigma},\mathbf{U}\rangle}{\langle\boldsymbol{\sigma},\mathbf{I}\rangle} - 2a\|\mathbf{U}\|\right).\end{aligned} \tag{6}$$

3 Loading

3.1 Isotropic Loading

First, we consider the case of the isotropic loading, that is, $\mathbf{U} = \mathbf{I}$ Then $\|\mathbf{U}\| = \sqrt{3}$ and $\langle \mathbf{U}, \mathbf{I} \rangle = 3$. In this case, Eq. (6) reduces to

$$\boldsymbol{\sigma}' = \langle \boldsymbol{\sigma}, \mathbf{I} \rangle \left(a^2 + \tfrac{a}{\sqrt{3}} + 3b\right)\mathbf{I} + \boldsymbol{\sigma}\left(1 - 2a\sqrt{3}\right). \tag{7}$$

The scalar product of (7) with $\mathbf{I}$ yields

$$\left\langle \boldsymbol{\sigma}', \mathbf{I} \right\rangle = \lambda \left\langle \boldsymbol{\sigma}, \mathbf{I} \right\rangle, \tag{8}$$

with $\lambda = 3a^2 - a\sqrt{3} + 1 + 9b$, where $3a^2 - a\sqrt{3} + 1 \geq 3/4$ and $b > 0$, therefore $\lambda > 0$. Hence $\langle \boldsymbol{\sigma}(s), \mathbf{I} \rangle = \langle \boldsymbol{\sigma}(0), \mathbf{I} \rangle \, \mathrm{e}^{\lambda s}$, and Eq. (7) can thus be written as

$$\boldsymbol{\sigma}' = -\mu \boldsymbol{\sigma} + R\mathrm{e}^{\lambda s}\mathbf{I},$$

where $\mu = 2a\sqrt{3} - 1$, $R = \langle \boldsymbol{\sigma}(0), \mathbf{I} \rangle \, (\lambda + \mu)/3 > 0$. The solution is

$$\boldsymbol{\sigma}(s) = \mathrm{e}^{-\mu s}\boldsymbol{\sigma}(0) + \frac{\langle \boldsymbol{\sigma}(0), \mathbf{I} \rangle}{3} \left(\mathrm{e}^{\lambda s} - \mathrm{e}^{-\mu s}\right)\mathbf{I}. \tag{9}$$

In other words, we have

$$\boldsymbol{\sigma}(s) - \frac{\langle \boldsymbol{\sigma}(0), \mathbf{I} \rangle}{3}\mathrm{e}^{\lambda s}\mathbf{I} = \mathrm{e}^{-\mu s}\left(\boldsymbol{\sigma}(0) - \frac{\langle \boldsymbol{\sigma}(0), \mathbf{I} \rangle}{3}\mathbf{I}\right). \tag{10}$$

The physically relevant case observed in experiments is $\mu > 0$, that is,

$$a > \frac{1}{2\sqrt{3}} \approx 0.289. \tag{11}$$

Then (10) means that the trajectory of $\boldsymbol{\sigma}(s)$ exponentially converges to the linear trajectory $\frac{\langle \boldsymbol{\sigma}(0), \mathbf{I} \rangle}{3}\mathrm{e}^{\lambda s}\mathbf{I}$ along the unit tensor $\mathbf{I}$ with initial condition given by the orthogonal projection of the initial condition $\boldsymbol{\sigma}(0)$ onto the line spanned by $\mathbf{I}$. The phenomenon that the influence of the initial condition is exponentially decreasing is typical for hypoplastic materials.

3.2 Anisotropic Loading

Let now $\mathbf{U} \in \mathbb{R}^{3\times 3}$ be arbitrary. As mentioned above, loading corresponds to $\langle \mathbf{U}, \mathbf{I} \rangle > 0$. It follows from (6) that the term $\langle \boldsymbol{\sigma}, \mathbf{U} \rangle \,/\, \langle \boldsymbol{\sigma}, \mathbf{I} \rangle$ satisfies a linear ODE

$$\left(\frac{\langle \boldsymbol{\sigma}, \mathbf{U} \rangle}{\langle \boldsymbol{\sigma}, \mathbf{I} \rangle} \right)' = A - \eta \frac{\langle \boldsymbol{\sigma}, \mathbf{U} \rangle}{\langle \boldsymbol{\sigma}, \mathbf{I} \rangle} \tag{12}$$

with

$$A = a^2 \|\mathbf{U}\|^2 + \frac{a}{3} \|\mathbf{U}\| \langle \mathbf{U}, \mathbf{I} \rangle + b \langle \mathbf{U}, \mathbf{I} \rangle^2 , \quad \eta = a\|\mathbf{U}\| + (a^2 + 3b) \langle \mathbf{U}, \mathbf{I} \rangle .$$

The solution of (12) is of the form

$$\frac{\langle \boldsymbol{\sigma}(s), \mathbf{U} \rangle}{\langle \boldsymbol{\sigma}(s), \mathbf{I} \rangle} = B + C\mathrm{e}^{-\eta s},$$

with

$$B = \frac{A}{\eta}, \quad C = \frac{\langle \boldsymbol{\sigma}(0), \mathbf{U} \rangle}{\langle \boldsymbol{\sigma}(0), \mathbf{I} \rangle} - B.$$

Equation (6) is therefore equivalent to a linear equation

$$\boldsymbol{\sigma}' = \langle \boldsymbol{\sigma}, \mathbf{I} \rangle \left(a^2 \mathbf{U} + \frac{a}{3} \|\mathbf{U}\| \mathbf{I} + b \langle \mathbf{U}, \mathbf{I} \rangle \mathbf{I} \right) + \boldsymbol{\sigma} \left(B - 2a\|\mathbf{U}\| + C\mathrm{e}^{-\eta s} \right). \tag{13}$$

To solve Eq. (13), we proceed as in the isotropic case taking the scalar product of (13) with $\mathbf{I}$, which yields $\langle \boldsymbol{\sigma}', \mathbf{I} \rangle = \left(D + C\mathrm{e}^{-\eta s} \right) \langle \boldsymbol{\sigma}, \mathbf{I} \rangle$, where

$$D = (a^2 + 3b) \langle \mathbf{U}, \mathbf{I} \rangle - a\|\mathbf{U}\| + B = \frac{1}{\eta} \left((a^2 + 3b)^2 \langle \mathbf{U}, \mathbf{I} \rangle^2 + \frac{a}{3} \|\mathbf{U}\| \langle \mathbf{U}, \mathbf{I} \rangle + b \langle \mathbf{U}, \mathbf{I} \rangle^2 \right) > 0.$$

Hence,

$$\langle \boldsymbol{\sigma}(s), \mathbf{I} \rangle = \langle \boldsymbol{\sigma}(0), \mathbf{I} \rangle \, \mathrm{e}^{f(s)}, \quad f(s) = Ds + \frac{C}{\eta}(1 - \mathrm{e}^{-\eta s}),$$

and we can rewrite (13) as

$$\boldsymbol{\sigma}'(s) = \langle \boldsymbol{\sigma}(0), \mathbf{I} \rangle \, \mathrm{e}^{f(s)} \mathbf{V} + g'(s) \boldsymbol{\sigma}(s), \tag{14}$$

where

$$\mathbf{V} = a^2 \mathbf{U} + \frac{a}{3} \|\mathbf{U}\| \mathbf{I} + b \langle \mathbf{U}, \mathbf{I} \rangle \mathbf{I}, \quad g(s) = (B - 2a\|\mathbf{U}\|)s + \frac{C}{\eta}(1 - \mathrm{e}^{-\eta s}).$$

From (14) it follows that

$$\left(\mathrm{e}^{-g(s)}\boldsymbol{\sigma}(s)\right)' = \langle\boldsymbol{\sigma}(0), \mathbf{I}\rangle\, \mathrm{e}^{f(s)-g(s)}\mathbf{V} = \langle\boldsymbol{\sigma}(0), \mathbf{I}\rangle\, \mathrm{e}^{(D+2a\|\mathbf{U}\|-B)s}\mathbf{V},$$

hence

$$\mathrm{e}^{-g(s)}\boldsymbol{\sigma}(s) = \boldsymbol{\sigma}(0) + \frac{\langle\boldsymbol{\sigma}(0), \mathbf{I}\rangle}{D + 2a\|\mathbf{U}\| - B}\mathrm{e}^{(D+2a\|\mathbf{U}\|-B)s} - 1)\mathbf{V},$$

provided $D + 2a\|\mathbf{U}\| - B \neq 0$. We note that

$$B = \frac{\langle\mathbf{V}, \mathbf{U}\rangle}{\langle\mathbf{V}, \mathbf{I}\rangle}, \quad \eta = \langle\mathbf{V}, \mathbf{I}\rangle, \quad C = \frac{\langle\boldsymbol{\sigma}(0), \mathbf{U}\rangle}{\langle\boldsymbol{\sigma}(0), \mathbf{I}\rangle} - \frac{\langle\mathbf{V}, \mathbf{U}\rangle}{\langle\mathbf{V}, \mathbf{I}\rangle},$$

and

$$D + 2a\|\mathbf{U}\| - B = a\|\mathbf{U}\| + a^2\langle\mathbf{U}, \mathbf{I}\rangle + 3b\langle\mathbf{U}, \mathbf{I}\rangle = \langle\mathbf{V}, \mathbf{I}\rangle = \eta. \tag{15}$$

The formula for $\boldsymbol{\sigma}(s)$ then reads

$$\boldsymbol{\sigma}(s) = \mathrm{e}^{g(s)}\boldsymbol{\sigma}(0) + \frac{\langle\boldsymbol{\sigma}(0), \mathbf{I}\rangle}{\langle\mathbf{V}, \mathbf{I}\rangle}\left(\mathrm{e}^{f(s)} - \mathrm{e}^{g(s)}\right)\mathbf{V}. \tag{16}$$

We are in the same situation as in (9) provided

$$\lim_{s\to\infty} g(s) = -\infty, \quad \lim_{s\to\infty} f(s) = +\infty. \tag{17}$$

This condition can be reformulated in terms of the parameters a, b, c, where

$$c = \frac{\langle\mathbf{U}, \mathbf{I}\rangle}{\sqrt{3}\,\|\mathbf{U}\|}$$

is the cosine of the angle between the loading direction $\mathbf{U}$ and the isotropic direction $\mathbf{I}$. It can be stated as follows.

Theorem 1 *The stability condition* (17) *is satisfied if and only if*

$$3b(c^2 - 2\sqrt{3}ac) < \left(2a^2 - \frac{1}{3}\right)\sqrt{3}ac + a^2. \tag{18}$$

Proof We have $D > 0$ according to (15), then the fact that $\lim_{s\to\infty} f(s) = +\infty$ follows immediately. It remains to find conditions on $\mathbf{U}$ and a under which

$$B < 2a\|\mathbf{U}\| \tag{19}$$

in order to obtain $\lim_{s\to\infty} g(s) = -\infty$. A straightforward computation yields that (19) is fulfilled if and only if

$$- a^2 + \frac{a \langle \mathbf{U}, \mathbf{I} \rangle}{3\|\mathbf{U}\|} - \frac{2a^3 \langle \mathbf{U}, \mathbf{I} \rangle}{\|\mathbf{U}\|} - \frac{6ab \langle \mathbf{U}, \mathbf{I} \rangle}{\|\mathbf{U}\|} + \frac{b \langle \mathbf{U}, \mathbf{I} \rangle^2}{\|\mathbf{U}\|^2} < 0\,, \tag{20}$$

which we wanted to prove. □

Rewriting (16) as

$$\boldsymbol{\sigma}(s) - \frac{\langle \boldsymbol{\sigma}(0), \mathbf{I} \rangle}{\langle \mathbf{V}, \mathbf{I} \rangle} \mathrm{e}^{f(s)} \mathbf{V} = \mathrm{e}^{g(s)} \left(\boldsymbol{\sigma}(0) - \frac{\langle \boldsymbol{\sigma}(0), \mathbf{I} \rangle}{\langle \mathbf{V}, \mathbf{I} \rangle} \mathbf{V} \right), \tag{21}$$

we see that the trajectory of $\boldsymbol{\sigma}(s)$ exponentially converges to the linear trajectory $\frac{\langle \boldsymbol{\sigma}(0), \mathbf{I} \rangle}{\langle \mathbf{V}, \mathbf{I} \rangle} \mathrm{e}^{f(s)} \mathbf{V}$ propagating along $\mathbf{V}$ with initial condition given by the projection of $\boldsymbol{\sigma}(0)$ orthogonal to $\mathbf{I}$ onto the line spanned by $\mathbf{V}$.

Remark 2 The stability condition in Theorem 1 admits a geometric interpretation in the parameter space. A more detailed discussion will be made in a forthcoming paper. Notice only that if inequality (11) holds, then (20) holds for every $\mathbf{U} \in \mathbb{R}^{3\times 3}$. Indeed, we have $c \le 1$, hence $c^2 - 2\sqrt{3}ac < 0$ for all a satisfying (11). This means in particular that the interval of the parameters a which ensure the stability of the proportional path is minimal in case of isotropic loading, while for the "almost volume-preserving loading" $\langle \mathbf{U}, \mathbf{I} \rangle / \|\mathbf{U}\| \to 0$ it becomes maximal.

4 Unloading

4.1 Isotropic Unloading

Isotropic loading can be described by $\mathbf{U} = -\mathbf{I}$. In comparison with the isotropic loading case, in Eq. (10) the coefficients $\lambda = -3a^2 - a\sqrt{3} - 9b - 1 < 0$ and $\mu = 1 + 2\sqrt{3}a > 0$. The condition $\mu > 0$ is satisfied for all $a > 0$.

4.2 Anisotropic Unloading

The case of the anisotropic loading is described by $\langle \mathbf{U}, \mathbf{I} \rangle < 0$. Asymptotic convergence to the asymptotic direction $\mathbf{V}^-$ is guaranteed by the condition $g(s) \to 0$ and $f(s) \to -\infty$ in the Eq. (16). It is satisfied when $D < 0, \eta > 0$, and $B - 2a\|\mathbf{U}\| < 0$. The last condition coincides with the case (18) of anisotropic loading. Here, however, additional conditions come into play, namely, $\eta > 0$, which is equivalent to

$$D + 2a\|\mathbf{U}\| - B > 0,$$

and in terms of c it can be rewritten as

$$\sqrt{3}c > -\frac{a}{a^2 + 3b}.$$

The condition $D < 0$ holds provided $3((a^2 + 3b)^2 + b)c^2 + ac/\sqrt{3} < 0$.

5 Conclusion

The modified model for constitutive behavior of granular materials proposed by Bauer [3] was studied in here for the particular case of proportional loading and unloading. We have determined the parameter range in which asymptotic stabilities of the proportional loading and unloading processes are guaranteed.

References

1. B.D. Annin, V.A. Kovtunenko, V.M. Sadovskii, Variational and hemivariational inequalities in mechanics of elastoplastic, granular media, and quasibrittle cracks, in *Analysis, Modelling, Optimization, and Numerical Techniques*, ed. by G.O. Tost, O. Vasilieva, Springer Proceedings in Mathematics & Statistics, vol. 121 (2015), pp. 49–56
2. P.J. Armstrong, C.O. Frederick, A mathematical representation of the multiaxial Bauschinger effect. C.E.G.B., Report RD/B/N 731 (1966)
3. E. Bauer, Modelling limit states within the framework of hypoplasticity, in *AIP Conference Proceedings*, ed by J. Goddard, P. Giovine, J.T. Jenkin, vol. 1227 (AIP, 2010), pp. 290–305
4. E. Bauer, V.A. Kovtunenko, P. Krejčí, L. Siváková, A. Zubkova, On Lyapunov stability in hypoplasticity, in *Proceedings of Equadiff Conference*, ed. by K. Mikula, D. Ševčovič, J. Urbán (Spektrum STU, 2017), pp. 107–116
5. E. Bauer, W. Wu, A hypoplastic model for granular soils under cyclic loading, in *Proceedings of the International Workshop Modern Approaches to Plasticity*, ed. by D. Kolymbas (Elsevier, 2010), pp. 247–258
6. M. Brokate, P. Krejčí, Wellposedness of kinematic hardening models in elastoplasticity, in *Mathematical Modelling and Numerical Analysis* (1998), pp. 177–209
7. G. Gudehus, *Physical Soil Mechanics* (Springer, Berlin, 2011)
8. W. Huang, E. Bauer, Numerical investigations of shear localization in a micro-polar hypoplastic material. Int. J. Numer. Anal. Meth. Geomech. **27**, 325–352 (2003)
9. A.M. Khludnev, V.A. Kovtunenko, *Analysis of Cracks in Solids* (WIT-Press, Southampton, 2000)
10. D. Kolymbas, *Introduction to Hypoplasticity* (A.A Balkema, Rotterdam, 2000)
11. V.A. Kovtunenko, A.V. Zubkova, Mathematical modeling of a discontinuous solution of the generalized Poisson-Nernst-Planck problem in a two-phase medium. Kinet. Relat. Mod. **11**, 119–135 (2018)
12. P. Krejčí, P. O'Kane, A. Pokrovskii, D. Rachinskii, Stability results for a soil model with singular hysteretic hydrology, in *Proceedings of the 5th International Workshop Multi-Rate Processes and Hysteresis*, ed. by A. Ivanyi, P. Ivanyi, D. Rachinskii, V. Sobolev. Journal of Physics: Conference Series, vol. 268 (2011), pp. 012016
13. A. Niemunis, Extended hypoplastic models for soils, Habilitation thesis, Ruhr University, Bochum, 2002
14. A. Niemunis, I. Herle, Hypoplastic model for cohesionless soils with elastic strain range. Mech. Cohes. Frict. Mat. **2**, 279–299 (1997)

15. B. Svendsen, K. Hutter, L. Laloui, Constitutive models for granular materials including quasi-static frictional behaviour: toward a thermodynamic theory of plasticity. Contin. Mech. Therm. **4**, 263–275 (1999)
16. W. Wu, E. Bauer, D. Kolymbas, Hypoplastic constitutive model with critical state for granular materials. Mech. Mater. **23**, 45–69 (1996)

Modeling of Excitatory Amino Acid Transporters

Denis Shchepakin, Leonid Kalachev and Michael Kavanaugh

Abstract A kinetic model for Excitatory Amino Acid Transporters (EAATs) was derived to analyze data from patch clamp experiments. The model was fitted to experimental data for EAAT1, and reliable parameter estimations were obtained. This allowed for inference and estimate of the turnover rate of EAAT1 and, potentially, for other EAATs, which would resolve longstanding discrepancies in the literature.

1 Introduction

Glutamate transporters mediate uptake of the neurotransmitter glutamate from the synaptic cleft and thereby limit the activation of glutamate receptors on the postsynaptic neuron. This is necessary to maintain the efficiency of information transfer by pulsatile synaptic release of glutamate and to prevent excitotoxic damage from excessive receptor activity. To date, five major subtypes of excitatory amino acid transporters have been identified in the mammalian CNS (EAAT1-5). These transporters are members of the solute carrier 1 (SLC1) gene family that also contains two neutral amino acid transporters (ASCT1-2); see [9]. Various experimental approaches have been utilized to establish and measure the kinetic characteristics of EAATs; see [5].

The present study focuses on EAAT1 (SLC1A3); a member of the SLC1 gene family that together with EAAT2 (SLC1A2) is widely expressed on astrocytes throughout the brain. Knowledge of the characteristic time required to transfer one molecule of glutamate by the transporters is crucial for understanding their physiological roles,

The authors thank Anastassios Tzingounis for data on Excitatory Amino Acid Transporter currents.

D. Shchepakin (✉) · L. Kalachev
Department of Mathematical Sciences, University of Montana, Missoula, MT, USA
e-mail: denis.shchepakin@umconnect.umt.edu

L. Kalachev
e-mail: kalachevl@mso.umt.edu

D. Shchepakin · M. Kavanaugh
Department of Biomedical Pharmaceutical Sciences, University of Montana, Missoula, MT, USA
e-mail: michael.kavanaugh@mso.umt.edu

A. Korobeinikov et al. (eds.), *Extended Abstracts Spring 2018*,
Trends in Mathematics 11, https://doi.org/10.1007/978-3-030-25261-8_31

particularly in shaping synaptic transmission. However, there is a great variety in turnover rate estimates of EAATs, ranging from few molecules per second (see [7]) to estimates in the hundreds per second (see [1]). In this work, we derived a model for EAATs and fitted it to the experimental data. We obtained a fitted set of parameters allowing us to reliably estimate the turnover rate of EAAT1 and conclude that is similar to that previously estimated for EAAT2 ($\approx$ 15/sec; see [7]). This establishes a kinetic framework for understanding the role of the molecular species underlying astrocytic glutamate transport in the brain.

2 EAAT Models

2.1 Existing Models

EAATs mediate transport of glutamate (Glu^-) with the cotransport of 3 Na^+ and 1 H^+, and the countertransport of 1 K^+, resulting in a net movement of two positive charges into the cell [3]. Previously developed Markov models take into account binding and unbinding of substrates, Fig. 1, e.g., [3]. Moreover, some states allow a thermodynamically uncoupled flow of chloride ions across the cell membrane [8]. As it is yet unknown which states can mediate this conductance, it is assumed to be possible for any state (which effectively expands the depicted model to twice as many states).

Most individual reaction rates are not yet established, and reported estimates of the rates obtained by fitting the full model to the experimental data are statistically unreliable due to overparameterization of the model with respect to the existing data. Under certain conditions (depending on the experimental design), it is possible to reduce the full model to a smaller one with fewer states, which can describe the experiment and predict outcomes while maintaining basic important characteristics of the initial system, e.g., turnover rate. In an excised outside-out patch voltage clamp experiment, one can regulate concentrations of substrates in both extracellular and intracellular spaces and rapidly change extracellular solutions. We consider an experimental scheme with saturating concentrations of sodium, hydrogen ions, and

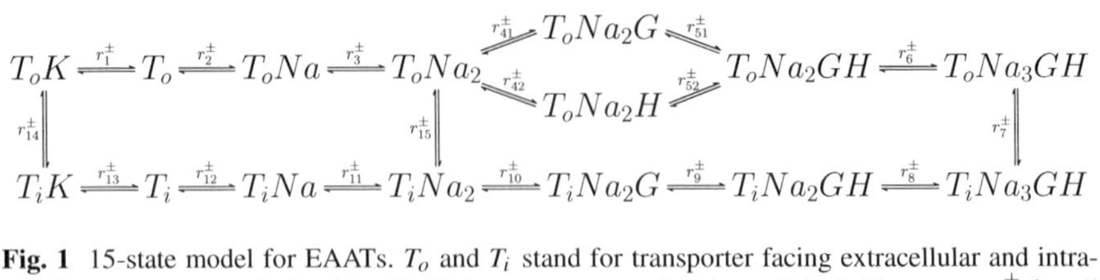

$$T_oK \underset{}{\overset{r_1^\pm}{\rightleftharpoons}} T_o \overset{r_2^\pm}{\rightleftharpoons} T_oNa \overset{r_3^\pm}{\rightleftharpoons} T_oNa_2 \quad \begin{matrix} \overset{r_{41}^\pm}{\nearrow} T_oNa_2G \overset{r_{51}^\pm}{\searrow} \\ \underset{r_{42}^\pm}{\searrow} T_oNa_2H \underset{r_{52}^\pm}{\nearrow} \end{matrix} \quad T_oNa_2GH \overset{r_6^\pm}{\rightleftharpoons} T_oNa_3GH$$

$$r_{14}^\pm \updownarrow T_oK/T_iK \qquad r_{15}^\pm \updownarrow T_oNa_2/T_iNa_2 \qquad r_7^\pm \updownarrow T_oNa_3GH/T_iNa_3GH$$

$$T_iK \overset{r_{13}^\pm}{\rightleftharpoons} T_i \overset{r_{12}^\pm}{\rightleftharpoons} T_iNa \overset{r_{11}^\pm}{\rightleftharpoons} T_iNa_2 \overset{r_{10}^\pm}{\rightleftharpoons} T_iNa_2G \overset{r_9^\pm}{\rightleftharpoons} T_iNa_2GH \overset{r_8^\pm}{\rightleftharpoons} T_iNa_3GH$$

Fig. 1 15-state model for EAATs. T_o and T_i stand for transporter facing extracellular and intracellular spaces, respectively; Na, K, H denote corresponding ions; G is L-Glutamate; $r_i^\pm$ for all i are reaction rate constants. r_i^+ and r_i^- correspond to clockwise and counterclockwise directions, respectively

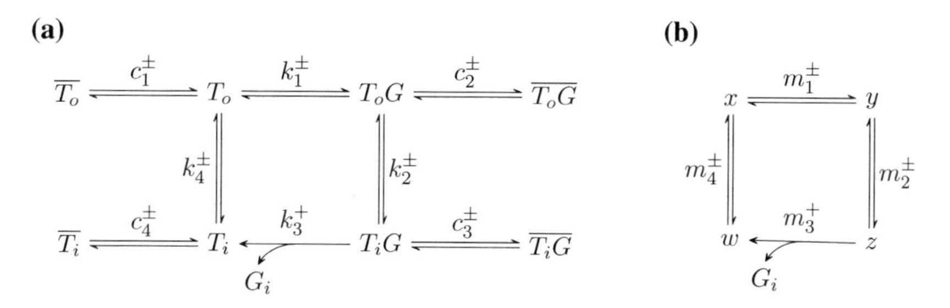

Fig. 2 **a** The simplified eight-state model for the patch clamp experiment. The states with bars are the corresponding conducting states, which allow the flow of chloride ions. **b** The chemical kinetic scheme that corresponds to the reduced model (1)

no potassium ions on the outside (and *vice versa* on the inside), with brief pulses of saturating glutamate applied to the outside membrane face with a piezoelectric switch. By selectively accelerating a subset of the state transitions in this scheme, we will be able to effectively observe only four states out of fifteen shown in Fig. 1. Adding four more states that correspond to chloride ion conductance, we obtain a model, depicted in Fig. 2a.

Similar models have been proposed before (e.g., [1]), but because these models are still overparameterized with respect to the data obtained in patch clamp experiments (see Fig. 3a; notice number of transitions), the conclusions drawn are not statistically reliable.

2.2 Reduced Model

In order to use the patch experiment data to identify unique parameter values, we need to simplify the model corresponding to the chemical kinetic scheme depicted in Fig. 2a even further. The model is a system of differential equations with eight variables, each corresponding to one state. Let us denote these variables using states notations, i.e., the variable $T_o(t)$ is a fraction of all transporters that are in T_o state, etc. According to [8], the transitions to conducting states are much faster compared to reactions which correspond to glutamate transportation. That allows us to define a small parameter $0 < \varepsilon \ll 1$:

$$c_i^{\pm} = \frac{\widetilde{c}_i^{\pm}}{\varepsilon}, \qquad \widetilde{c}_i^{\pm} \sim O(1), \qquad k_i^{\pm} \sim O(1), \qquad i = \overline{1,4}.$$

Using boundary function method [6], we can reduce the model. The leading order approximations can be written in the following form:

$$
\begin{aligned}
T_o(t) &= \tfrac{c_1^+}{c_1^+ + c_1^-} x(t) + O(\varepsilon), & \overline{T_o}(t) &= \tfrac{c_1^-}{c_1^+ + c_1^-} x(t) + O(\varepsilon), \\
T_oG(t) &= \tfrac{c_2^-}{c_2^- + c_2^+} y(t) + O(\varepsilon), & \overline{T_oG}(t) &= \tfrac{c_2^+}{c_2^- + c_2^+} y(t) + O(\varepsilon), \\
T_iG(t) &= \tfrac{c_3^-}{c_3^- + c_3^+} z(t) + O(\varepsilon), & \overline{T_iG}(t) &= \tfrac{c_3^+}{c_3^- + c_3^+} z(t) + O(\varepsilon), \\
T_i(t) &= \tfrac{c_4^+}{c_4^+ + c_4^-} w(t) + O(\varepsilon), & \overline{T_i}(t) &= \tfrac{c_4^-}{c_4^+ + c_4^-} w(t) + O(\varepsilon),
\end{aligned}
$$

where functions x, y, z, and w are the solutions of the system

$$
\begin{aligned}
\frac{dx}{dt} &= -m_1^+ x + m_1^- y + m_4^+ w - m_4^- x, \\
\frac{dy}{dt} &= m_1^+ x - m_1^- y - m_2^+ y + m_2^- z, \\
\frac{dz}{dt} &= m_2^+ y - m_2^- z - m_3^+ z, \\
\frac{dw}{dt} &= m_3^+ z - m_4^+ w + m_4^- x,
\end{aligned}
\tag{1}
$$

with

$$
\begin{aligned}
m_1^+ &= \frac{c_1^+ k_1^+}{c_1^+ + c_1^-}, & m_1^- &= \frac{c_2^- k_1^-}{c_2^- + c_2^+}, & m_2^+ &= \frac{c_2^- k_2^+}{c_2^- + c_2^+}, & m_2^- &= \frac{c_3^- k_2^-}{c_3^- + c_3^+}, \\
m_3^+ &= \frac{c_3^- k_3^+}{c_3^- + c_3^+}, & & & m_4^+ &= \frac{c_4^+ k_4^+}{c_4^+ + c_4^-}, & m_4^- &= \frac{c_1^+ k_4^-}{c_1^+ + c_1^-}.
\end{aligned}
$$

The system (1) can be depicted as a chemical kinetic scheme on its own, see Fig. 2b. The corresponding initial conditions are

$$
\begin{aligned}
x(0) &= T_o(0) + \overline{T_o}(0), & y(0) &= T_oG(0) + \overline{T_oG}(0), \\
z(0) &= T_iG(0) + \overline{T_iG}(0), & w(0) &= T_i(0) + \overline{T_i}(0).
\end{aligned}
$$

As stated above, the functions represent fractions of all transporters in particular states; therefore, the constant sum of all variables is 1. This stays true for both systems depicted in Fig. 2. Since in the absence of glutamate $k_1^+ = m_1^+ = 0$, if the system is allowed to reach a steady state before the first pulse, we have

$$
x(0) = \frac{m_4^+}{m_4^+ + m_4^-}, \qquad y(0) = \frac{m_4^-}{m_4^- + m_4^+}, \qquad z(0) = 0, \qquad w(0) = 0. \tag{2}
$$

The total current recorded during the experiment is a sum of the conductive current due to the flow of chloride ions (proportional to conductive states), stoichiometric current (coupled flux of glutamate molecules and ions across the membrane), and some constant leak current; see [4]. The voltage dependence of transport suggests that the transporter also mediates a capacitive charge transfer (gating charge movement) resulting in two positive charges into the cell as a glutamate molecule crosses the cell membrane, and electrically neutral countertransport of the potassium ion (rather than three positive charges into the cell with glutamate and one positive charge outside from the cell with potassium); see [3]. The resulting current formula is

$$I = -A \cdot \overline{T_o} - B \cdot \overline{T_o G} - C \cdot \overline{T_i G} - D \cdot \overline{T_i} - E\left(k_2^+ T_o G - k_2^- T_i G\right) + I_{leak}.$$

Assuming the system is at steady state prior to the first pulse of glutamate in the patch clamp experiment, we can set the steady state current to zero and find the formula for I_{leak}. The leading order approximation of the current will depend on the functions x, y, z, and w from the system (1). Finally, one can express any one variable using the other three. We can rewrite the current formula in the following way:

$$I(t) = -\mathcal{A} \cdot x(t) - \mathcal{B} \cdot y(t) - \mathcal{D} \cdot w(t) + \frac{\mathcal{A} m_4^+ + \mathcal{D} m_4^-}{m_4^+ + m_4^-} + O(\varepsilon), \qquad (3)$$

where

$$\mathcal{A} = \frac{c_1^- A}{c_1^+ + c_1^-} + \frac{c_3^+ C - c_3^- k_2^- E}{c_3^- + c_3^+},$$
$$\mathcal{B} = \frac{c_2^+ B + c_2^- k_2^+ E}{c_2^- + c_2^+} + \frac{c_3^+ C - c_3^- k_2^- E}{c_3^- + c_3^+},$$
$$\mathcal{D} = \frac{c_4^- D}{c_4^+ + c_4^-} + \frac{c_3^+ C - c_3^- k_2^- E}{c_3^- + c_3^+}.$$

The turnover rate is given by the influx of internal glutamate after all the other states reached the steady state:

$$\Theta = \lim_{t\to\infty} \frac{dG_i}{dt} = \lim_{t\to\infty} k_3^+ T_i G(t) =$$
$$\frac{m_1^+ m_2^+ m_3^+ m_4^+}{m_1^+ m_2^+ (m_3^+ + m_4^+) + m_1^+ m_4^+ (m_2^- + m_3^+) + (m_4^+ + m_4^-)(m_1^- m_2^- + m_1^- m_3^+ + m_2^+ m_3^+)} + O(\varepsilon). \qquad (4)$$

3 Experiments and Model Fitting

Patch clamp experiments were performed under the assumptions stated in Sect. 2.1. The experimental technique allows one to record the current corresponding to the flow of ions through a cell membrane; see [4]. After the current reaches steady state in the absence of glutamate, two consecutive short pulses of glutamate were applied. The current activates quickly and then decays to a steady state. After each pulse, the system returns to its initial steady state. The dynamics of the current

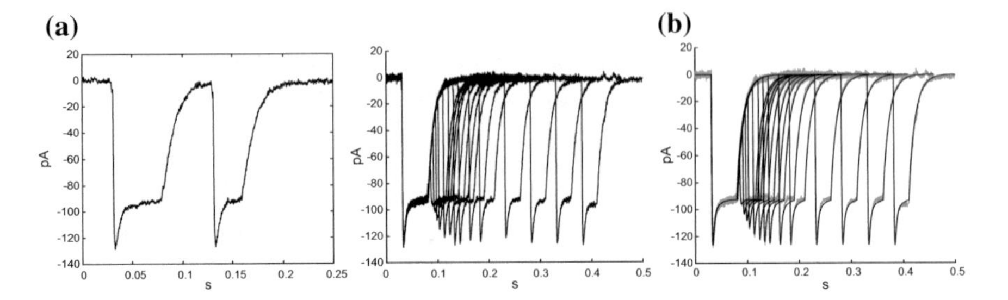

Fig. 3 **a** Current recorded during patch clamp experiment. Two consecutive pulses of glutamate caused inward (negative) current. Single pair (left) and all recorded pairs overlapped (right); **b** The data (gray) were fitted using the model (black)

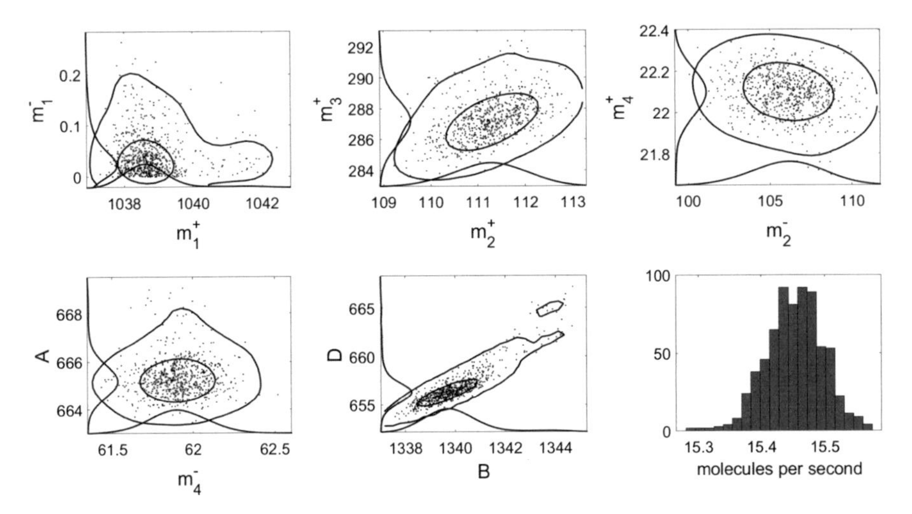

Fig. 4 95 and 99% confidence regions (inner and outer contours, respectively) yielded by MCMC method. The last picture shows the distribution of turnover rate value 15.45s^{-1}, 99% confidence interval: [15.31, 15.57]

were highly sensitive to the delay between pulses: the difference in current peaks between the first and the second pulses disappears as delay increases (see Fig. 3a). The data were fitted with the model (1), (2), and (3) using delayed rejection adaptive Metropolis Markov Chain Monte Carlo (MCMC) method; see [2]. The fitted model solutions practically coincide with the data curves, see Fig. 3b. Also, Fig. 4 shows chain confidence regions and turnover rate estimate and inference, obtained using the resulting chain and formula (4).

References

1. C. Grewer, N. Watzke, M. Wiessner, T. Rauen, Glutamate translocation of the neuronal glutamate transporter EAAC1 occurs within milliseconds. Proc. Natl. Acad. Sci. USA **97**, 9706–9711 (2000)
2. H. Haario, M. Laine, A. Mira, E. Saksman, DRAM: efficient adaptive MCMC. Stat. Comput. **16**, 339–354 (2006)
3. H.P. Larsson, A.V. Tzingounis, H.P. Koch, M.P. Kavanaugh, Fluorometric measurements of conformational changes in glutamate transporters. Proc. Natl. Acad. Sci. USA **101**, 3951–3956 (2004)
4. T.S. Otis, M.P. Kavanaugh, Isolation of current components and partial reaction cycles in the glial glutamate transporter EAAT2. J. Neurosci. **20**, 2749–2757 (2000)
5. A.V. Tzingounis, J.I. Wadiche, Glutamate transporters: confining runaway excitation by shaping synaptic transmission. Nat. Rev. Neurosci. **8**, 935–947 (2007)
6. A.B. Vasileva, V.F. Butuzov, L.V. Kalachev, *The Boundary Function Method for Singularly Perturbed Problems* (SIAM, 1995)
7. J.I. Wadiche, J.L. Arriza, S.Gm. Amara, and M.P. Kavanaugh, Kinetics of a human glutamate transporter. Neuron **14**, 1019–1027 (1995)
8. J.L. Wadiche, M.P. Kavanaugh, Macroscopic and microscopic properties of a cloned glutamate transporter/chloride channel. J. Neurosci. **18**, 7650–7661 (1998)
9. Y. Zhou, N.C. Danbolt, Glutamate as a neurotransmitter in the healthy brain. J. Neural Transm. **121**, 799817 (2014)

Limitations in Computational Analysis of Retrovirus Evolution

Lidia Nefedova

Abstract Retroviruses and retrotransposons with long terminal repeats (a type of transposable elements in eukaryotic genome) are very similar in their structure and life cycle that strongly indicates their common origin. Obviously, one of the structures transformed into others in the process of evolution. However, it is not clear which of the structures appeared earlier. There are two quite convincing scenarios of evolution: a scenario describing the ways of transformation of retrotransposons into retroviruses and the reverse scenario. The *Drosophila melanogaster* genome provides an excellent opportunity to analyze both possible scenarios for the evolution of retroelements, since it, unlike, for example, the human genome, is filled with diverse families of functionally active retrotransposons, including retrotransposons–retroviruses with infectious properties. The construction of evolutionary models—evolutionary trees—requires alignment of conserved amino acid sequences encoded by both types of retroelements. For phylogenetic trees construction, a variety of algorithms are developed. The most reliable approach is based on the principle of maximum likelihood. However, in the process of computational analysis, we meet with several problems that algorithms for constructing phylogenetic trees usually ignore. On the example of the model object, Drosophila, we consider the main limitations of modeling the evolution of retroelements: a tendency to accumulation of repeats, a horizontal transfer of sequences, and a rate of viral sequence evolution.

The basis of the evolutionary process is the variability of organisms, which is provided by mutations—changes in the nucleotide structure of genes. Because genomes evolve through the gradual accumulation of mutations, the number of differences in nucleotide sequences between a pair of genomes of different species should provide information about how long the organisms have divided as species. Two genomes of organisms, whose evolutionary lines have diverged recently, should have smaller

This work is supported by RFBR, pr. N 17-04-01250.

L. Nefedova (✉)
Faculty of Biology, Department of Genetics, Lomonosov Moscow State University, 119899 Moscow, Russia
e-mail: lidia_nefedova@mail.ru

A. Korobeinikov et al. (eds.), *Extended Abstracts Spring 2018*,
Trends in Mathematics 11, https://doi.org/10.1007/978-3-030-25261-8_32

differences than in organisms whose common ancestor existed a long time ago. It is the way to build evolutionary trees (molecular trees) which usually reflect the previously obtained morphological trees. The phylogeny of species is based on a comparison of the sequences of individual protein-coding genes. Compared genomes have thousands of genes, but only sets of several conservative genes are used for the tree building. Here, we are waiting for the first problem: different genes evolve at different and inconstant rates, so we *cannot definitely calibrate molecular timescale of evolution*. There is a countless range of possible combinations of mutation rate and time, and with access to only percentage data, the researchers will not be able to determine which combination is correct; see [2].

Mutations in protein-coding genes can be synonymous substitutions of nucleotides (K_s) that do not change a specific protein structure and non-synonymous substitutions (K_a) that change a specific protein structure. To estimate the selection, non-synonymous/synonymous mutation ratio (K_a/K_s ratio) is used; see [8]. Ks value is assumed to be neutral. A ratio greater than 1 caught positive, or Darwinian, selection; less than 1 implies purifying, or stabilizing, selection (acting against change); and a ratio of exactly 1 indicates neutral (i.e., no) selection. This approach allows observing the consequence of the most important evolutionary process—selection—at the level of individual genes. But K_a/K_s *estimation is applicable only to the analysis of the protein-coding part of the gene* and does not estimate evolutionary changes in the regulatory noncoding regions of the gene, which affects the level, timing or location of the gene expression. In addition, *the spectrum of mutations in the genome is not limited to only nucleotide substitutions in genes*: processes of genetic recombination often occur. Genetic recombination is a powerful tool of evolution. Some computer algorithms take into account the possibility of recombination. But *the level of recombination is not constant over time and depends on the influence of the environmental factors*. Recombination is stimulated by chromosome breaks, which can occur in the result of radiation, UV, and other exposure. We cannot say with certainty what factors acted millions of years ago. Thus, K_a/K_s *estimation reflects microevolutionary processes* occurring within one species and genus, but weakly reflects macroevolutionary processes and far less reflects the phylogenetic relationships between different species.

In most cases, the alignment of the compared nucleotide or amino acid sequences is a necessary stage in the reconstruction of phylogenetic trees. Pairwise sequence alignment methods are used to find the best-matching (local or global) alignments of two query sequences. Aligned sequences are represented as rows within a matrix. Gaps are inserted between the residues so that identical or similar characters are aligned in successive columns. The technique of dynamic programming is applied to produce local alignments via the Smith–Waterman algorithm, and global alignments via the Needleman–Wunsch algorithm. But pair alignment is only the first step. For the construction of phylogenetic relationships in phylogenetic trees, it is necessary to compare a number of sequences, i.e., build multiple alignments. Finding the optimal multileveling by the dynamic programming method has too much time complexity, so multiple alignments are built on the basis of different heuristics. Clustal (cluster alignment)—one of the most widely used computer programs for multiple alignments of nucleotide and amino acid sequences, based on progressive programming that uses a heuristic algorithm, was developed in 1984 and has since undergone many improvements; see [3]. But, at the same time, there are other alternative approaches to constructing multiple alignments, which means that there is no universal method.

It seems that viruses are most convenient for comparative genome analysis and the construction of phylogenetic trees. They have very small genomes, sometimes in eight orders smaller than the genomes of their hosts. But, despite this, they are the most problematic organisms for evolutionary analysis. Viruses are a noncellular form of life and are obligate cellular parasites, and their replication outside the cell is impossible. The genetic diversity of viruses significantly exceeds the genetic diversity of all the other organisms combined. Due to such a high genetic diversity, comparative phylogenetic studies of viruses are difficult; attempts to reconstruct the evolutionary relationships between viruses with different types of replication and to build traditional bifurcation phylogenetic trees with one common ancestor are highly speculative. The origin of viruses is one of the most controversial issues of biology. There are three main hypotheses for the origin of the virus domain: the regressive hypothesis, the hypothesis of cellular origin, and the hypothesis of coevolution. Each hypothesis has its own "pluses" and "minuses".

From an evolutionary point of view, retroviruses are the most interesting virus as they use in their life cycle the stage of integration of DNA copies of their RNA genome into the host genome. The integrated form remains in the genome forever. Thus, the retrovirus does not pass without a trace in the genome. Such "traces" can be used for phylogenetic analysis. Remarkably, some transposable elements of multicellular organisms—retrotransposons with long terminal repeats—and integrated DNA copies of retroviruses have the same structure. Thus, retroviruses and retrotransposons have the same evolutionary history. The fact that the transposable elements present in the genomes of all cellular organisms says that retrotransposons and viruses are related. Nevertheless, it is impossible to understand which structures originated in the evolution earlier—transposable elements or viruses. When did retroviruses arise? Numerous comparative genomic studies show that, apparently, retroviruses should have appeared at least 250–300 million years ago together with the first vertebrates of the land. However, this can be argued, if we assume that only vertebrates have retroviruses. Then how to be with the fruit fly Drosophila, invertebrate, which also contains retroviruses. It should be noted that all eukaryotes have retroelements, including plants and fungi.

Evolution of the virus cannot be considered in isolation from the evolution of the host genome. Because the viruses and their hosts coexist for a long time, their evolutionary relations are balanced. The "Red Queen" hypothesis is used to describe an idea based on coevolution of host and parasite; see [7]. Under this view, the species had to "run" (evolve) in order to stay in the same place (extant). The high rate of evolution of exogenous retroviruses and the relatively slow rate of evolution of genomic proviral DNA (approximately 4 orders of magnitude slower) should be manifested by serious differences in the nucleotide composition of exogenous and endogenous retroviruses.

Fruit fly Drosophila is an excellent model for studying the processes of evolution of retroviruses and their phylogenetic relationships with transposable elements. Transposable elements, most of which are represented by a variety of retrotransposons, account for up to 20% of the Drosophila genome. A variety of forms of retrotransposons allows reconstructing phylogenetic relationships between them and building a logical chain of the retroelement evolution. Using the methods of phylogenetic analysis, we classified all retrovirus-like transposable elements of Drosophila, which showed that assembly of retrovirus in host genome is possible, but reverse process degradation of retrovirus sequences and loss of genes, not necessary for the existence of retrotransposon, also occur; see [4].

The rate of sequence evolution is calculated by the level of mutations. Evidently, the rate of evolution (the speed of accumulation of mutations, the level of variability) will be lower in sequences of those proteins whose functions are associated with replication, and higher in those sequences that are more often attacked by host defense systems. It is noteworthy that *phylogenetic tree for retrotransposons and retroviruses* (and, consequently, for the concept of evolution) *may be different in dependence on what sequence comparison it was constructed* (the virus capsid protein or the sequence of the proteins responsible for the virus replication).

For retroviruses, an evolution rate often estimate for their long terminal repeats (LTRs) has been developed since left and right repeat sequences should be identical (according to the mechanism of the retrovirus replication). Using the Ka/Ks ratio, Drosophila retroelements' LTR divergence was calculated. The average age of the elements was found to be $137,000 \pm 89,000$ years; see [1]. This calculation was based on two assumptions. The first assumption is that LTR sequences are not under selection in the host genome. The second assumption is that the rate of mutations was unchangeable during all period of evolution. But these are only assumptions and cannot reflect the real evolutionary processes.

We have found cases of interspecific transfer in representatives of the Drosophila genus. We consider the cases of interspecific transfer to be the presence of almost identical sequences of retroviruses in different species of Drosophila, including quite distant ones. Thus, *the problem of evolutionary analysis of retroviruses is that they can be transmitted both horizontally* (between individuals of one or several species) *and vertically* (in generations, acting as transposable elements). This fact makes their evolutionary analysis very difficult.

We showed that new genes can arise in the Drosophila genome as a result of the retrotransposon genes fixing; see [5]. We found that the retrovirus capsid gene gag was fixed in Drosophila genome and evolved under strong stabilizing selection, but several sites of the gene were under positive selection. Thus, our example demonstrates complexity of the evolutionary processes inside the gene sequence. Thus, it should be taken into account that *different parts of individual genes can evolve at different rates*.

Clustal algorithm is actively used to build phylogenetic trees, despite the author's warnings that *unedited alignments should not be used in building trees*; see [6]. If one of aligned sequences has "excess" information, it will lead to gaps in other aligned sequences, which are highly undesirable for tree building. It means that any repeats should be deleted from the set of aligned sequences. But DNA motif accumulation is one of the ways of retroelement evolution. For example, in Drosophila genome, there are several copies of retroelement *Tirant* with different number of tandem repeats in its regulatory sequence. The number of repeats correlates with traspositional activity of this element. Consequently, *Tirant* evolved through the accumulation of certain repeats in its sequence to change mechanism of the interaction with the host genome. Thus, we have to consider such sequence modifications in the analysis of this element evolution.

Mathematical models and algorithms describing evolution are based on parametrization of many acting factors. The main problem is to detect all the factors. These factors have to be determined only by experimental investigations. On the other hand, mathematical models have to be confirmed experimentally for living systems.

References

1. N.J. Bowen, J.F. McDonald, Drosophila euchromatic LTR retrotransposons are much younger than the host species in which they reside. Genome Res. **11**(9), 1527–1540 (2001)
2. S. Ho, The molecular clock and estimating species divergence. Nat. Educ. **1**(1), 168 (2008)
3. P. Hogeweg, B. Hesper, The alignment of sets of sequences and the construction of phyletic trees: an integrated method. J. Mol. Evol. **20**(2), 175–186 (1984)
4. L.N. Nefedova, A.I. Kim, Molecular phylogeny and systematics of Drosophila retrotransposons and retroviruses. Mol. Biol. (Mosk.) **43**(5), 807–817 (2009)
5. L.N. Nefedova, I.V. Kuzmin, P.A. Makhnovskii, A.I. Kim, Domesticated retroviral GAG gene in Drosophila: new functions for an old gene. Virology **450–451**, 196–204 (2014)
6. J.D. Thompson, T.J. Gibson, D.G. Higgins, Multiple sequence alignment using ClustalW and ClustalX. Currl. Protoc. Bioinform. (2002) (Chapter 2:Unit 2.3)
7. D. Vergara, J. Jokela, C.M. Lively, Infection dynamics in coexisting sexual and asexual host populations: support for the red queen hypothesis. Am. Nat. **184**(Suppl. 1), S22–30 (2014)
8. J. Zhang, R. Nielsen, Z. Yang, Evaluation of an improved branch-site likelihood method for detecting positive selection at the molecular level. Mol. Biol. Evol. **22**(12), 2472–2479 (2005)

A Non-local Formulation of the One-Phase Stefan Problem Based on Extended Irreversible Thermodynamics

M. Calvo-Schwarzwälder

Abstract Non-local effects are introduced into a mathematical description of a solidification process based on Fourier's law with a size-dependent thermal conductivity. An asymptotic solution based on a large Stefan number is proposed. The agreement with the numerical solution is excellent for any Nusselt number.

1 Introduction

It is widely known that heat transport in nanostructures cannot be described by the classical equations [2, 3]. There exist a wide range of theoretical models which extend the classical Fourier law to account for non-local effects which become dominant on length scales which are comparable to the phonon mean free path, such as, for example, the hydrodynamic or the thermomass models [5, 6, 9].

Alvarez and Jou [1] propose including non-local effects into the Fourier law by considering a size-dependent thermal conductivity. Based on the extended irreversible thermodynamics framework [11], they derive an expression where the key parameter is the ratio between the phonon mean free path and the size of the system, known as the Knudsen number. Furthermore, dependence of the thermal conductivity on the size of the device has been observed experimentally [12] and analytically [4, 5]. Hennessy et al. [10] found that size-dependent thermal conductivities in solidification problems based on the Guyer–Krumhansl equation. Recently, Font [7] included the expression proposed by Alvarez and Jou into the formulation of the Stefan problem with a fixed temperature boundary condition at the origin.

The author acknowledges that the research leading to these results has been funded by 'La Caixa' Foundation and by the CERCA programme of the Generalitat de Catalunya.

M. Calvo-Schwarzwälder (✉)
Centre de Recerca Matemàtica, Campus de Bellaterra, Edifici C, 08193 Bellaterra, Barcelona, Spain
e-mail: mcalvo@crm.cat

Departament de Matemàtiques, Universitat Politècnica de Catalunya, 08028 Barcelona, Spain

A. Korobeinikov et al. (eds.), *Extended Abstracts Spring 2018*,
Trends in Mathematics 11, https://doi.org/10.1007/978-3-030-25261-8_33

Analytical solutions to phase-change problems are rare and numerics are non-trivial due to the moving boundary. A common approach is exploiting the fact that the Stefan number is expected to be large [7, 8, 13, 14] to obtain approximate solutions that typically are in good agreement with the numerical simulations.

2 Mathematical Model

We consider a one-dimensional, liquid bath, which is initially at the freezing temperature T_f. Due to the cold temperature T_e of the environment in contact with the bath at $x = 0$, the liquid bath starts solidifying and a solid phase begins to grow into the bath, occupying the space $[0, s(t)]$; see Fig. 1. The liquid is assumed to be at the freezing temperature, we only need to determine the temperature profile in the solid, which then determines the position of the interface $s(t)$ through an energy balance known as Stefan condition. In addition, at the boundary that is in contact with the cold environment we assume that the temperature exchange is described by the temperature difference at either side of the surface in contact with the cold environment. This is known as Newton cooling condition and it is a physically more realistic boundary condition that has been used previously in similar studies [13, 14].

In non-dimensional formulation, the problem is described by the following set of equations [7]:

$$\frac{\partial T}{\partial t} = f(s)\frac{\partial^2 T}{\partial x^2}, \qquad 0 \le x \le s(t), \tag{1a}$$

$$f(s)\frac{\partial T}{\partial x} = \mathrm{Nu}(1 + T), \qquad x = 0, \tag{1b}$$

$$T = 0, \qquad x = s(t), \tag{1c}$$

$$\beta\frac{\mathrm{d}s}{\mathrm{d}t} = f(s)\frac{\partial T}{\partial x}, \qquad x = s(t), \tag{1d}$$

$$s = 0, \qquad t = 0, \tag{1e}$$

where $f(s) = 2s(\sqrt{s^2 + 1} - s)$ is the non-dimensional form of the size-dependent thermal conductivity derived in [1], β is the Stefan number and Nu is the Nusselt number. The limit $\mathrm{Nu} \to \infty$ corresponds to the fixed temperature condition considered in [7]. For simplicity, we are going to avoid writing explicitly the dependence of f on s.

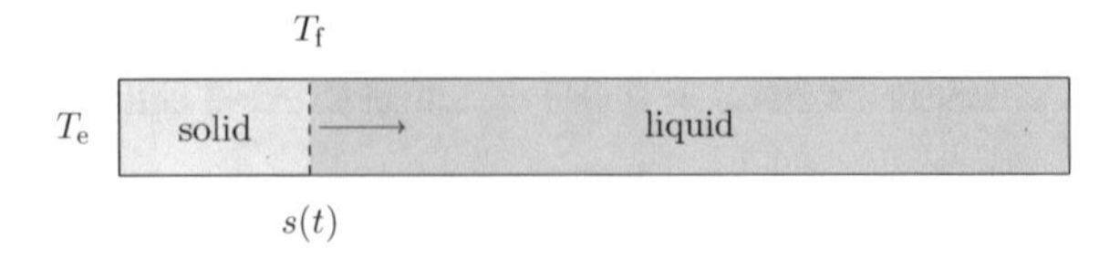

Fig. 1 Semi-infinite bath, initially at the freezing temperature, that is solidifying from $x = 0$ due to a temperature $T_e < T_f$. The solid–liquid interface is located at $x = s(t)$

3 Perturbation Solution

In many thermal problems, it is likely to expect $\beta \gg 1$; see [7, 13]. We can benefit from this assumption by introducing the time variable $\tau = t/\beta$ and expanding T in powers of β^{-1}. The solution, to first order, is

$$T(x,\tau) \approx A + Bx + \frac{1}{\beta}\left[\frac{A'}{2f}x^2 + \frac{B'}{6f}x^3 - \left(\frac{A'}{2f}s^2 + \frac{B'}{6f}s^3\right)\frac{\mathrm{Nu}x - f}{\mathrm{Nu}s + f}\right]\frac{\mathrm{d}s}{\mathrm{d}\tau}, \tag{2}$$

where

$$A(s) = -\frac{\mathrm{Nu}s}{f(s) + \mathrm{Nu}s}, \qquad B(s) = \frac{\mathrm{Nu}}{f(s) + \mathrm{Nu}s}. \tag{3}$$

After substituting (2) into (1d) and rearranging terms, we obtain

$$\frac{\mathrm{d}s}{\mathrm{d}\tau} = \frac{\beta f B}{\beta - A's - \frac{1}{2}B's^2 + \mathrm{Nu}\left(\frac{1}{2}A's^2 + \frac{1}{6}B's^3\right)(\mathrm{Nu}s + f)^{-1}}, \tag{4}$$

which, subject to (1e), is trivial to solve numerically.

4 Results and Conclusion

The approximate solution based on the assumption is now tested against the numerical solution of (1) for different values of β and Nu. For a detailed description of the numerical scheme, we refer to [7, 8, 14].

In Fig. 2, we have plotted the evolution of the solid–liquid interface s in time for different values of β and Nu and the dynamics in the cases of a constant and a variable thermal conductivity are compared. In the latter case, we also compare the perturbation method presented in Sect. 3 against the numerical solution.

The solidification rate is clearly affected by both parameters at the same time: if either β or Nu decrease, the speed of the interface decreases as well. In addition, the Nu has direct impact on the behaviour of the solution, in Fig. 2a and d (small Nu) we observe $s \sim t$ whereas in Fig. 2c and f (large Nu) we observe a transition from a $s \sim t$ to $s \sim \sqrt{t}$. Since the Nusselt number determines the heat flux at $x = 0$, which drives the whole process, this effect is not directly related to the fact of considering a size-dependent thermal conductivity.

The main effect that the non-local formulation has on the dynamics is a decrease in the solidification rate, as shown in Fig. 2. Furthermore, this effect is strengthened as the Stefan number increases, as it can be observed in Fig. 2c and f, for example. This can be understood by looking at (1d), which states that the solidification rate is proportional to β^{-1} and therefore a large Stefan number slows the process down, making non-local effects to be dominant for a larger period of time, which will reduce the solidification rate even more due to the presence of f in (1d).

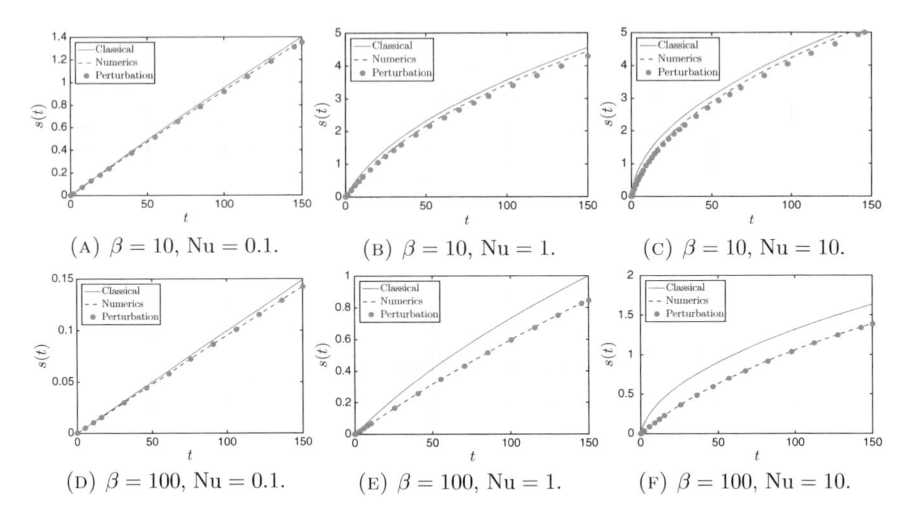

(A) $\beta = 10$, $\mathrm{Nu} = 0.1$. (B) $\beta = 10$, $\mathrm{Nu} = 1$. (C) $\beta = 10$, $\mathrm{Nu} = 10$.

(D) $\beta = 100$, $\mathrm{Nu} = 0.1$. (E) $\beta = 100$, $\mathrm{Nu} = 1$. (F) $\beta = 100$, $\mathrm{Nu} = 10$.

Fig. 2 Evolution of the solid–liquid interface for different values of β and Nu

Finally, we observe that the asymptotic solution is in very good agreement with the numerical solution for any value of Nu and β and the accuracy increases as β increases from 10 to 100. In the latter, both numerical and asymptotic solutions are identical.

To conclude, the present study shows that the changes in the solidification dynamics when a size-dependent conductivity is considered are only expected for certain range of values of the key parameters. In particular, when the solidification process is slow due to a large Stefan number, i.e. a small temperature difference between the initial liquid and the environment, non-local effects reduce the solidification rate even more. In any case, the perturbation method is in excellent agreement with the numerical solution.

References

1. F.X. Alvarez, D. Jou, Size and frequency dependence of effective thermal conductivity. J. Appl. Phys. **103**(9), 094321 (2008)
2. D.G. Cahill, P.V. Braun, G. Chen, D.R. Clarke, S. Fan, K.E. Goodson, P. Keblinski, W.P. King, G.D. Mahan, A. Majumdar, H.J. Maris, S.R. Phillpot, E. Pop, L. Shi, Nanoscale thermal transport II. 2003–2012. Appl. Phys. Rev. **1**(1), 011305 (2014)
3. D.G. Cahill, W.K. Ford, K.E. Goodson, G.D. Mahan, A. Majumdar, H.J. Maris, R. Merlin, S.R. Phillpot, Nanoscale thermal transport. J. Appl. Phys. **93**(2), 793–818 (2003)
4. M. Calvo-Schwarzwälder, M.G. Hennessy, P. Torres, T.G. Myers, F.X. Álvarez, A slip-based model for the size-dependent effective thermal conductivity of nanowires. Int. Commun. Heat Mass Transf. **91**, 57–63 (2018)
5. M. Calvo-Schwarzwälder, M.G. Hennessy, P. Torres, T.G. Myers, F.X. Álvarez, Effective thermal conductivity of rectangular nanowires based on phonon hydrodynamics. Int. Commun. Heat Mass Transf. **126**, 1120–1128 (2018)

6. B.Y. Cao, Z.Y. Guo, Equation of motion of a phonon gas and non-Fourier heat conduction. J. Appl. Phys. **102**(5), 053503 (2007)
7. F. Font, A one-phase Stefan problem with size-dependent thermal conductivity. Appl. Math. Modell. **63**, 172–178 (2018)
8. F. Font, T.G. Myers, S.L. Mitchell, A mathematical model for nanoparticle melting with density change. Microfluid. Nanofluidics **18**(2), 233–243 (2015)
9. Y. Guo, M. Wang, Phonon hydrodynamics and its applications in nanoscale heat transpor. Phys. Rep. **595**, 1–44 (2015)
10. M.G. Hennessy, M. Calvo-Schwarzwälder, T.G. Myers, Asymptotic analysis of the Guyer-Krumhansl-Stefan model for nanoscale solidification. Appl. Math. Modell. **61**, 1–17 (2018)
11. D. Jou, J. Casas-Vázquez, G. Lebon, Extended irreversible thermodynamics, 2nd edn. (Springer, Berlin, 1996)
12. D. Li, Y. Wu, P. Kim, L. Shi, P. Yang, A. Majumdar, Thermal conductivity of individual silicon nanowires. Appl. Phys. Lett. **83**(14), 2934–2936 (2003)
13. T.G. Myers, Mathematical modelling of phase change at the nanoscale. Int. Commun. Heat Mass Transf. **76**, 59–62 (2016)
14. H. Ribera, T.G. Myers, A mathematical model for nanoparticle melting with size-dependent latent heat and melt temperature. Microfluid. Nanofluidics **20**(11), 147 (2016)

Experimental Investigation of Viscoelastic Hysteresis in a Flex Sensor

Maxim Demenkov

Abstract We consider flex (or bending) polymer-based soft sensor that measures its own bending in terms of electrical resistance. When it is used for precise positioning as a part of fast control feedback, e.g. in soft robotic manipulators, sensor measurements show hysteresis-like rate-dependent behaviour typical for viscoelastic materials. We have constructed a simple electromechanical device for its investigation.

1 Introduction

Various devices using resistive flex sensors [8] have been made available in different areas such as biometric measurements for medical purposes or interfacing virtual reality. Nevertheless, these are slow-motion, imprecise applications. The advancements in soft robotics [4, 11], where such sensors could be printed directly on a robot body or used in other capacity as a part of control feedback, pose a serious question about reliability of their readings. Since the sensors are made of polymers, it is natural to assume that the well-known viscoelastic behaviour of polymers (see, e.g. [2, 3, 10]), including rate-dependent hysteresis, can affect the sensor readings.

Nonlinear rate-dependent hysteresis was not extensively studied from the mathematical viewpoint, and so far only a few publications (see, e.g. [1]) are available on the subject. Well-known classical models of hysteresis, such as the Prandtl–Ishlinskii and Krasnoselskii–Pokrovskii model, do not incorporate rate dependence [7, 9]. Even the application of the term 'hysteresis' in a rate-dependent setting is questionable for some researchers in mathematics. It is therefore important to create a simple and cheap device to analyse such behaviour in a typical university environment without complex laboratory setup, to facilitate the development of its mathematical models.

The viscoelastic materials have a relationship between stress and strain that depends on time or frequency. In engineering, linear viscoelastic material models can be represented by an arbitrary composition of linear springs and dampers; see

M. Demenkov (✉)
Institute of Control Sciences, Profsoyuznaya str. 65, Moscow 117997, Russia
e-mail: max.demenkov@gmail.com

A. Korobeinikov et al. (eds.), *Extended Abstracts Spring 2018*,
Trends in Mathematics 11, https://doi.org/10.1007/978-3-030-25261-8_34

Fig. 1 Linear viscoelastic model

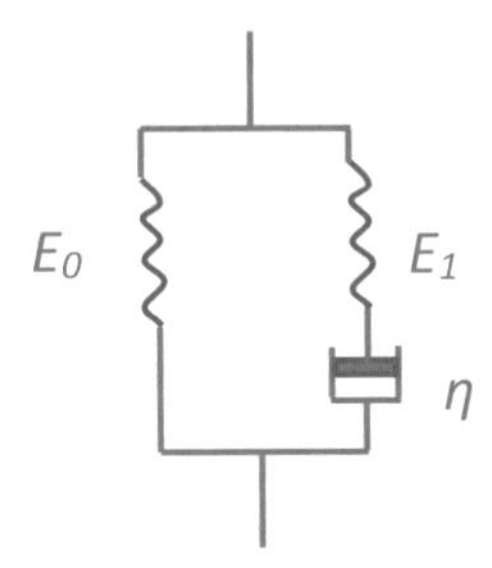

Fig. 1. The simplest model for solids that is able to show all phenomena related to viscoelasticity is the following three-parameter model (also called the standard linear solid or the Zener model, see, e.g. [10]):

$$\sigma + \frac{\eta}{E_1}\dot{\sigma} = E_0\varepsilon + \eta\frac{E_0 + E_1}{E_1}\dot{\varepsilon},$$

where σ is stress, ε is strain, η is viscosity, E_0 and E_1 are Young's modulus and the parameters of the relaxation function, respectively:

$$E(t) = E_0 + E_1 \exp(-t/\tau_R), t > 0,$$

and $\tau_R = \eta/E_1$ denotes the relaxation time. In the nonlinear case, no 'standard' model is available.

2 Experimental Results

In our experimental device, a small oval-shaped plastic arm is attached to an Arduino-controlled servomotor and is kinematically linked with the Flex Sensor® 4.5" available from Spectra Symbol, which is connected to the same Arduino module as the servomotor. Arduino is a popular robotics hobbyist platform, which can be also used for data acquisition in experiments. One side of the sensor is printed with a polymer ink that has conductive particles embedded in it. When the sensor is bent away from the ink, the conductive particles move further apart, increasing its resistance.

The positional servomotor SG90 can approximate a trajectory of its angular motion using a number of reference points. The delay before setting the next reference point is related to its angular speed (maximum is 500 deg/sec). Angular speed for the sensor tip is linearly related to the angular speed of the motor, which is six times higher.

This device (see Fig. 2) is a modification (simply the replacement of a fan with servomotor) of Flexy [5], initially created at the Slovak University of Technology in

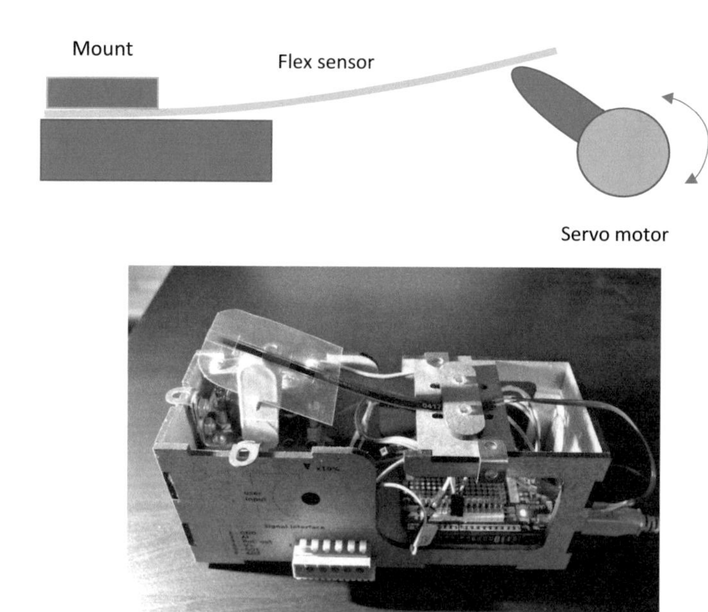

Fig. 2 Scheme of our device and its actual implementation

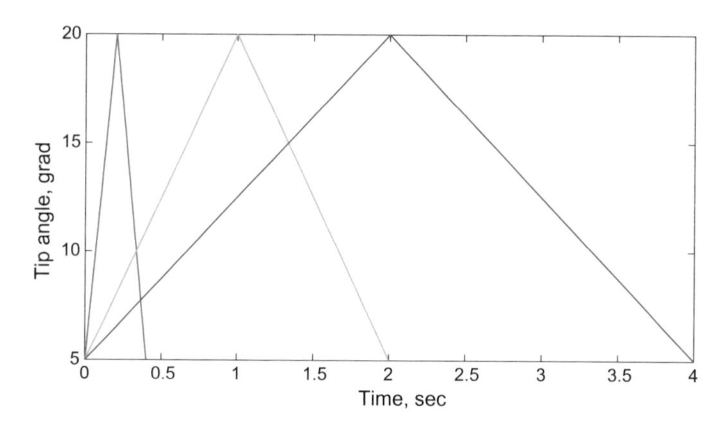

Fig. 3 Piecewise-linear angular input (blue colour corresponds to the motor speed of 45 deg/sec, green—90 deg/sec, magenta—450 deg/sec)

Bratislava for teaching students the basics of automatic control. Its electrical schemes and blueprints for laser cutting are freely available [6].

To test the hypothesis of sensor viscoelastic behaviour, we have conducted a number of experiments with different piecewise-linear angular inputs, see Fig. 3. In each case, we vary the sensor tip angle from 5 to 20 degrees (measured by a protractor). We divide the whole angle interval into 40 reference points and then vary the delay

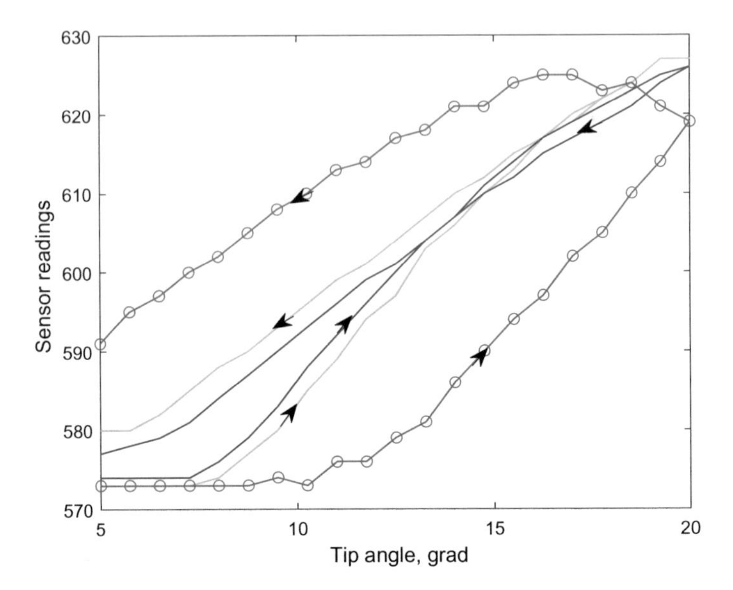

Fig. 4 Sensor readings under piecewise-linear angular input (magenta colour corresponds to the fastest motion with circles representing the reference points)

before setting a new reference point for the servo. As a result, with decreasing of the delay we have obtained different curves for loading (angle increasing) and unloading (angle decreasing), see Fig. 3. In the fastest case depicted in Fig. 4, a possible error in sensor readings can be as large as 60–70%.

As one can see, under slow angular motion (blue in Fig. 3) the hysteresis curve (also blue in Fig. 4) can cross itself. It is also not a loop: probably due to the viscosity of the sensor, it never returns exactly to the same reading as started, but it can get there after some time if unloaded. It appears that the hysteresis curve is asymmetrical and its width increases with increasing of the angular speed.

3 Conclusion

We have developed a simple electromechanical device based on freely available design, which can be used for investigation of rate-dependent viscoelastic hysteresis. At the present time, the results are inconclusive: the servomotor has no position feedback and cannot be used for correct system identification. Our future goal is to improve our device which, hopefully, can help in deriving mathematical model of the sensor.

References

1. R.S. Anderssen, I.G. Götz, K.H. Hoffmann, The global behavior of elastoplastic and viscoelastic materials with hysteresis-type state equations. SIAM J. Appl. Math. **58**(2), 703–723 (1998)
2. H.T. Banks, A brief review of some approaches to hysteresis in viscoelastic polymers. Nonlinear Anal. **69**(3), 807–815 (2008)
3. G. Bles, W.K. Nowacki, A. Tourabi, Experimental study of the cyclic visco-elasto-plastic behaviour of a polyamide fibre strap. Int. J. Solids Struct. **46**, 2693–2705 (2009)
4. K. Elgeneidy, N. Lohse, M. Jackson, Bending angle prediction and control of soft pneumatic actuators with embedded flex sensors–a data-driven approach. Mechatronics **50**, 234–247 (2018)
5. M. Kalúz, L. Čirka, M. Fikar, Flexy: an open-source device for control education, in *13th APCA International Conference on Control and Soft Computing (CONTROLO)* (University of the Azores, 2018)
6. M. Kalúz, Flexy (2018) (preprint). https://www.uiam.sk/~kaluz/opensourceprojects/contents/flexy.html
7. M.A. Krasnosel'skii, A.V. Pokrovskii, *Systems with Hysteresis* (Springer, Berlin, 1989)
8. G. Saggio, F. Riillo, L. Sbernini, Quitadamo LR Resistive flex sensors: a survey. Smart Mater. Struct. **25**, 1–30 (2016)
9. A. Visintin, *Differential Models of Hysteresis* (Springer, Berlin, 1994)
10. A.S. Wineman, K.R. Rajagopal, *Mechanical Response of Polymers* (Cambridge University Press, Cambridge, 2000)
11. J. Zhang, K. Iyer, A. Simeonov, M.C. Yip, Modeling and inverse compensation of hysteresis in supercoiled polymer artificial muscles. IEEE Robot. Autom. Lett. **2**(2), 773–780 (2017)

How Does the Hopf Bifurcation Appear in the Hydrogen Atom in a Circularly Polarized (CP) Microwave Field?

Mercè Ollé and Juan R. Pacha

Abstract The dynamics of the Rydberg atom in a rotating electric field can be described by a two degree of freedom Hamiltonian depending on a parameter K. This Hamiltonian has two equilibrium points L_1 and L_2. While L_1 is a saddle-center for all values of K, the hyperbolic character of L_2 changes from a double center to a complex saddle as the parameter K crosses a critical threshold, giving rise to the Hopf bifurcation. Here we analyze the dynamics close to this transition. We finally remark that the full details (proofs, figures, and further discussion) referred in this extended abstract can be found in Ollé and Pacha (Commun Nonlinear Sci Numer Simul 62:27–60, 2018, [9]).

1 Introduction

Throughout this note, we shall consider the hydrogen—or, more generally, the Rydberg—atom interacting with a circularly polarized microwave field. In this context, we shall refer this problem as the *circularly polarized* (CP) problem. In a suitable rotating reference frame, and assuming planar motion for the electron, the dynamics of the CP problem is described by a two degree of freedom Hamiltonian that depends on a positive parameter K. Some aspects of this problem have been studied by several authors; see [1, 3, 6]. However, some others remain to be well understood. Particularly, chaotic regions are expected to appear since we know that the system is non-integrable; see [4].

It is well known (see [1]) that the CP problem has two equilibrium points for any positive value of the parameter K, say L_1 and L_2. The first one, L_1, is a saddle-center for all values of $K > 0$. The second, L_2, is a center×center for $K < K_{crit}$ and changes to a complex saddle as $K > K_{crit}$. For the critical value, $K_{crit} = 3^{-4/3}/2 \approx 0.11556021$, the equilibrium point L_2 corresponds a $1:1$ non-semisimple resonance

M. Ollé (✉) · J. R. Pacha
Departament de Matemàtiques, Universitat Politècnica de Catalunya, Barcelona, Spain
e-mail: merce.olle@upc.edu

J. R. Pacha
e-mail: juan.ramon.pacha@upc.edu

A. Korobeinikov et al. (eds.), *Extended Abstracts Spring 2018*,
Trends in Mathematics 11, https://doi.org/10.1007/978-3-030-25261-8_35

corresponding to a pairwise collision of the characteristic exponents on the imaginary axis.

As far as we know, most of the papers are devoted to study the system for small $0 < K < K_{crit}$. The case $K > K_{crit}$ has been tackled in [7] (though the approach there is perturbative without computing the current normal form) and more recently in [9]. In this text, we give some highlights of this last paper.

2 The Hamiltonian for the Hydrogen Atom in a Circularly Polarized Microwave Field

The Hamiltonian for the hydrogen atom subjected to a CP microwave field, in atomic units ($m_e = \bar{h} = e = 1$) and assuming that the electron moves in the horizontal (x, y) plane, is the following:

$$H(x, y, p_x, p_y, t) = \frac{1}{2}(p_x^2 + p_y^2) - \frac{1}{r} + F(x \cos \tilde{\omega} t + y \sin \tilde{\omega} t), \qquad (1)$$

where (x, y) are the positions, $(p_x, p_y) = (x' - y, y' + x)$ their conjugate momenta, $r^2 = x^2 + y^2$, $\tilde{\omega}$ is the angular frequency of the microwave field, and $F > 0$ is the field strength; see [3]. The generating function,

$$S(X, Y, p_x, p_y, t) = \left[X \cos(\tilde{\omega} t) - Y \sin(\tilde{\omega} t)\right] p_x + \left[X \sin(\tilde{\omega} t) + Y \cos(\tilde{\omega} t)\right] p_y$$

gives the change to a coordinate frame rotating with the same angular frequency $\tilde{\omega}$. The transformed Hamiltonian is

$$\begin{aligned}\mathcal{H}(X, Y, P_X, P_Y, t) &= H\left(\partial_{p_x} S\left(X, Y, p_x, p_y, t\right), \partial_{p_y} S\left(X, Y, p_x, p_y, t\right), p_x, p_y, t\right) \\ &\quad - \partial_t S\left(X, Y, p_x, p_y, t\right) \\ &= \frac{1}{2}\left(P_X^2 + P_Y^2\right) - \tilde{\omega}\left(X P_Y - Y P_X\right) - \frac{1}{\sqrt{X^2 + Y^2}} + F X,\end{aligned}$$

where p_x and p_y in the formula above are expressed in terms of the "new" positions (X, Y) and momenta (P_X, P_Y), by solving in the "old" momenta (p_x, p_y) the equations $P_X = \partial_X S(X, Y, p_x, p_y, t)$ and $P_Y = \partial_Y S(X, Y, p_x, p_y, t)$. Note that, for the current case, these equations are linear in (p_x, p_y), so this can be done explicitly. We remark that the resulting Hamiltonian, $\mathcal{H}$, does not depend on t. Moreover, it is checked at once that the application of the scaling (in both time and coordinates): $t = \tau/\tilde{\omega}$, $(X, Y) = \tilde{\omega}^{-2/3}(x, y)$, $(P_X, P_Y) = \tilde{\omega}^{1/3}\left(p_x, p_y\right)$, to the Hamiltonian equations associated to $\mathcal{H}$, yields a new Hamiltonian system whose corresponding Hamiltonian function casts

$$H\left(x, y, p_x, p_y\right) = \frac{1}{2}(p_x^2 + p_y^2) - x p_y + y p_x - \frac{1}{r} + K x, \qquad (2)$$

with $K = F/\tilde{\omega}^{4/3} > 0$ and where, for the sake of simplicity, we use the same names for the transformed Hamiltonian (2), and for the new position and momenta as in the initial Hamiltonian (1). Finally, we stress that only one parameter (here, K) is necessary to describe the dynamics.

3 Normal Form Around L_2

Reduction to normal form (NF) around the equilibrium point L_2 of Hamiltonian (2) involves several steps, roughly: shifting the origin to the L_2 equilibrium point, reduction of the quadratic part, complexification, and nonlinear reduction (for example, by means of some Lie series method) up to the required degree. Details for this case in point can be found in [9]. Up to order four, the (real) NF turns out to be

$$\begin{aligned}\widetilde{Z}^{(4)}(\mu, \tilde{x}, \tilde{y}) &= \tilde{\alpha}\mu + \tilde{\beta}\mu^2 + \frac{1}{2}\left(\tilde{y}_1^2 + \tilde{y}_2^2\right) + \omega(\mu)\,(\tilde{y}_1\tilde{x}_2 - \tilde{y}_2\tilde{x}_1) - \frac{\varepsilon(\mu)}{2}\left(\tilde{x}_1^2 + \tilde{x}_2^2\right) \\ &\quad + A\left(\tilde{x}_1^2 + \tilde{x}_2^2\right)^2 + B\left(\tilde{x}_1^2 + \tilde{x}_2^2\right)(\tilde{y}_1\tilde{x}_2 - \tilde{y}_2\tilde{x}_1) + C\,(\tilde{y}_1\tilde{x}_2 - \tilde{y}_2\tilde{x}_1)^2,\end{aligned} \tag{3}$$

$\tilde{x} = (\tilde{x}_1, \tilde{x}_2)$, $\tilde{y} = (\tilde{y}_1, \tilde{y}_2)$, and the coefficients $\tilde{\alpha}$, $\tilde{\beta}$, $\omega(\mu)$, $\varepsilon(\mu)$, A, B, C are given by

$$\begin{aligned} &\tilde{\alpha} = \frac{25}{8}, && \tilde{\beta} = \frac{27}{32}\,3^{2/3}, && \tilde{\gamma} = \frac{14}{25}\,3^{2/3}\sqrt{5}, \\ &\varpi = \frac{\sqrt{5}}{3}, && \omega(\mu) = \varpi + \tilde{\gamma}\mu, && \varepsilon(\mu) = \frac{6}{5}\,3^{2/3}\mu, \\ &A = \frac{7}{1250}\,3^{2/3}, && B = -\frac{6}{625}\,3^{2/3}\sqrt{5}, && C = -\frac{88}{28125}\,3^{2/3}. \end{aligned}$$

Here, μ is a parameter related with K through

$$K := \frac{-1 + \delta_0^3 + 3\delta_0\mu - 3\delta_0^2\mu^2 + 3\delta_0^3\mu^3}{\delta_0^2\,(1 - \delta_0\mu)},$$

$\mu < 1/\delta_0$ and $\delta_0 := 3^{2/3}/2$ is the distance of L_2 to the origin $(x, y) = (0, 0)$ of the configuration space. It turns out that $K < K_{crit}$ for $\mu < 0$, $K = K_{crit}$ for $\mu = 0$, and $K > K_{crit}$ when $0 < \mu < 1/\delta_0$. Thus, Proposition 1 follows from the Hamiltonian Hopf bifurcation theorems; see [2, 8, 10].

Proposition 1 (from [9]) *The equilibrium point L_2 of Hamiltonian* (2) *is stable for $K \le K_{crit}$ and unstable for $K > K_{crit}$. For $K < K_{crit}$, there exist two Lyapunov families of elliptic periodic orbits that contain the equilibrium point. Both families become one unique family of elliptic periodic orbits for $K = K_{crit}$. For $K > K_{crit}$, the unique family of elliptic periodic orbits persists but no longer holds the equilibrium. Moreover, if for some K close to K_{crit} one singles out an elliptic periodic* *orbit, then the excitations in the elliptic directions yield the unfolding of a Cantor family of Lagrangian 2D-tori having that periodic orbit as its fiber.*

3.1 Dynamics of the Hamiltonian NF of Order 4

It is convenient to introduce polar coordinates Q, $\theta \in \mathbb{T}^1$ for the positions, and their canonically conjugate momentum P and action J (see [5]),

$$\tilde{x}_1 = Q\cos\theta, \qquad \tilde{y}_1 = P\cos\theta - \frac{J}{Q}\sin\theta,$$

$$\tilde{x}_2 = -Q\sin\theta, \qquad \tilde{y}_2 = -P\sin\theta - \frac{J}{Q}\cos\theta.$$

In these new coordinates (θ, Q, J, P), the NF (3) transforms to a new (integrable) Hamiltonian, Γ, which is written as

$$\Gamma(\mu, Q, J, P) = \Gamma_0(\mu) + \omega(\mu)J + CJ^2 + \frac{P^2}{2} + \frac{J^2}{2Q^2} + \left(BJ - \frac{\varepsilon(\mu)}{2}\right)Q^2 + AQ^4, \tag{4}$$

with $\Gamma_0(\mu) = \tilde{\alpha}\mu + \tilde{\beta}\mu^2$. Note that (4) does not depend explicitly on the angle θ, so J is a first integral of Γ. Hence, (4) can be regarded as a one degree of freedom Hamiltonian depending on parameters μ and J. From this point of view, it is clear that the difference $\widetilde{\Gamma} := \Gamma - \Gamma_0(\mu) = P^2/2 + V(\mu, Q, J)/2$, with

$$V(\mu, Q, J) := 2\omega(\mu)J + 2CJ^2 + \frac{J^2}{Q^2} + (2BJ - \varepsilon(\mu))\,Q^2 + 2AQ^4$$

is a *natural* Hamiltonian, for it is a sum of a kinetic term, $P^2/2$, and a potential function, $V(\mu, Q, J)/2$, that depends only on the position. Then, for μ (and so, for K) fixed, and for a given level of the energy E, the phase portrait of $\widetilde{\Gamma}$ consists of the curves

$$(Q, P) = \left(Q, \pm\sqrt{E - V(\mu, Q, J)}\right), \tag{5}$$

parametrized by the action J.

In Fig. 1, we plot these phase portraits for $\mu > 0$. Figure 1a corresponds to the level set $E = 0$. The 2D Lagrangian tori with $J < 0$ are drawn in thin continuous lines. For $J = 0$, the thick black line follows the stable and unstable manifolds of L_2 (at the origin in the current coordinates). In the approximation given by the NF, these manifolds coincide. The 2D Lagrangian tori around the bifurcated periodic orbit (that is marked with a cross in the figures) have $J > 0$ and are drawn in dashed lines. In Fig. 1b, we sketch the analogous phase portrait for $E < 0$—the same line

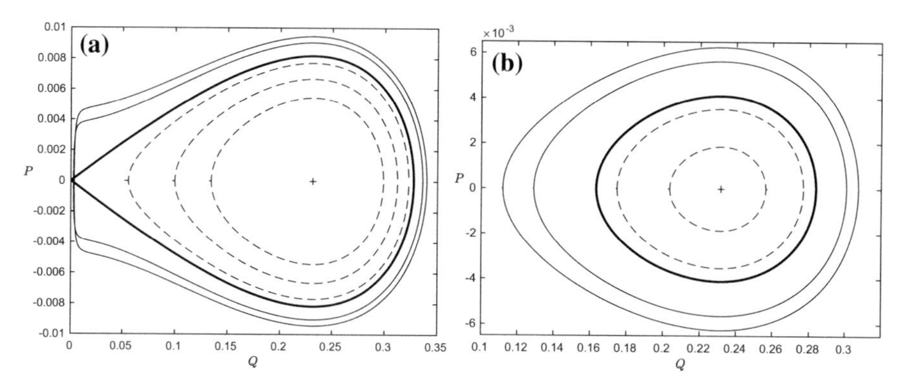

Fig. 1 Phase portraits of the Hamiltonian system $\tilde{\Gamma}$ for $\mu > 0$. $E = 0$ in (**a**), and $E < 0$ in (**b**). See the text for details

code is used continuous lines for orbits with $J < 0$, thick continuous line for the one with $J = 0$, and dashed lines for those with $J > 0$.

Of course, the main difference between the two phase portraits in Fig. 1 is that the stable and unstable invariant manifolds of L_2 (recall that $\mu > 0$ is assumed), $W^{\pm}(L_2)$, "live" in the level set $E = 0$, so they do not appear in Fig. 1b. In the current polar coordinates (Q, P), their parametrization follows immediately setting $E = 0$ and $J = 0$ in (5). Hence,

$$W^{\pm}(L_2) = \left\{(\theta, Q, J, P) = \left(\theta, Q, 0, \pm\sqrt{-V(\mu, Q, 0)}\right),\ \theta \in \mathbb{T}^1, 0 < Q \leq \sqrt{\varepsilon(\mu)/(2A)}\right\},$$

where the plus sign describes the unstable (outgoing) manifold, whereas the minus sign describes the stable (incoming) one.

If $\mu < 0$, L_2 is a center×center with two pure imaginary characteristic exponents $\pm \mathrm{i}\omega_1$, $\pm \mathrm{i}\omega_2$ ($\omega_1 > 0$, $\omega_2 > 0$ depending on μ). For $E = 0$, one sees just the equilibrium point and a family of 2D Lagrangian tori. For each $E > 0$ ($E < 0$) fixed, we have, for a suitable value of J, a stable periodic orbit that is represented on the (Q, P)-plane by a point on the Q axis. This orbit is surrounded by a family of closed curves, parametrized by J, that corresponds—in the whole phase space—to a family of Lagrangian 2D tori. Variation of $E > 0$ (similarly of $E < 0$) leads to a family of periodic orbits, characterized by the solutions of the equations $\partial_Q V(\mu, J, Q) = 0$, $V(\mu, J, Q) = E$, that are stable. This yields two families of periodic orbits, parametrized by the energy, in accordance with the Lyapunov Center Theorem.

Finally, L_2 is a degenerate center×center for $\mu = 0$ with characteristic exponents $\pm \mathrm{i}\omega_1 = \pm \mathrm{i}\omega_2 = \pm \mathrm{i}\omega$ ($\omega > 0$). The two Lyapunov families of periodic orbits parametrized by the energy E become one family in the sense that, as E tends to 0, the period of the orbits of both families tends to the same value $2\pi/\omega$. Moreover, as in the previous case, for a fixed $E > 0$ ($E < 0$), we have a stable periodic orbit surrounded by 2D-invariant Lagrangian tori.

4 Numerical Computations

Next, numerical simulations are conducted to study the resulting Hopf bifurcation beyond the description given by the NF. In order to show the dynamics of the CP problem from a global point of view, we consider the so-called Poincaré Section Plot (PSP): given K, and for an energy level h fixed, we take, for some range of time, the intersection of a solution (or a set of solutions) with the Poincaré section $x' = 0$, $y' < 0$, and we plot the (x, y) projection of such intersection points. We recall that a detailed description of the dynamics and the consequences of the Hopf bifurcation are explained in [9]. In this extended abstract, we just remark two comments.

On one hand, concerning the invariant manifolds of L_2. We recall that the NF (3) is integrable, so it is not useful to detect phenomena related to chaos, such as splitting. However, the original system is non-integrable (see [4]) and we expect to have such splitting between the manifolds $W^{\pm}(L_2)$. For example, Fig. 2 shows the projection on the (x, y) plane of the Poincaré section plot (PSP) of $W^{\pm}(L_2)$, where the splitting is clearly seen (compare with Fig. 1b, where splitting does not show up at all).

On the other hand, focusing on the global picture of the dynamics, from our numerical approach—coined on the basis of the different relevant invariant objects involved—we conclude the following.

4.1 For $K > K_{crit}$

In a region near the origin: A first global property to remark, and independent of the Hopf bifurcation phenomenon, is the existence of the retrograde periodic stable orbit close to the origin; this orbit together with the invariant surrounding 2d tori confine a clear region for the electron. This was already observed in [1] for very small values of $K < K_{crit}$.

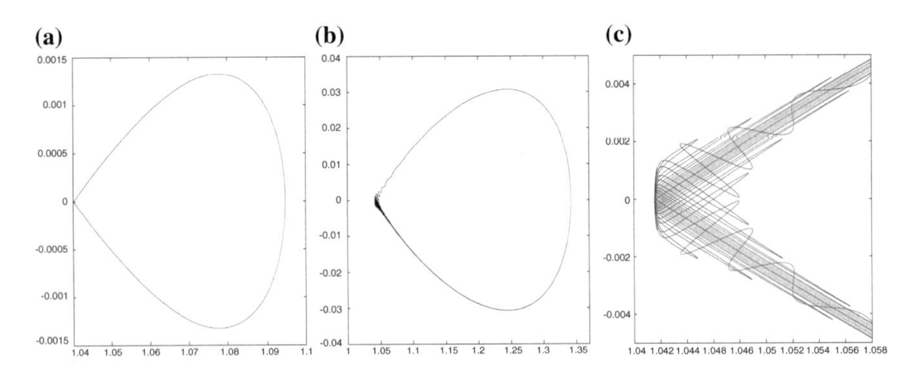

Fig. 2 (x, y) projection of the PSP for $W^u(L_2)$. **a** $K = 0.1157$; **b** $K = 0.12$; **c** a zoom close to L_2 with both $W^{\pm}(L_2)$

The invariant manifolds of the Lyapunov periodic orbit, ol_1: Regardless the value of K, for $h = h(L_2)$, there is the unstable periodic orbit around L_1, ol_1, and their invariant manifolds which play a clear role on the dynamics. These manifolds visit a small (x, y) region for $0 < K < K_{crit}$ small, but we have just shown that they become more complex (and their homoclinic tangles as well) as K grows visiting large regions in the (x, y) plane.

The region influenced by the Hopf bifurcation: When K is bigger and close to K_{crit}, and for $h = h(L_2)$, the dynamics of the Hopf bifurcation is very local in the sense that $W^+(L_2)$, $W^-(L_2)$, the bifurcated stable periodic orbit and the surrounding invariant 2D tori are confined by the external KAM tori, so they do not play a significant role in the global dynamics. Nevertheless, as far as K increases, these KAM tori disappear and both invariant manifolds—those of ol_1 and those of L_2—are mixed, giving rise to many different kinds of orbits and chaos. Relevant to say (and as mentioned above) is that, among this chaos, the stable bifurcated periodic orbit and associated 2D tori confine a clear region among this chaotic sea. However, for bigger values of K such that the periodic bifurcated orbit is unstable, we obtain the same kind of PSP but there is no confinement around the periodic orbit at all.

Fast escaping orbits and erratic orbits: We finally mention that apart from chaotic orbits, other solutions, which may escape very fast, or which are erratic and finally escape (or not) typically appear.

4.2 For $K < K_{crit}$

Let us take $K = 0.115$, very close to K_{crit} and h near $h(L_2)$. As in Case $K > K_{crit}$, there is the *big* stable region close to the origin due to the retrograde o_r orbit and the intricate invariant manifolds associated with the periodic orbit ol_1 that cover a big

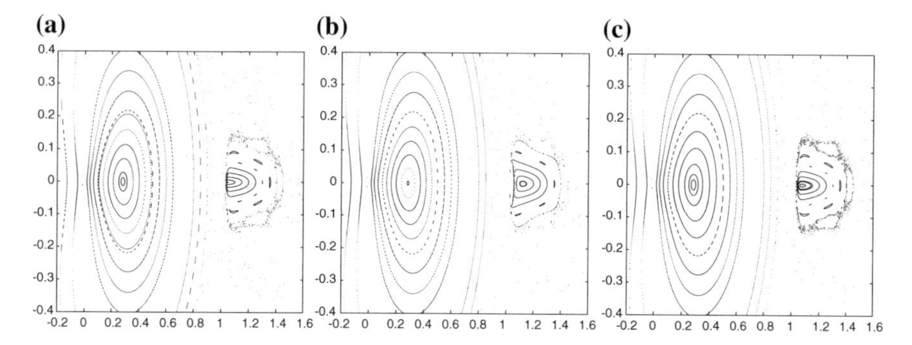

Fig. 3 $K = 0.115 < K_{crit}$. PSP: the *big* stable region close to the origin is clearly seen. We remark the stable equilibrium point for $h = h(L_2)$ in (**a**), the Lyapunov stable periodic orbit for $h < h(L_2)$ in (**b**), and the one for $h > h(L_2)$ in (**c**)

chaotic region in the (x, y) plane. For $h = h(L_2)$, we have the equilibrium point L_2 which is stable, whereas for increasing/decreasing h we obtain the corresponding family of stable Lyapunov periodic orbits surrounded by 2D tori. This is shown in Fig. 3. Of course, this dynamics is in accordance with the theoretical results obtained from the NF analysis; see Sect. 3.

References

1. E. Barrabés, M. Ollé, F. Borondo, D. Farrelly, J.M. Mondelo, Phase space structure of the hydrogen atom in a circularly, polarized microwave field. Phys. D **241**(4), 333–349 (2012)
2. H.W. Broer, H. Hanßmann, J. Hoo, The quasi-periodic Hamiltonian Hopf bifurcation. Nonlinearity **20**(2), 417–460 (2007)
3. A.F. Brunello, T. Uzer, D. Farrelly, Hydrogen atom in circularly polarized microwaves: Chaotic ionization via core scattering. Phys. Rev. **55**(5), 3730–3745 (1997)
4. L. Guirao, M. López, J. Vera, C^1 non integrability of a hydrogen atom in a circularly polarized microwave field. Cent. Eur. J. Phys. **10**(4), 742–748 (2012)
5. D.C. Heggie, Bifurcation at complex instability. Celest. Mech. **35**(4), 357–382 (1985)
6. C. Jaffé, D. Farrelly, T. Uzer, Transition state theory without time-reversal symmetry: Chaotic ionization of the hydrogen atom. Phys. Rev. Lett. **84**(4), 610–613 (2000)
7. A. Lahiri, M. Roy, The Hamiltonian Hopf bifurcation: an elementary perturbative approach. Int. J. Non-Linear Mech. **36**, 787–802 (2001)
8. K.R. Meyer, G.R. Hall, D. Offin, *Introduction to Hamiltonian Dynamical Systems and the N-Body Problem*, Applied Mathematical Sciences, vol. 90 (Springer, New York, 2009)
9. M. Ollé, J.R. Pacha, Hopf bifurcation for the hydrogen atom in a circularly polarized microwave field. Commun. Nonlinear Sci. Numer. Simul. **62**, 27–60 (2018)
10. J.C. van der Meer, *The Hamiltonian Hopf Bifurcation*, vol. 1160, Lecture Notes in Mathematics (Springer, Berlin, 1985)

Stabilization of Unstable Periodic Solutions for Inverted Pendulum Under Hysteretic Control: The Magnitskii Approach

Mikhail E. Semenov, Peter A. Meleshenko, Igor N. Ishchuk, Valeriy N. Tyapkin and Zainib Hatif Abbas

Abstract In this paper, a mathematical model of the stabilization of the inverted pendulum with vertically oscillating suspension under hysteretic control is constructed. The stabilization of unstable periodic solutions for such a system is considered using the Magnitskii approach.

1 Introduction

The problem of inverted pendulum has a long history [5, 6, 18] and remains relevant even in the present days [1, 3, 16]. As it is well known, the model of inverted pendulum plays the central role in the control theory [4, 7]. It is well-established benchmark problem that provides many challenging problems to control design. Because of their nonlinear nature, pendulums have maintained their usefulness and they are now used to illustrate many of the ideas emerging in the field of nonlinear control [2]. Typical examples are feedback stabilization, variable structure control, passivity-based

This work was supported by the RFBR (Grants 16-08-00312-a, 17-01-00251-a, 18-47-310003 and 18-08-00053-a).

M. E. Semenov (✉) · P. A. Meleshenko
Voronezh State University, Universitetskaya sq.1, 394006 Voronezh, Russia
e-mail: mkl150@mail.ru

P. A. Meleshenko
e-mail: melechp@yandex.ru

M. E. Semenov
Geophysical Survey of Russia Academy of Science, Lenina av. 189, 249035 Obninsk, Russia

M. E. Semenov · Z. H. Abbas
Voronezh State Technical University, XX-letiya Oktyabrya st. 84, 394006 Voronezh, Russia

M. E. Semenov · P. A. Meleshenko · I. N. Ishchuk
Zhukovsky–Gagarin Air Force Academy, Starykh Bolshevikov st. 54 "A", 394064 Voronezh, Russia

I. N. Ishchuk · V. N. Tyapkin
Siberian Federal University, Svobodny 79, 660041 Krasnoyarsk, Russia

A. Korobeinikov et al. (eds.), *Extended Abstracts Spring 2018*,
Trends in Mathematics 11, https://doi.org/10.1007/978-3-030-25261-8_36

control, back-stepping and forwarding, nonlinear observers, friction compensation, and nonlinear model reduction. A special role plays the systems containing oscillating parts taking into account the hysteretic properties of the internal links or some parts of such systems. One of the most interesting phenomena in the oscillating systems is a resonance behavior of the dynamical parameters. Investigation of the resonance phenomena in a system containing hysteretic nonlinearity was made in [9].

The model of inverted pendulum with oscillating suspension point (see panel *a* in Fig. 1) was studied in detail by Kapitza [5, 6]. Let us recall that the equation of motion of pendulum has the following form:

$$\ddot{\phi} - \frac{1}{l}[g + \ddot{f}(t)] \sin \phi = 0, \tag{1}$$

where ϕ is the angle of vertical deviation of the pendulum, l is the pendulum's length, g is the gravitational acceleration, and $f(t)$ is the law of motion of the suspension point (of course, this equation should be considered together with the corresponding initial conditions). As it is known, if motion of the suspension point is of harmonic character, then Eq. (1) reduces to the Mathieu equation.

In order to make an adequately description of the dynamics of real-life physical and mechanical systems, it is necessary to take into account effects of hysteretic nature such as "backlash", "stops", etc. The mathematical models of such nonlinearities, according to the classical patterns of Krasnosel'skii and Pokrovskii [8], reduce to operators, which are treated as converters in an appropriate function spaces. The dynamics of such converters are described by the relation of "input-state" and "state-output". Let us note that the hysteretic systems are of interest in various fields of modern physical, mathematical, and technical sciences. Interesting role plays the hystertic phenomena in biology [12–14].

The majority of real-life physical and technical systems contain parts that can be represented as a cylinder with a piston. Inevitably, the backlashes appear in such systems during its long operation due to the "aging" of materials. Such backlashes

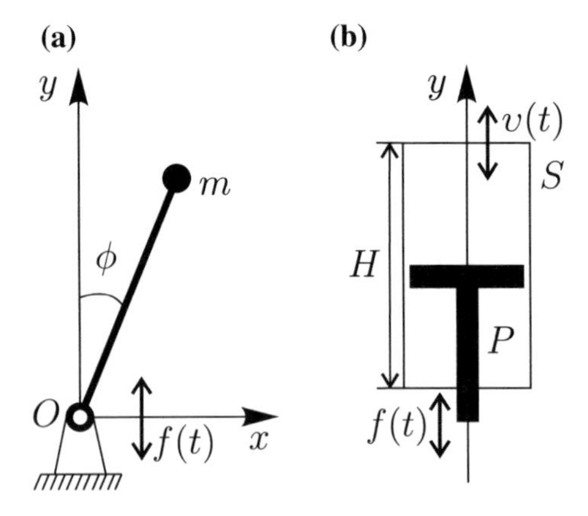

Fig. 1 Geometry of the problem. **a** general view of the inverted pendulum. **b** the suspension point (cylinder and piston)

are of hysteretic nature and the analysis of such nonlinearities is quite important and relevant problem. In this paper, we investigate the problem of the inverted pendulum under hysteretic control in the form of backlash. We pay specific attention to stabilization of unstable periodic solutions for such a system using the Magnitskii approach.

2 Mathematical Model

Let us consider a system where the base of the pendulum is a physical system (P, S) formed by a cylinder of length H and the piston P (both the cylinder and the piston are supposed to be ideal and absolutely rigid). Both the cylinder and the piston can move in the direction of the vertical axis as it is shown in panel b of Fig. 1.

We determine the piston's position by the coordinate $f(t)$ and the cylinder's position by coordinate $v(t)$. The system (P, S) can be considered as a converter Γ with the input signal $f(t)$ (piston's position) and the output signal $v(t)$ (cylinder's position). Such a converter is called *backlash*. The set of its possible states is $f(t) \leq v(t) \leq f(t) + H$ $(-\infty < f(t) < \infty)$. The cylinder's position $v(t)$ at $t > t_0$ is defined by $v(t) = \Gamma[t_0, v(t_0)]f(t)$, where $\Gamma[t_0, v(t_0)]$ is the operator defined for each $v_0 = v(t_0)$ on the set of continuous inputs $f(t)$ $(t > t_0)$ for which $v_0 - H < f(t) < v_0$; see [8].

We suppose that the piston's acceleration periodically changes from $-a\omega^2$ to $a\omega^2$ with the frequency ω. This assumption implies that the linearized equation of motion of such a pendulum can be written in the following form:

$$\begin{aligned} &\ddot{\phi} - \tfrac{1}{l}[g + a\omega^2 G(t, H)w(t)]\phi = 0, \\ &w(t) = -\text{sign}[\sin(\omega t)], \\ &\phi(0) = \phi_{10}, \ \dot{\phi}(0) = \phi_{20}, \end{aligned} \tag{2}$$

where $\text{sign}(z)$ is the usual signum function, $G(t, H)w(t)$ is the acceleration of the suspension point and

$$G(t, H) = \begin{cases} 0, \ t \in (t^*, t^* + \Delta t), \\ 1, \ t \text{ out of } (t^*, t^* + \Delta t), \end{cases}$$

where t^* are the moments after which the acceleration's sign change takes place, $\Delta t = \sqrt{2H/a\omega^2}$ is the time for which the piston passes through the cylinder.

3 Stability Zones

Passing to dimensionless units the equation of motion can be rewritten in the form of the corresponding equivalent system:

$$\begin{cases} \dot{x}_1 = x_2, \\ \dot{x}_2 = p(\tau)x_1, \end{cases} \\ x_1(0) = x_{10}, \ x_2(0) = x_{20}. \tag{3}$$

The matrix of this system has the following form:

$$\mathbf{P}(\tau) = \begin{pmatrix} 0 & 1 \\ p(\tau) & 0 \end{pmatrix},$$

where $p(\tau) = k - sG(\tau, H)\text{sign}(\sin \tau)$. The matrix $\mathbf{P}(\tau)$ is a periodic function of time with the period 2π, namely, $\mathbf{P}(\tau + 2\pi) \equiv \mathbf{P}(\tau)$.

Following the results of Floquet [15], the investigation of the stability of such systems reduces to the problem of finding of the fundamental matrix of the solutions at the moment 2π (the so-called *monodromy matrix*) and evaluation of its eigenvalues (the so-called *multipliers*). For the stability of the periodic system, it is necessary and sufficient that the following condition takes place: $|\varrho| < 1$ (all the multipliers are placed inside the unit circle).

Due to the fact that the matrix $\mathbf{P}(\tau)$ is a piecewise constant, the fundamental system of solutions and the monodromy matrix can be constructed in the closed analytic form (details can be found in [17]). The characteristic equation for the fundamental matrix $\mathbf{A}$ has the following form:

$$\|\mathbf{A} - \varrho\mathbf{E}\| = \begin{vmatrix} a_{11} - \varrho & a_{12} \\ a_{21} & a_{22} - \varrho \end{vmatrix} = \varrho^2 + \alpha\varrho + \beta = 0, \tag{4}$$

where $\beta = (-1)^2 \exp\left(\int_0^T \text{Sp}[\mathbf{P}(\tau)]\mathrm{d}\tau\right) = 1$ (see [11]) and $\alpha = -(a_{11} + a_{22})$.

The product of roots ϱ_1 and ϱ_2 of Eq. (4) is equal to unity, so the motion is stable at $|\alpha| < 2$ only. This condition can be written in the explicit form as given below:

$$\Big|\cos(k_2\gamma)\Big[2\cosh(2\sqrt{k}\Delta\tau)\cosh(k_1\gamma) + \sinh(2\sqrt{k}\Delta\tau)\sinh(k_1\gamma)\left(\tfrac{\sqrt{k}}{k_1} + \tfrac{k_1}{\sqrt{k}}\right)\Big] + \\ \sin(k_2\gamma)\Big[\sinh(2\sqrt{k}\Delta\tau)\cosh(k_1\gamma)\left(\tfrac{\sqrt{k}}{k_2} - \tfrac{k_2}{\sqrt{k}}\right) + \cosh^2(\sqrt{k}\Delta\tau)\sinh(k_1\gamma)\left(\tfrac{k_1}{k_2} - \tfrac{k_2}{k_1}\right) + \\ \sinh^2(\sqrt{k}\Delta\tau)\sinh(k_1\gamma)\left(\tfrac{k}{k_1k_2} - \tfrac{k_1k_2}{k}\right)\Big]\Big| < 2. \tag{5}$$

Thus, the *stability zone* of the system (3) in the space of parameters is defined by the inequality (5).

4 Periodic Regimes

Analysis of the monodromy matrix allows to get the following result: *periodic behavior of the pendulum corresponds to the edges of the stability zone*. On the left edge (at $a_{11} + a_{22} = -2$), we have a periodic regime with period $T_1 = 2\pi/\omega$. On the right edge (at $a_{11} + a_{22} = 2$), we have a periodic regime with period $T_2 = 4\pi/\omega$. However, *not for all of the nonzero initial values the periodic solutions exist*. The periodic regime with period T_1 exists, if the following relations for initial conditions take place:

$$\phi_{10} = \frac{a_{12}}{a_{11} - 1}\phi_{20}, \quad \phi_{20} = \frac{a_{21}}{a_{22} - 1}\phi_{10}, \tag{6}$$

In a similar manner, the periodic regime with period T_2 exists for the initial conditions that satisfy the following relations:

$$\phi_{10} = \frac{a_{12}}{1 + a_{11}}\phi_{20}, \quad \phi_{20} = \frac{a_{21}}{1 + a_{22}}\phi_{10}. \tag{7}$$

Here, we would like to note that all the periodic solutions are unstable.

4.1 Stabilization of the Periodic Regimes

Following Magnitskii [10], we consider the following approach to stabilization of periodic solutions. We introduce the so-called expanded system describing not the point's motion, but the motion of a cycle. The relationship between the basic and expanded systems can be described in the following sentences:

(i) to any *periodic solution* of the basic system corresponds a *stationary solution* of the expanded system;
(ii) to any *stationary solution* of the expanded system corresponds a *periodic solution* of the basic system;
(iii) there is a *bijection* between the stability of the periodic solution of the basic system and the stability of the stationary solution of the expanded system.

Using this technique, the stabilizing system for our mechanical system becomes

$$\begin{aligned}
\dot{x}_1 &= x_2 - x_{1\tau}\frac{x_{1\tau}x_2 + x_{2\tau}\left[k - sG(\tau, H)\text{sign}(\sin x)\right]}{x_{1\tau}^2 + x_{2\tau}^2} + \varepsilon_{x_1} q,\\
\dot{x}_2 &= \left[k - sG(\tau, H)\text{sign}(\sin x)\right]x_1 - x_{2\tau}\frac{x_{1\tau}x_2 + x_{2\tau}\left[k - sG(\tau, H)\text{sign}(\sin x)\right]}{x_{1\tau}^2 + x_{2\tau}^2} + \varepsilon_{x_2} q,\\
\dot{q} &= a_x\left(x_2 - x_{1\tau}\frac{x_{1\tau}x_2 + x_{2\tau}\left[k - sG(\tau, H)\text{sign}(\sin x)\right]}{x_{1\tau}^2 + x_{2\tau}^2}\right) +\\
&+ a_y\left(\left[k - sG(\tau, H)\text{sign}(\sin x)\right]x_1 - x_{2\tau}\frac{x_{1\tau}x_2 + x_{2\tau}\left[k - sG(\tau, H)\text{sign}(\sin x)\right]}{x_{1\tau}^2 + x_{2\tau}^2}\right) + \beta q,\\
&x_1|_{\tau=0} = x_1|_{\tau=2\pi}, x_2|_{\tau=0} = x_2|_{\tau=2\pi}, q|_{\tau=0} = q|_{\tau=2\pi},\\
&x_{1\tau}|_{\tau=0} = x_{1\tau}|_{\tau=2\pi}, x_{2\tau}|_{\tau=0} = x_{2\tau}|_{\tau=2\pi}, q_\tau|_{\tau=0} = q_\tau|_{\tau=2\pi}.
\end{aligned}$$

The numerical results for the periodic solutions of the considered mechanical system are presented in Fig. 2.

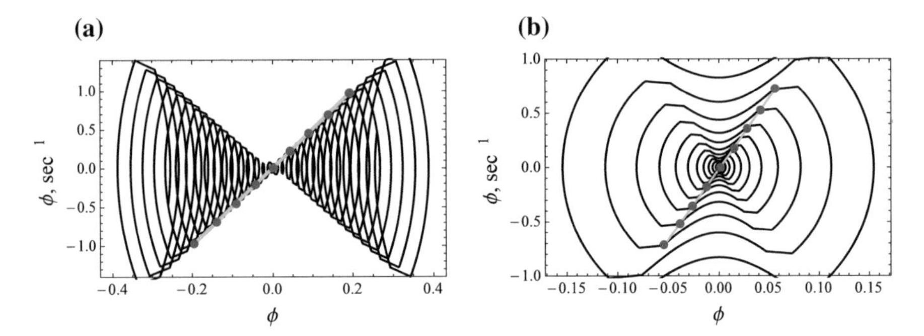

Fig. 2 Stabilized periodic solutions. **a** $T_1 = 2\pi/\omega$; **b** $T_2 = 4\pi/\omega$

5 Conclusions

In this paper, we have constructed the model of inverted pendulum with the oscillating suspension point under hysteretic control in the form of a backlash. More specifically, the explicit condition for the stability of such a system has been obtained using the monodromy matrix technique. The periodic solutions in such a system are also analyzed and the corresponding relations for the parameters a and ω are obtained. Using the Magnitskii approach, we presented a method for stabilization of the unstable periodic solutions for such a mechanical system.

References

1. A. Arinstein, M. Gitterman, Inverted spring pendulum driven by a periodic force: linear versus nonlinear analysis. Eur. J. Phys. **29**, 385–392 (2008)
2. K.J. Åström, K. Furuta, Swinging up a pendulum by energy control. Automatica **36**(2), 287–295 (2000)
3. E.I. Butikov, Oscillations of a simple pendulum with extremely large amplitudes. Eur. J. Phys. **33**(6), 1555–1563 (2012)
4. J. Huang, F. Ding, T. Fukuda, T. Matsuno, Modeling and velocity control for a novel narrow vehicle based on mobile wheeled inverted pendulum. IEEE Trans. Control Syst. Tech. **21**(5), 1607–1617 (2013)
5. P.L. Kapitza, Sov. Phys. JETP **21**, 588–592 (1951)
6. P.L. Kapitza, Usp. Fiz. Nauk **44**, 7–15 (1951)
7. K.D. Kim, P. Kumar, Real-time middleware for networked control systems and application to an unstable system. IEEE Trans. Control Syst. Technol. **21**, 1898–1906 (2013)
8. M.A. Krasnosel'skii, A.V. Pokrovskii, *Systems with Hysteresis* (Springer, Berlin, 1989)
9. P. Krejčí, Resonance in preisach systems. Appl. Math. **45**, 439–468 (2000)
10. N.A. Magnitskii, S.V. Sidorov, *New Methods for Chaotic Dynamics* (World Scientific, 2006)
11. D.R. Merkin, *Introduction to the Theory of Stability* (Springer, Berlin, 1997)
12. A. Pimenov, T.C. Kelly, A. Korobeinikov, M.J.A. O'Callaghan, A.V. Pokrovskii, *Mathematical Modeling, Clustering Algorithms and Applications* (Nova Science Pub Inc., 2010), pp. 247–279
13. A. Pimenov, T.C. Kelly, A. Korobeinikov, M.J.A. O'Callaghan, A.V. Pokrovskii, D. Rachinskii, Math. Model. Nat. Phenom. **7**, 204–226 (2012)

14. A. Pimenov, T.C. Kelly, A. Korobeinikov, M.J.A. O'Callaghan, D. Rachinskii, Memory and adaptive behavior in population dynamics: anti-predator behavior as a case study. J. Math. Biol. **74**, 1533–1559 (2017)
15. V.A. Pliss, *Nonlocal Problems of the Theory of Oscillations* (Academic Press, 1966)
16. M.E. Semenov, A.M. Solovyov, M.A. Popov, P.A. Meleshenko, Coupled inverted pendulums: stabilization problem. Arch. Appl. Mech. **88**(4), 517–524 (2018)
17. M.E. Semenov, P.A. Meleshenko, A.M. Solovyov, A.M. Semenov, Springer Proc. Phys. **168**, 463–506 (2015)
18. A. Stephenson, On induced stability. Phil. Mag. **15**, 233 (1908)

Nonideal Relay with Random Parameters

Mikhail E. Semenov, Peter A. Meleshenko, Igor N. Ishchuk, Dmitry D. Dmitriev, Sergey V. Borzunov and Nataliya N. Nekrasova

Abstract In this work, we introduce a novel class of hysteretic operators with random parameters. A definition of these operators is made in terms of the "input–output" relations. Properties of such operators are considered in an example of a nonideal relay with random parameters.

1 Introduction

Hysteresis phenomena are of interest from both the fundamental and the applied points of view. This interest is caused by high incidence of these phenomena in various technical systems (robotic, mechanical, electromechanical systems, management systems for aircraft tracking, etc.) Also, these phenomena determine some unusual elastoplastic properties of modern nanomaterials based on fullerene films. Moreover, the hysteretic phenomena are widely known in biology, chemistry, economics, etc.

This work was supported by the RFBR (Grants 17-01-00251-a, 18-47-310003, 18-08-00053-a and 19-08-00158-a).

M. E. Semenov (✉) · P. A. Meleshenko · S. V. Borzunov
Voronezh State University, Universitetskaya sq.1, 394006 Voronezh, Russia
e-mail: mkl150@mail.ru

P. A. Meleshenko
e-mail: melechp@yandex.ru

M. E. Semenov
Geophysical Survey of Russia Academy of Science, Lenina av. 189, 249035 Obninsk, Russia

M. E. Semenov · N. N. Nekrasova
Voronezh State Technical University, XX-letiya Oktyabrya st. 84, 394006 Voronezh, Russia

M. E. Semenov · P. A. Meleshenko · I. N. Ishchuk
Zhukovsky–Gagarin Air Force Academy, Starykh Bolshevikov st. 54 "A", 394064 Voronezh, Russia

I. N. Ishchuk · D. D. Dmitriev
Siberian Federal University, Svobodny 79, 660041 Krasnoyarsk, Russia

A. Korobeinikov et al. (eds.), *Extended Abstracts Spring 2018*,
Trends in Mathematics 11, https://doi.org/10.1007/978-3-030-25261-8_37

The hysteretic behavior of such systems is caused by either their internal structure, or the presence of separate blocks with hysteretic characteristics.

Currently used models of hysteretic phenomena both constructive (nonideal relay, Preisach and Ishlinskii–Prandtl models, etc. [6]), and phenomenological (Bouc–Wen model, Duhem model, etc. [3]) assume the stability of the parameters that identify the hysteretic properties of the corresponding operators. However, the stability of parameters in real engineering systems (e.g., in the systems described in [7]) does not always take place. Thus, such operators are the natural model in the situation where the parameters of a hysteresis carrier are under influence of stochastic uncontrollable affections. These facts make it necessary to develop the extended models of hysteretic effects, taking into account the stochastic changes of the parameters determining the hysteretic operators. Here we would like to note that the stochastic and unusual properties of the systems with hysteresis were considered in [2, 5]. Let us note that the equations with random parameters were considered in [1, 10, 11]. However, principally, equations considered in these and other publications are linear. The strongly nonlinear differential equations, containing the operator nonlinearity with random properties, have not been considered in the literature. Thus, the hysteretic operators (e.g., the nonideal relay) with random parameters seem novel and promising objects.

2 Nonideal Relay with Random Parameters

Let us consider a nonideal relay (a detailed description of this and other hysteretic converters is given in the classical book [4]) with the switching numbers that are not fixed and treated as random variables with absolutely continuous distribution function. Let us make the following assumption: the probability density of each of the switching numbers are supposed to be finite with non-intersecting supports. We denote these switching numbers as $\varphi_\alpha(u)$ and $\varphi_\beta(u)$. We consider a case where the supports of functions $\varphi_\alpha(u)$ and $\varphi_\beta(u)$ are contained in the intervals $[u_\alpha^-, u_\alpha^+]$ and $[u_\beta^-, u_\beta^+]$, respectively.

The dynamics of the input–output relations (see [4]) for the operator of a nonideal relay with random switching numbers is determined by two relations, namely, "input-state", and "state-output". We suppose that all permissible for the converter R continuous inputs are given on the nonnegative semi-axis ($t > 0$) (the input–output relation for this converter has the form $x(t) = R\,[t_0, x_0, \alpha, \beta]\,u(t)$, $(t \geq t_0)$). The space of possible states of such an operator is defined as $\Omega = \Omega(\omega, p, u)$ ($\omega = 0, 1$, $0 \leq p \leq 1$, $-\infty < u < +\infty$).

The variable state of the converter $R\left[(1;\, p_0);\, x_0;\, \varphi_\alpha(u);\, \varphi_\beta(u)\right] u(t)$ is a random value that takes the value 0 with probability $(1 - p(t))$ and a value of 1 with probability $p(t)$. In other words, it can be presented as a pair $\{1;\, p(t)\}$ (here the second output component corresponds to the probability that at the time t the first component is 1). The output of this converter is a random function $x(t)$ (Markovian process) taking a value of 1 with probability $p(t)$. The rule that determines the value of probability $p(t)$ will be given below.

2.1 Definition

The definition of the input–output relation can be made by means of a three-step construction (see [4]):

(i) at the first step, we define the input–output relation on the monotonic inputs only;
(ii) at the second step, using the semi-group identity, the input–output relation is defined for all piecewise monotonic inputs;
(iii) at the third step, using the special limit construction, the corresponding converter will be defined for all monotonic inputs.

We define the operator R on the monotonic inputs. Let us assume that at the initial time t_0 (to simplify, we assume that $t_0 = 0$) the operator R is in the state $1; p_0; u_0 \in \Omega$, $(u(0) = u_0)$. Let the input $u(t)$ be a monotonic increase, then for the time $t > 0$ the output is $x(t) = \{1; p(t)\}$, where

$$p(t) = \max\left\{p_0; \int_{-\infty}^{u(t)} \varphi_\beta(u)du\right\}. \tag{1}$$

Let t_1 be an arbitrary moment of time satisfying the inequality $0 < t_1 < t$, then the semi-group identity for the operator of a nonideal relay has the form (following from the definition):

$$R\left[t_0; p_0; u_0; \varphi_\alpha; \varphi_\beta\right]u(t) = R\left[t_1; R\left[t_0; p_0; u_0; \varphi_\alpha; \varphi_\beta\right]u(t_1); u(t_1); \varphi_\alpha; \varphi_\beta\right]u(t). \tag{2}$$

To define an operator on the piecewise monotonic inputs (in the case of a finite interval $[0, T]$), we break this interval by points $t_1, t_2, \ldots, t_n$ into intervals of monotonicity. On each of them, we define the corresponding operator as an operator on a strictly monotonic input whose initial state will be defined as the state at the instant corresponding to the "last" change in the behavior of the input.

To determine the operator R on continuous inputs, we use the following limit construction. Let $u(t)$, $t \in [0, T]$, be an arbitrary continuous input. Let us consider an arbitrary sequence of piecewise monotonic inputs $u_n(t)$, $(n = 1, 2, \ldots)$ that converges uniformly to each element of this sequence $u(t)$. A single-variable state $p_n(t)$ $(n = 1, 2, \ldots)$ forms a sequence of state variables $p_n(t)$, $(n = 1, 2, \ldots)$. Let us prove that the sequence $p_n(t)$ $(n = 1, 2, \ldots)$ converges uniformly. We estimate the absolute value of the difference:

$$|p_n(t) - p_m(t)| \leq \max_t \left| \int_{-\infty}^{u_n(t)} \varphi_\alpha(u)du - \int_{-\infty}^{u_m(t)} \varphi_\alpha(u)du \right| = \max_t \left| \int_{u_n(t}^{u_m(t)} \varphi_\alpha(u)du \right|. \tag{3}$$

Since the function $\varphi_\alpha(u)$ is continuous and because of uniform convergence

$$\lim_{n,m\to\infty} \max_t |u_n(t) - u_m(t)| = 0,$$

as well as, using the mean value theorem:

$$\max_t \left| \int_{u_n(t}^{u_m(t)} \varphi_\alpha(u)du \right| \leq \max_t \varphi_\alpha(t) [u_n(t) - u_m(t)]$$

the right-hand side of the inequality (3) tends to zero. Thus, the sequence $p_n(t)$ is fundamental (the continuity is obvious). Then, there is $\lim_{n\to\infty} p_n(t) = p(t)$, which is comparable to an arbitrary continuous input $u(t)$.

2.2 *Monotonicity*

Let us consider the monotonicity property for the constructed converter. We determine the monotonicity with respect to the initial state of the nonideal relay, namely, if $\{u(t_0, x_0\}$, $\{v(t_0, y_0\} \in \Omega(\alpha, \beta)$, $x_0 \leq y_0$, and $u(t) \leq v(t)$ $(t \geq t_0)$, then the following inequality takes place: $R[t_0, x_0, \alpha, \beta]u(t) \leq R[t_0, x_0, \alpha, \beta]v(t)$ $(t \geq t_0)$.

With respect to the modified operator of a nonideal relay with random parameters, the analog of monotonicity can be presented in the form of the following theorem.

Theorem 1 *Let* $p\{x_{01} = 1\} \geq p\{x_{02} = 1\}$ *and* $x_1(t) \geq x_2(t)$. *Then, for any* t, $p\{x_1 = 1\} \geq p\{x_2 = 1\}$.

3 Dynamics of a System Under NonIdeal Relay with Random Parameters

To demonstrate the action of the developed operator on the real-life physical system, let us consider a simple oscillating system under hysteretic force with random parameters. Such a system was considered in [8, 9] with the external force in the form of a nonideal relay with an inversion of the switching numbers. One of the main results of these papers is the existence of unlimited solutions with the growth rate of amplitude as the square root of time. Let us consider an analogous system with a nonideal relay where switching numbers distributed according to an even-dimensional law. The equation of motion together with the corresponding initial conditions has the following form:

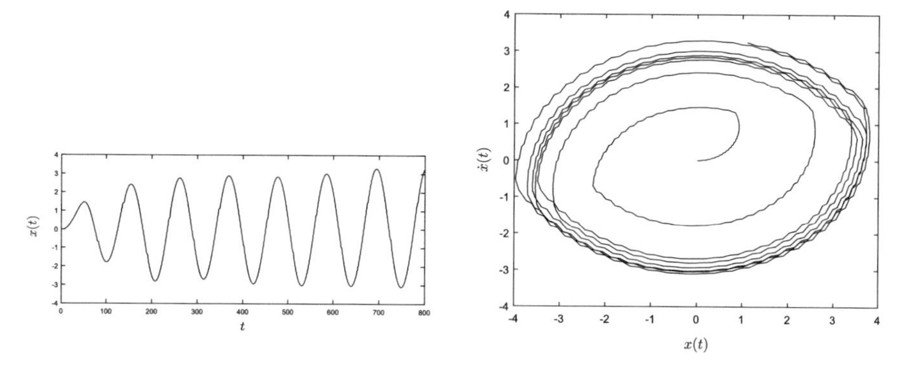

Fig. 1 A solution (left-hand panel) and phase portrait (right-hand panel) for the system (4)

$$\begin{aligned} \ddot{x}(t) + \omega^2 x(t) &= R[t_0;\ p_0;\ u_0;\ \varphi_\alpha;\ \varphi_\beta]x(t), \\ x(0) = x_0,\ \dot{x}(0) &= x_1. \end{aligned} \tag{4}$$

To find a numerical solution of the system (4), it is necessary to generate a set of random values corresponding to the switching numbers. Let us consider a case when $\varphi_\alpha(u)$ and $\varphi_\beta(u)$ corresponds to the uniform distribution law for α and β. For definiteness, we assume that these functions correspond to uniform distributions in the intervals $[-1.5, -0.5]$ and $[0.5, 1.5]$, respectively. At each period, switching numbers are selected from the corresponding distributions with the initial conditions for the next realization corresponding to the values of the phase coordinates obtained at the previous step. Using this algorithm, a solution to system (4) is obtained and the corresponding law of motion together with the phase portrait is shown in Fig. 1.

The following theorem characterizes the dynamics of system (4).

Theorem 2 *Let us suppose that the supports of function φ_α and φ_β do not intersect. Then, $\overline{\lim_{t\to\infty} x(t)} = \infty$, that is the amplitude tends to infinity with probability* 1.

The proof of this theorem follows from the fact that the area of the minimal hysteretic loop is positive $S_{min} > 0$. As a consequence, an amplitude value at each cycle satisfies the inequality: $A_n^2(t) \geq n S_{min}$.

We also note that under the conditions of the theorem, the rate of growth of the amplitude with probability 1 is proportional to the square root of time.

4 Conclusions

The paper presents a generalization of the classical hysteretic converter in the form of nonideal relay to the case when its switching numbers are randomly distributed according to a corresponding law. The properties of this converter are established, as well as the dynamics of the simple mechanical system in the form of oscillator under hysteretic force determined by a nonideal relay with random parameters is considered.

References

1. Ø. Bernt, *Stochastic Differential Equations. An Introduction with Applications* (Springer, Berlin, 2003), p. 379
2. P. Gurevich, D. Rachinskii, Pattern formation in parabolic equations containing hysteresis with diffusive thresholds. J. Math. Anal. Appl. **424**(2), 1103–1124 (2015)
3. F. Ikhouane, V. Mañosa, J. Rodellar, Dynamic properties of the hysteretic Bouc-Wen model. Syst. Control Lett. **56**(3), 197–205 (2007)
4. M.A. Krasnosel'skii, A.V. Pokrovskii, *Systems with Hysteresis* (Springer, Berlin, 1989), p. 410
5. D. Rachinskii, M. Ruderman, Convergence of direct recursive algorithm for identification of Preisach hysteresis model with stochastic input. SIAM J. Appl. Math. **76**(4), 1270–1295 (2016)
6. M.E. Semenov et al., MATEC Web Conference **83**, 01008(1)–01008(5) (2016)
7. M.E. Semenov, A.M. Solovyov, M.A. Popov, P.A. Meleshenko, Coupled inverted pendulums: stabilization problem. Arch. Appl. Mech. **88**(4), 517–524 (2018)
8. A.M. Solovyov, M.E. Semenov, P.A. Meleshenko, A.I. Barsukov, Bouc-Wen model of hysteretic damping. Procedia Eng. **201**, 549–555 (2017)
9. A.M. Solovyov, M.E. Semenov, P.A. Meleshenko, O.O. Reshetova, M.A. Popov, E.G. Kabulova, Hysteretic nonlinearity and unbounded solutions in oscillating systems. Procedia Eng. **201**, 578–583 (2017)
10. V.G. Zadorozhniy, Linear chaotic resonance in vortex motion. Comp. Math. Math. Phys. **53**(4), 486–502 (2013)
11. V.G. Zadorozhniy, S.S. Khrebtova, First moment functions of the solution to the heat equation with random coefficients. Comp. Math. Math. Phys. **49**, 1853–1868 (2009)

Critical Phenomena in a Dynamical System Under Random Perturbations

Natalia Firstova and Elena Shchepakina

Abstract The effect of Gaussian white noise on a canard cycle in dynamical model of an electrochemical reaction is analyzed. A critical noise intensity, at which the small-amplitude oscillations are transformed into mixed-mode oscillations, is obtained.

1 Introduction

It is well known that random perturbations can decisively affect the long-term behavior of dynamical systems. It should be noted that all realistic systems are subject to noise. For example, in a chemical system, the role of random perturbations can be played by various impurities, thermal vibrations, and many other external factors.

In this paper, we analyze the influence of an external noise on a canard cycle using an electrochemical reactor model as an example. The analysis is based on the stochastic sensitivity functions technique [1, 2]. We investigate the stochastic sensitivity of the equilibrium and the limit cycle of the model. We demonstrate transitions induced by the noise and find out the critical value of the noise intensity corresponding to the beginning of the transitions.

This work was funded by RFBR and Samara Region (project 16-41-630529-p) and the Ministry of Education and Science of the Russian Federation under the Competitiveness Enhancement Program of Samara University (2013–2020).

N. Firstova (✉)
Department of Technical Cybernetics, Samara National Research University, Moskovskoye shosse 34, Samara 443086, Russian Federation
e-mail: firstova.natalia@yandex.ru

E. Shchepakina
Department of Differential Equations and Control Theory, Samara National Research University, Moskovskoye shosse 34, Samara 443086, Russian Federation
e-mail: shchepakina@ssau.ru

A. Korobeinikov et al. (eds.), *Extended Abstracts Spring 2018*,
Trends in Mathematics 11, https://doi.org/10.1007/978-3-030-25261-8_38

2 Stochastic Model of Electrochemical Reactor

Consider a model of an electrochemical reaction of the Koper–Sluyters type [6] with allowance for random perturbations. It is assumed that the system is affected by a Gaussian white noise of low intensity. In this case, the model can be represented by the following system:

$$\frac{du}{dt} = -k_a e^{\gamma\theta/2} u(1-\theta) + k_d e^{-\gamma\theta/2}\theta + 1 - u + \epsilon w_1 = f(u,\theta), \tag{1}$$

$$\beta\frac{d\theta}{dt} = k_a e^{\gamma\theta/2} u(1-\theta) - k_d e^{-\gamma\theta/2}\theta - k_e e^{\alpha_0\zeta E}\theta + \epsilon w_2 = g(u,\theta), \tag{2}$$

where u is the dimensionless interfacial concentration of electrolyte, θ is the dimensionless amount of electrolyte that is adsorbed on the electrode surface, E is the electrode potential, β is the coverage ratio of the adsorbate, α_0 is the symmetry factor for the electron transfer, w_1 and w_2 are (in)dependent Wiener processes, ϵ reflects the noise intensity, and the current density is given in dimensionless form by $J = k_e e^{\alpha_0\zeta E}\theta$; $\zeta = F/(RT)$, where R is the universal gas constant, F is Faraday's constant, and T is the temperature. The parameter γ is interpreted an interaction parameter. Positive γ signifies attractive and negative γ signifies repulsive adsorbate interactions.

A detailed analysis of the deterministic model was carried out in [4, 7] using the theory of invariant manifolds. A critical regime corresponding to the canard cycle (see, for example, [8, 9] and references therein) was discovered. It was shown that the critical regime plays the role of a border between two main types of the reaction modes: a nonperiodic slow regime and relaxation oscillations.

In this paper, we investigate the influence of an external noise on the canard cycle [3, 5]. We start with the analysis of the stochastic sensitivity of the equilibrium of the system.

3 Theoretical Sensitivity to Random Perturbations

The stochastic sensitivity function method [1, 2] is applied to analyze the sensitivity of a stochastic equilibrium of a dynamical system to random perturbations. This method is based on the calculation of a stochastic sensitivity matrix W. The positively definite symmetric matrix W characterizes the spread of random trajectories of the system around the equilibrium position. The eigenvalues of W are the so-called theoretical characteristics of noise sensitivity.

The matrix W is found from the solution of the matrix equation

$$FW + WF^T + S = 0, \tag{3}$$

where

$$F = \begin{pmatrix} \frac{\partial f}{\partial u} & \frac{\partial f}{\partial \theta} \\ \frac{\partial g}{\partial u} & \frac{\partial g}{\partial \theta} \end{pmatrix}_{(\bar{u},\bar{\theta})}, \quad S = \begin{pmatrix} 1 & 0 \\ 0 & 1 \end{pmatrix}, \quad W = \begin{pmatrix} w_{11} & w_{12} \\ w_{12} & w_{22}. \end{pmatrix}.$$

From (3), we can find the elements of the matrix W:

$$w_{11} = \frac{-1 - 2f_\theta w_{12}}{2f_u}, \quad w_{22} = \frac{-1 - 2g_u w_{12}}{2g_\theta}, \quad w_{12} = \frac{f_u f_\theta + g_u g_\theta}{2(f_\theta^2 g_\theta + g_\theta^2 f_u - f_u f_\theta g_u - f_u g_u g_\theta)},$$

and the eigenvalues:

$$\lambda_{1,2} = \frac{w_{11} + w_{22} \pm \sqrt{(w_{11} + w_{22})^2 - 4(w_{11}w_{22} - w_{12}^2)}}{2}. \tag{4}$$

Here,

$$f_u = \frac{\partial f}{\partial u}(\bar{u}, \bar{\theta}), \quad f_\theta = \frac{\partial f}{\partial \theta}(\bar{u}, \bar{\theta}), \quad g_u = \frac{\partial g}{\partial u}(\bar{u}, \bar{\theta}), \quad g_\theta = \frac{\partial g}{\partial \theta}(\bar{u}, \bar{\theta}).$$

Figure 1a demonstrates the stochastic sensitivity of the equilibrium with respect to parameter k_e. Without loss of generality, the parameters' values are chosen to be $\epsilon = 0.2$, $\gamma = 8.99$, $k_a = 10$, $k_d = 100$, $\alpha_0 = 0.05$, $f = 38.7$, $E = 0.207564$ unless other values are specified in figure captions. Note that one of the eigenvalues (4) is sufficiently small (see the red curve), so the degree of stochastic sensitivity is determined by the highest eigenvalue. This figure shows that the equilibrium becomes more sensitive to random perturbations when the value of the control parameter k_e is higher.

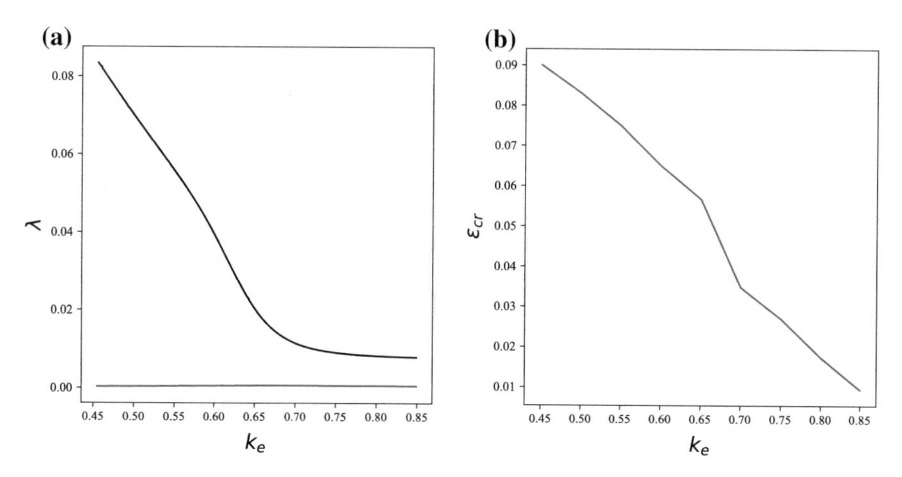

Fig. 1 **a** Theoretical sensitivity to the random perturbations and **b** the critical noise intensity value as a function of the control parameter k_e

4 Noise-Induced Transitions

Qualitative changes are possible in the stochastic model under the noise influence: when a certain critical value of the noise intensity ϵ_{cr} is reached, a transition from one deterministic attractor (stable point) to another (limit cycle) occurs. Random trajectories leave the pool of attraction of the deterministic attractor and wind up the limit cycle. Such qualitative changes in the system are called noise-induced transitions. Consider the change in the stochastic phase portrait depending on the intensity of the noise.

For weak noise, the randomly forced system (1) and (2) exhibits the small-amplitude stochastic oscillations near its equilibrium. Rare transitions occur through the unstable cycle to the limit cycle and back with increasing noise intensity. In that case, the oscillations of mixed type are observed, see Fig. 2.

However, as noise intensity increases, the large-amplitude stochastic oscillations appear, see Fig. 3. Transitions become more frequent with further increase of noise intensity. Thus, using the stochastic sensitivity function, we can predict the value of the noise intensity ϵ_{cr} corresponding to the beginning of the transitions.

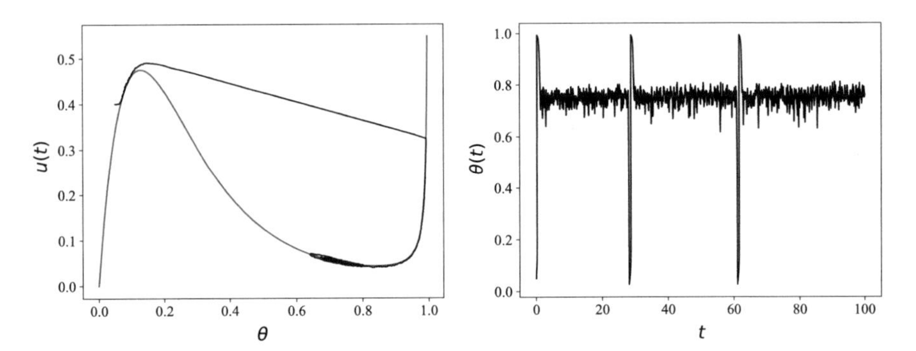

Fig. 2 Noise-induced transitions for $k_e = 0.85$, $\epsilon = 0.0098$

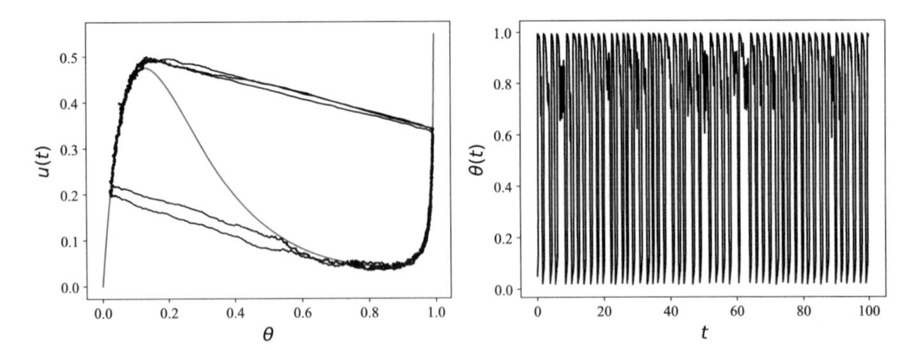

Fig. 3 Noise-induced transitions for $k_e = 0.85$, $\epsilon = 0.02$

We demonstrate transitions induced by noise for the control parameter $k_e = 0.85$ and find out that the critical value of the noise intensity approximately equals to $\epsilon_{cr} \approx 0.009495$. After searching for the critical values of the noise intensity for the value of the parameter k_e from the stable zone, we obtain the dependence of the ϵ_{cr} from the control parameter.

As it can be seen from Fig. 1b, the increase in control parameter value leads to the decrease in the noise intensity value, at which the transitions between attractors begin to appear.

References

1. I.A. Bashkirtseva, L.B. Ryashko, Sensitivity analysis of the stochastically and periodically forced brusselator. Physica A **278**, 126–139 (2000)
2. I.A. Bashkirtseva, Stochastic sensitivity analysis: theory and numerical algorithms. IOP Conf. Ser. Mater. Sci. Eng. **192**, 012024 (2017)
3. N. Berglund, B. Gentz, C. Kuehn, Hunting french ducks in a noisy environment. J. Differ. Equ. **252**(9), 4786–4841 (2012)
4. N. Firstova, E. Shchepakina, Conditions for the critical phenomena in a dynamic model of an electrocatalytic reaction. IOP Conf. Ser. J. Phys. Conf. Ser. **811**, 012002 (2017)
5. J. Grasman, Asymptotic analysis of nonlinear systems with small stochastic perturbations. Math. Comput. Simul. **31**(1–2), 41–54 (1989)
6. M.T.M. Koper, J.H. Sluyters, Instabilities and oscillations in simple models of electrocatalytic surface reactions. J. Electroanal. Chem. **371**(1), 149–159 (1994)
7. E.A. Shchepakina, N.M. Firstova, Study of oscillatory processes in the one model of electrochemical reactor. in *CEUR Workshop Proceedings*, vol. 1638 (2016), pp. 731–741
8. E. Shchepakina, V. Sobolev, Black swans and canards in laser and combustion models, in: *Singular Perturbations and Hysteresis*, ed. by M. Mortell, R. O'Malley, A. Pokrovskii, V. Sobolev (SIAM, 2005), pp. 207–256
9. E. Shchepakina, V. Sobolev, M.P. Mortell, *Singular Perturbations: Introduction to System Order Reduction Methods with Applications*. Lecture Notes in Mathematics, vol. 2114 (Springer, Berlin, 2014)

Burgers-Type Equations with Nonlinear Amplification: Front Motion and Blow-Up

Nikolay Nefedov

Abstract Recent results for some classes of initial boundary value problem for some classes of Burgers-type equations, for which we investigate moving fronts by using the developed comparison technique, are presented. There are also presented our recent results on singularly perturbed reaction–advection–diffusion problems, which are based on a further development of the asymptotic comparison principle. An asymptotic approximation of solutions with a moving front is constructed in the case of modular and quadratic nonlinearity and nonlinear amplification. The influence exerted by nonlinear amplification on front propagation and collapse is determined. The front localization and the collapse time are estimated.

1 Introduction

Recent results for some classes of initial boundary value problem for some classes of Burgers-type equations, for which we investigate moving fronts by using the developed comparison technique, are presented. There are also presented our recent results on singularly perturbed reaction–advection–diffusion problems, which are based on a further development of the asymptotic comparison principle (see [1–4, 6]). For these initial boundary value problems, the existence of moving fronts and its asymptotic approximation were investigated. These results were illustrated by the problem

$$\varepsilon\frac{\partial^2 u}{\partial x^2} - A(u,x,t)\frac{\partial u}{\partial x} - \frac{\partial u}{\partial t} = f(u,x,t,\varepsilon), \quad x \in (0,1), 0 < t \le T, \qquad (1)$$

$$u(0,t,\varepsilon) = u^0(t), \quad u(1,t,\varepsilon) = u^1(t), \quad t \in [0,T],$$
$$u(x,0,\varepsilon) = u_{init}(x,\varepsilon), \quad x \in [0,1].$$

This work was supported by the Russian Science Foundation (project N 18–11–00042).

N. Nefedov (✉)
Faculty of Physics, Department of Mathematics, Lomonosov Moscow State University, 119899 Moscow, Russia
e-mail: nefedov@phys.msu.ru

A. Korobeinikov et al. (eds.), *Extended Abstracts Spring 2018*,
Trends in Mathematics 11, https://doi.org/10.1007/978-3-030-25261-8_39

An asymptotic approximation of solutions with a moving front for specific forms of equation (1) in the case of modular and quadratic nonlinearity and nonlinear amplification is constructed. Such problems are typical in numerous applications of nonlinear wave theory; see [5] and references therein. Note that the applications make use of a more natural formulation of problem (1) in which the coordinate and time are swapped ("wave formulation"). Then, the equation describes quadratically nonlinear or modular waves propagating in a nondispersive medium with cubically nonlinear amplification. The influence exerted by nonlinear amplification on front propagation and collapse is determined. The front localization and the collapse time are estimated.

2 Initial Boundary Value Problems with Fronts: Motion and Blow-Up

The author considers the following problems:

(i) the reaction–diffusion equation

$$\varepsilon^2 \frac{\partial^2 u}{\partial x^2} - \frac{\partial u}{\partial t} = f(u, x, \varepsilon), \quad x \in (0,1), t > 0;$$

(ii) the reaction–advection–diffusion equation

$$\varepsilon \frac{\partial^2 u}{\partial x^2} - A(u, x)\frac{\partial u}{\partial x} - \frac{\partial u}{\partial t} = f(u, x, \varepsilon), \quad x \in (0,1), t > 0;$$

(iii) the reaction–diffusion system

$$\begin{aligned} \varepsilon^q \frac{\partial^2 u}{\partial x^2} - \frac{\partial v}{\partial t} &= f(u, v, x, \varepsilon), \\ \varepsilon^p \frac{\partial^2 u}{\partial x^2} - \frac{\partial u}{\partial t} &= g(u, v, x, \varepsilon), \quad x \in (0,1), t > 0. \end{aligned}$$

For these initial boundary value problems, it was proven the existence of fronts and their asymptotic approximation was obtained. In particular, it was shown that the principal term describing the location of the moving front is determined by the initial value problem

$$\frac{dx_0}{dt} = V(x_0), \quad x_0(0) = x_{00}, \tag{2}$$

where x_{00} is the initial location of the front, and $V(x_0)$ is a known function, defined by the input data. It was proved that the Lyapunov stability of steady states of equation (2) determine the Lyapunov stability of stationary solutions with the interior layer of the IBVP. There are also proved that under some conditions the blow-up of the solution to problem (2) determines the blow-up of the interior layer solution of the IBVP.

2.1 Reaction–Advection–Diffusion Equations

We illustrate our results by the problem

$$\varepsilon\frac{\partial^2 u}{\partial x^2} - A(u,x)\frac{\partial u}{\partial x} - \frac{\partial u}{\partial t} = f(u,x,\varepsilon), \quad x \in (0,1), t > 0,$$

$$u(0,t,\varepsilon) = u^0, \quad u(1,t,\varepsilon) = u^1, \quad t \in [0,T],$$
$$u(x,0,\varepsilon) = u_{init}(x,\varepsilon), \quad x \in [0,1].$$

We consider a formulation similar to that for the Burgers' equation. The initial function represents a front formed with given outer branches. The task is to describe the motion of the front over time.

Let also assume the following:

(H_1) The equation $A(u,x)du/dx + B(u,x) = 0$, with the initial condition $u(0) = u^0$, has the solution $u = \varphi^l(x)$, and with the initial condition $u(1) = u^1$ has the solution $u = \varphi^r(x)$. Moreover, $\varphi^l(x) < \varphi^r(x)$, $x \in [0,1]$, and $A(\varphi^l(x),x) > 0$, $A(\varphi^r(x),x) < 0$, $x \in [0,1]$;

(H_2) $I(x) := \int_{\varphi^l(x)}^{\varphi^r(x)} A(u,x)du > 0$;

(H_3) The initial value problem

$$\frac{dx_0}{dt} = \frac{I(x_0)}{\varphi^r(x_0) - \varphi^l(x_0)} \equiv V(x_0), \quad x_0(0) = x_{00},$$

where x_{00} is the initial location of the front, has the solution $x_0(t) : [0;T] \to [0,1]$ such that $x_0(t) \in (0,1)$ for $t \in [0,T]$.

The main result for this problem is the theorem of existence and asymptotic approximation of the moving front with the principal term of the front location $x_0(t)$.

2.2 Burgers-Type Equation with Blow-Up of the Solution

Now suppose that assumption (H_1) is changed to the following one:

(H_1') the equation $A(u,x)du/dx + B(u,x) = 0$ with the initial condition $u(0) = u^0$ has the solution $u = \varphi^l(x)$ with blow-up near some point x_c: $\varphi^l(x) \to -\infty$ for $x \to x_c$, and with the initial condition $u(1) = u^1$ has the solution $u = \varphi^r(x)$. Moreover, $\varphi^l(x) < \varphi^r(x), x \in [0, x_c)$, and $A(\varphi^l(x), x) > 0, x \in [0, x_c)$, $A(\varphi^r(x), x) < 0, x \in [0, 1]$.

It was shown that the Burgers-type equation can exhibit the blow-up of the front-type solution with a jump at the front tending to infinity. The solution has blow-up near the point x_c, which is the point of the blow-up for the solution of the problem for φ^l for time $t \to T_c$, where $T_c = \int_{x_{00}}^{x_c} dx/V(x)$. In particular, we have shown that the following Burgers-type equation with cubic amplification exhibits the blow-up of the front-type solution:

$$\varepsilon\frac{\partial^2 u}{\partial x^2} - \frac{\partial u}{\partial t} = -u\frac{\partial u}{\partial x} - u^3, \quad x \in (0, 1), \quad t \in (0, 0.3],$$
$$u(0, t) = -2, \quad u(1, t) = \frac{1}{3},$$
$$u(x, 0) = \frac{7}{6}\tanh\frac{x - \frac{1}{4}}{\varepsilon} - \frac{5}{6}.$$

One can easily check that all conditions for this case are satisfied. The equation of front motion of fronts from H_3 has the following form:

$$\frac{dx_0}{dt} = -\frac{\varphi^r(x_0) + \varphi^l(x_0)}{2}, \qquad x_0(0) = x_{00}, \quad x_{00} \in (0, 1),$$

where $\varphi^l = 1/(x + 1/u^0)$ and $\varphi^r = 1/(x - 1 + 1/u^1)$. Therefore, the blow-up point is $x_c = 1/u^0$ and for this case $x_c = 1/2$.

We present some examples of calculations; see Fig. 1.

2.3 Case of Modular Nonlinearity and Cubic Amplification

As it was noted in Sect. 1, nonlinear waves can be described by the Burgers equation with the coordinate and time formally swapped. The typical situation for the nonlinear waves is the situation where both φ^l and φ^r blow-up. This is illustrated by the following problem:

$$\begin{aligned} &\varepsilon\frac{\partial^2 u}{\partial t^2} - \frac{\partial u}{\partial x} = -\frac{\partial |u|}{\partial t} - u^3, \quad x \in (0, d), \quad t \in [t_0, t_1], \\ &u(x, t_0, \varepsilon) = u^0 < 0, \quad u(x, t_1, \varepsilon) = u^1 > 0, \quad x \in [0, d], \\ &u(0, t, \varepsilon) = u_{init}(t, \varepsilon), \quad t \in [t_0, t_1]. \end{aligned} \tag{3}$$

The functions φ^l and φ^r have the form $\varphi^l = -(-1/(2t + c_0))^{1/2}$ and $\varphi^r = (1/(2t + c_1))^{1/2}$, where $c_0 = -1/(u_0)^2$ and $c_1 = 1/(u_1)^2$. Blow-up points for φ^l and φ^r are $t_{cl} = -c_0/2 > 0$ and $t_{cr} = -c_1/2 < 0$, respectively. The equation of front motion is

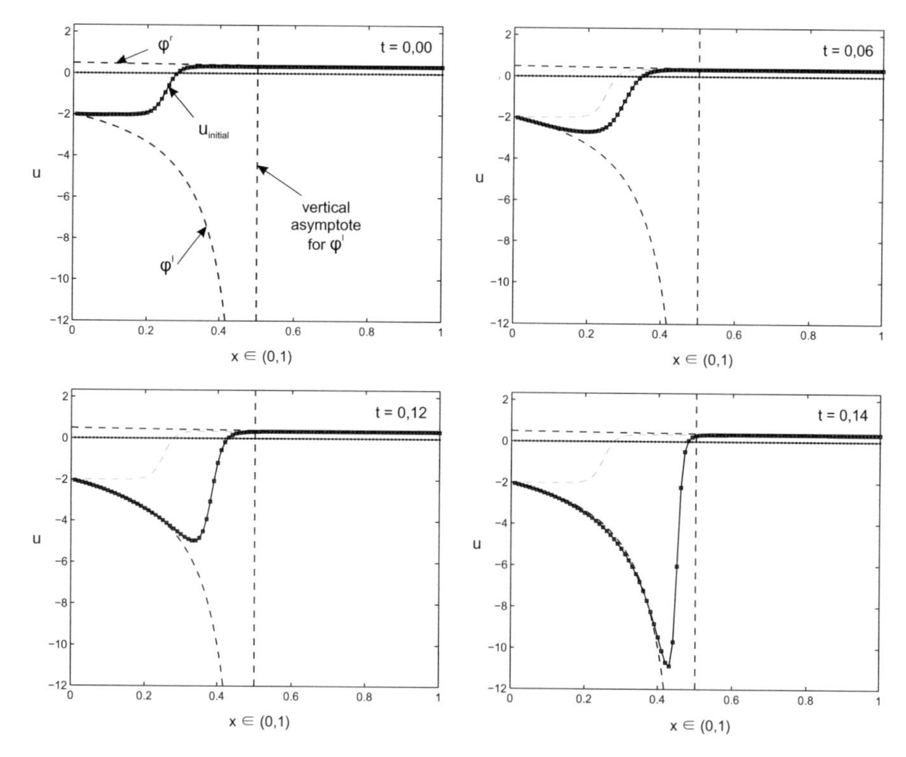

Fig. 1 Some typical example of the direct problem's solution $u(x, t)$ (some refining of the mesh in a neighborhood of the transition point and the bounds has been performed)

$$\frac{dt_0}{dx} = -\frac{\varphi^r(t_0) + \varphi^l(t_0)}{\varphi^r(t_0) - \varphi^l(t_0)} = V(t_0), \qquad t_0(0) = t_{00}, \quad t_{00} \in (t_{cr}, t_{cl}). \tag{4}$$

Thus, the problem (3) has a solution with a sharp front and its observation time depends on the coordinate and is determined by the problem (4).

References

1. V.F. Butuzov, N.N. Nefedov, L. Recke, K.R. Schneider, Periodic solutions with a boundary layer of reaction-diffusion equations with singularly perturbed Neumann boundary conditions. Int. J. Bifurc. Chaos **24**, 1440019–14400198 (2014)
2. N. Nefedov, Comparison principle for reaction-diffusion-advection problems with boundary and internal layers. Lect. Notes Comput. Sci. **8236**, 62–72 (2013)
3. N.N. Nefedov, E.I. Nikulin, Existence and stability of periodic contrast structures in the reaction-advection-diffusion problem. Russ. J. Math. Phys. **22**(2), 215–226 (2015)
4. N.N. Nefedov, L. Recke, K.R. Schnieder, Existence and asymptotic stability of periodic solutions with an interior layer of reaction-advection-diffusion equations. J. Math. Anal. Appl. **405**, 90–103 (2013)

5. N.N. Nefedov, O.V. Rudenko, On front motion in a Burgers-type equation with quadratic and modular nonlinearity and nonlinear amplification. Dokl. Math. **97**, 99–103 (2018)
6. A.B. Vasileva, V.F. Butuzov, N.N. Nefedov, Singularly perturbed problems with boundary and internal layers, in *Proceedings of the Steklov Institute of Mathematics*, vol. 268 (2010), pp. 258–273

On Uncertainty Quantification for Models Involving Hysteresis Operators

Olaf Klein

Abstract Parameters within hysteresis operators modeling real-world objects have to be identified from measurements and are therefore subject to error in measurement. To investigate the influence of these errors, the methods of *Uncertainty Quantification (UQ)* are applied.

1 Uncertainties in Models with Hysteresis Operators and Uncertainty Quantification

Considering, e.g., magnetization, piezo-electric effects, elastoplastic behavior, or magnetostrictive materials, one has to take into account hysteresis effects. Many models involve therefore *hysteresis operators*. The parameters in the models are identified using results from *measurements*, sometimes performed only for some *sample specimens* but also used also for other specimens.

The parameters in the hysteresis operators are therefore also subject to uncertainties. We apply the methods of *Uncertainty Quantification (UQ)*, see, e.g., [6, 7], to deal with these uncertainties, i.e., we describe them by introducing appropriate random variables modeling the corresponding information/assumptions/beliefs and use probability theory to describe and determine the influence of the uncertainties.

In this paper, we present results of *Forward UQ*, i.e., we consider the model output as random variable and compute properties like expected value, variation, and probabilities for outputs entering some interval, credible intervals, and other *Quantities of Interest (QoI)*. Moreover, we will also present a brief example of *Inverse UQ*, i.e., of using (further) data and measurements, to determine (reduce/adapt) the uncertainty of the parameter, i.e., to determine a (new) random variable taking into account the (new) information, and use the (new) random variable to represent the parameter afterward.

O. Klein (✉)
WIAS, Mohrenstr. 39, 10117 Berlin, Germany
e-mail: olaf.klein@wias-berlin.de

A. Korobeinikov et al. (eds.), *Extended Abstracts Spring 2018*,
Trends in Mathematics 11, https://doi.org/10.1007/978-3-030-25261-8_40

2 The Play Operator

2.1 The Play Operator with Deterministic Data

Considering some *yield limit* $r \geq 0$ and some *initial state* z_0, the *play operator* $\mathcal{P}_r[z_0, \cdot]$ maps $u \in C[0, T]$ being piecewise monotone to $\mathcal{P}_r[z_0, u] \in C[0, T]$ being piecewise monotone and it holds, see, e.g., [1, 4, 5, 8],

$$\mathcal{P}_r[z_0, u](0) = \max\left(u(0) - r, \min\left(u(0) + r, z_0\right)\right)$$

$$\mathcal{P}_r[z_0, u](t) = \begin{cases} \max\left(\mathcal{P}_r[z_0, u](t_0), u(t) - r\right), & \text{if } u \text{ is} \\ \qquad \text{increasing on } [t_0, t], & \\ \min\left(\mathcal{P}_r[z_0, u](t_0), u(t) + r\right), & \text{if } u \text{ is} \\ \qquad \text{decreasing on } [t_0, t], & \end{cases}$$

for all $t_0, t \in [0, T]$ with $t_0 < t$ such that u *is monotone on* $[t_0, t]$.

Now, we want to consider a situation, wherein the true value of the yield limit is not known, but we think that its values are near to 2. Hence, interpret the yield limit $r \geq 0$ as value of the random variable R generated from $N(2, \sqrt{0.5}^2)$ by ignoring $(-\infty, 0]$ and rescaling, leading to the following *probability density function* ρ_R of R:

$$\rho_R(r) = \begin{cases} \frac{1}{C} \mathrm{e}^{-\frac{(r-2)^2}{2 \cdot 0.5}} & \text{if } r \geq 0, \\ 0 & \text{if } r < 0, \end{cases} \qquad \text{with} \quad C = \int_0^\infty \mathrm{e}^{-\frac{(r-2)^2}{2 \cdot 0.5}}\, \mathrm{d}r\,. \tag{1}$$

The mapping $[0, \infty) \ni r \mapsto \mathcal{P}_r[w, u](t)$ is continuous, see, e.g., [5, Prop. 2.5]. Hence, it follows that the composition of this mapping with R generates a random variable, denoted by $\mathcal{P}_R[w, u](t)$.

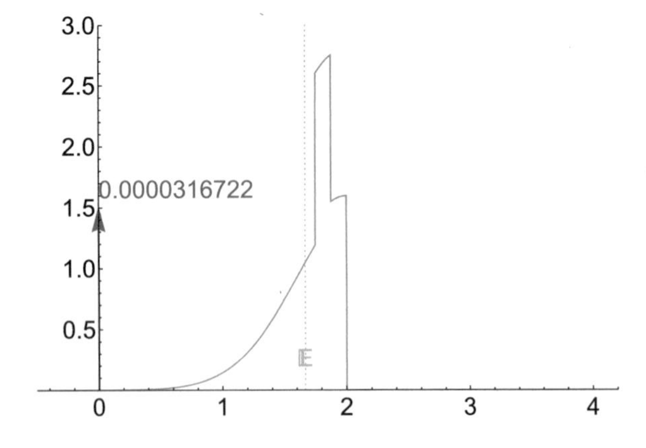

Fig. 1 Density function and position and weight for Dirac measure generating the measure representing the random variable $\mathcal{P}_R[0, u](t_5)$

We consider a piecewise monotone input function u such that the value of the function increases from 0 to 4, decreases afterward to 0, increases afterward to 3.5, and decreases then to 0.5 until $t_5 > 0$. The probability measure for the random variable $\mathcal{P}_R[0, u](t_5)$ is the sum of a measure with density function shown in Fig. 1 and a Dirac measure at 0 weighted by $\int_4^\infty \rho_R(r)\,\mathrm{d}r \approx 0.0000316722$.

3 Identification

3.1 Considered Situation

In [3], a magnetostrictive Terfenol actuator is investigated and the hysteresis between the current generating the magnetic field and the resulting displacement is considered, and the data of the *First-Order-Reversal-Curves (FORC)* are used to determine a Preisach operator and a generalized Prandtl–Ishlinskiĭ operator.

A further generalized Prandtl–Ishlinskiĭ operator has been considered in [2, Sects. 3, 5] and is defined by mapping $H \in C[0,T]$ to $\mathcal{G}_{c_1,c_2,c_3}[H] \in C[0,T]$ defined by

$$\mathcal{G}_{c_1,c_2,c_3}[H](t) = \int_0^\infty c_1 \mathrm{e}^{-r/c_2} \mathcal{P}_r[\lambda_0(r), \tanh(c_3 u)](t)\,\mathrm{d}r\,, \tag{2}$$

with c_1, c_2, c_3 being positive parameters and $\lambda_0\colon [0,\infty) \to \mathbb{R}$ being an appropriate function.

In a joined work with Daniele Davino and Ciro Visone of the Università del Sannio, Benevento, Italy, the data measured to prepare the FORC diagram in [3] are used to generate functions approximated by the initial loading curves. These functions should be approximated by the initial loading curve corresponding to $\mathcal{G}_{c_1,c_2,c_3}[H](t)$ for appropriate values. Moreover, information about the uncertainty for c_1 and c_2 should be determined.

3.2 *Identification of c_3*

To determine c_3, we consider the generated approximations of the initial loading curve. They are obtained after applying the transformation $x \mapsto \tanh(c_3 x)$ to the input parameter and the value for c_3 is determined by requesting that the sum over the squared L^2 difference between these generated approximations is minimized. We ends up with the optimal value $c_3 = c_{\tanh} = 1.2465$.

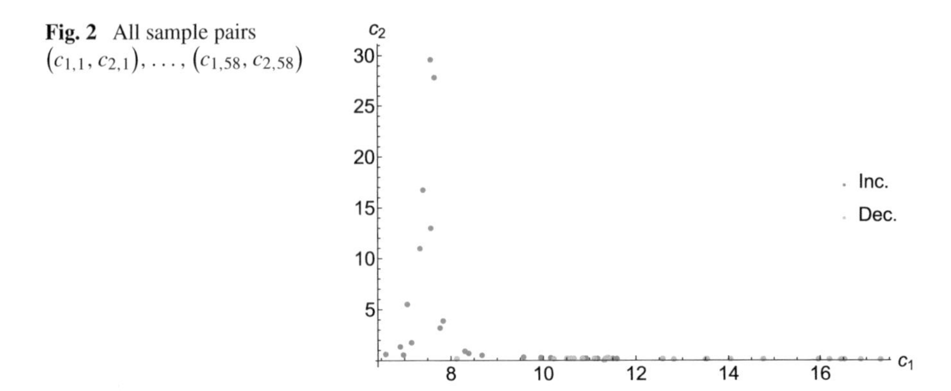

Fig. 2 All sample pairs $(c_{1,1}, c_{2,1}), \ldots, (c_{1,58}, c_{2,58})$

3.3 *Identification of* $c_{1,i}$, $c_{2,i}$

For each approximation of the initial loading cure derived from the measurements, we determine $(c_{1,i}, c_{2,i}) \in (0, \infty) \times (0, \infty)$ by minimizing L^2 difference between the approximations for the initial loading and the loading curve corresponding to the Prandtl–Ishlinskiĭ operator in $\mathcal{G}_{c_1,c_2,c_3}[H](t)$.

If an appropriate subset of the computed parameter pairs is considered, computing the *mean* and the *standard deviation* of the corresponding values for $c_{1,i}$ and $c_{2,i}$ leads to $\text{mean}(c_{1,i}) = 12.7$, $\text{mean}(c_{2,i}) = 0.13$, $\text{std}(c_{1,i}) = 2.6$, and $\text{std}(c_{2,i}) = 0.04$.

One could now assume that the value of c_i is represented by a random variable C_i that is a normal distributed random variables with the corresponding mean and standard deviation, maybe restricted to (δ, ∞) with some small $\delta > 0$.

If all computed parameter pairs are considered, computing the mean and the standard deviation for all values for c_1 and c_2, we get $\text{mean}(c_{1,i}) = 10.9$, $\text{mean}(c_{2,i}) = 2.13$, $\text{std}(c_{1,i}) = 2.7$, and $\text{std}(c_{2,i}) = 5.95$. In view of these values, it is obvious that one cannot use them to derive a satisfying assumption for the random variables representing the positive parameter c_1 and c_2.

Considering all computed parameter pairs, see Fig. 2, and taking into account that the correlation between $(c_{1,i})_i$ and $(c_{2,i})_i$ is -0.418327, it is obvious that we cannot get these pairs as samples if we assume that the parameter can be represented by two independent normal distributed random variables.

4 Further Research

We need to look for other random variables C_1^* and C_2^* such that $((c_{1,i}, c_{2,i}))_i$ can be considered as appropriate samples of the random variable (C_1^*, C_2^*).

To perform this inverse UQ will be subject of further research, in addition to some issues concerning the modeling.

References

1. M. Brokate, J. Sprekels, *Hysteresis and Phase Transitions* (Springer, Berlin, 1996)
2. D. Davino, P. Krejčí, C. Visone, Fully coupled modeling of magneto-mechanical hysteresis through 'thermodynamic' compatibility. Smart Mater. Struct. **22**(9) (2013)
3. M. Al Janaideh, C. Visone, D. Davino, P. Krejčí, The generalized Prandtl-Ishlinskii model: relation with the Preisach nonlinearity and inverse compensation error, in *2014 American Control Conference (ACC)* (2014)
4. M. Krasnosel'skiĭ, A. Pokrovskii, *Systems with Hysteresis*, Russian edn. (Springer, Berlin, 1989) (Nauka, Moscow, 1983)
5. P. Krejčí, Hysteresis, convexity and dissipation in hyperbolic equations. Gakuto Int. Ser. Math. Sci. Appl., vol. 8 (Gakkōtosho, Tokyo, 1996)
6. R.C. Smith, Uncertainty quantification: theory, implementation, and applications. Comput. Sci. Eng. Soc. Ind. Appl. Math. (SIAM), vol. 12 (Philadelphia PA, 2014)
7. T.J. Sullivan, *Introduction to Uncertainty Quantification*. Texts in Applied Mathematics, vol. 63 (Springer, Berlin, 2015)
8. A. Visintin, *Differential Models of Hysteresis*. Applied Mathematical Sciences, vol. 111 (Springer, Berlin, 1994)

Unusual Elastic–Plastic Properties of Fullerene Films: Dynamical Hysteretic Model

Peter A. Meleshenko, Andrey M. Semenov, Andrey I. Barsukov, Leonid V. Stenyukhin and Valentina P. Kuznetsova

Abstract In this paper, a model of an unusual elastic–plastic hysteresis is constructed and discussed following the recent progress in investigation of the fullerene films. The constructive model is based on the operator technique of hysteretic nonlinearities. To describe the input–output relations, we use the Ishlinkii's operator technique together with the probability model based on the Kolmogorov–Chapman equation.

1 Introduction

The hysteretic effects take place in various areas of material science (at both macro- and microlevels). Depending on purposes of investigation, both the phenomenological and constructive (based on the first principles) models can be used and there are many literature sources on this subject (see, e.g., [4, 5] and related references). As a rule, in the constructive models that are described by the relations "input-state" and "state-output" [8, 9], the dynamical properties of the hysteresis carrier were not taken into account.

As it is known, the mechanical properties almost all materials (namely, the elastic–plastic hysteresis, or hysteretic properties of the material) remain unchanged (the

This work was supported by the RFBR (Grants 16-08-00312-a, 17-01-00251-a, and 18-08-00053-a).

P. A. Meleshenko (✉)
Voronezh State University, Universitetskaya sq.1, 394006 Voronezh, Russia
e-mail: melechp@yandex.ru

Zhukovsky–Gagarin Air Force Academy, Starykh Bolshevikov st. 54A, 394064 Voronezh, Russia

A. M. Semenov
Institute of macromolecular compounds, Bolshoy pr. 31, 199004 Saint-Petersburg, Russia

A. I. Barsukov · L. V. Stenyukhin
Voronezh State Technical University, XX-letiya Oktyabrya st. 84, 394006 Voronezh, Russia

V. P. Kuznetsova
Voronezh State Medical University, Studencheskaya st. 10, 394036 Voronezh, Russia

A. Korobeinikov et al. (eds.), *Extended Abstracts Spring 2018*,
Trends in Mathematics 11, https://doi.org/10.1007/978-3-030-25261-8_41

hysteretic dependence of elastic–plastic materials does not depend on the speed of mechanical affection). However, the results of recent experiments with the fullerene nanofilms [6] show that the shape of hysteretic curve in the coordinates "force–displacement" depends on the speed of force application.

In this work, we propose a dynamical probability model of hysteresis for description of elastic–plastic properties of nanoscale fullerene film taking into account the electromagnetic nature of the fullerene clusters binding. This model is based on the fact that the decay law for fullerene supercluster $[C_{60}]_n$ (especially for $n = 2$) depends on external conditions (temperature, pressure, etc.) as well as is of probabilistic nature. A description of the decay law at the macroscopic level can be made using the well-known theory of random processes (a basic object in this field is the Kolmogorov–Chapman equation).

2 Hysteresis in Nanoscale Films

In recent years, the self-regenerating materials and covers are intensively investigated [2, 3, 6, 11]. Such covers regenerate when on its surface a little injury takes place. Usually, such covers contain the capsules with the "regenerating agent." When the damage takes place, the capsules break and, as a result, there are chemical reactions that lead to vanishing of the injury. In this work, we consider the covers that have self-regenerating properties, but this effect is provided by the hysteretic properties of the cover's material. This cover is coated by two beams of buckyball, namely, the molecular (the PVD technology) and ion (the magnetron technology). The basic "object" in the regeneration effect is the unusual elastic–plastic hysteresis which is caused by the depolymerization of fullerene. As it was shown in [6], on the surface of nanofilm there are some "liftings" at small mechanical affection by the probe with the diameter less than 200 nm. It is also shown that the relief changing does not connect with the adhesion of the film to the probe.

As it is known, the elastic–plastic hysteresis manifests in macro-level in such a way that the hysteresis loop gets over clockwise. Herewith, as a rule, a form and other characteristics of the loop do not depend on the speed of mechanical affections. However, for the material under consideration, such a dependence takes place (namely, the form of the hysteretic loop depends on the speed of force affection).

3 Dynamical Hysteretic Model

Here, we present a model of the observable effect. The dependence of the loop's form on time means that the presented model should be nonstationary.

It is well known that the physical properties of nanofilms depend on the structure of the materials. Main processes in such nanocovers occur due to the polymerization and depolymerization processes. These processes can be initialized by the temperature, light, or mechanical affections. It should be also pointed out that the polymerization together with the depolymerization occurs according to the probability laws. The main assumption of this model consists in the fact that the depolymerization process is turning on when the cover is under temporal excess pressure.

The state of the cover can be described by the pair of parameters $(\omega_1(t), \omega_2(t))$ that are the fraction of the domains under polymerization and depolymerization, respectively. The dynamics of these parameters can be described by the Kolmogorov–Chapman equation (here the dot displays the time derivative):

$$\begin{cases} \dot{\omega}_1 = -\lambda_1\omega_1 + \lambda_2\omega_2, \\ \omega_1 + \omega_2 = 1, \end{cases} \tag{1}$$

with the initial conditions $\omega_1(0) = \omega_{01}, \omega_2(0) = \omega_{02}$. Linear volumes, x_1 and x_2, are connected to these states, respectively. At the same time, the dependence of these linear volumes on the external force u, we can define as

$$x_1 = x_1(u), \ \ x_2 = x_2(u), \tag{2}$$

using the Ishlinskii's operator which will be described below. Finally, the dependence of a displacement on the external force can be determined by the following relation:

$$l = \omega_1 x_1 + \omega_2 x_2. \tag{3}$$

Equations (1)–(3) are the base of the proposed model.

At the same time, the intensities of transitions λ_1 and λ_2 should also depend on the external force and are driven by the relations $\lambda_1 = \lambda_1(u)$ and $\lambda_2 = \lambda_2(u)$. Identification of these dependencies from the known experimental data is a complicated problem. Namely, there are certain facts which indicate that these functions are monotonically increasing. In this work, we suppose that these functions can be chosen in the form

$$\lambda_1(u) = \lambda_{01} + c_1 u, \ \ \lambda_2(u) = \lambda_{02} + c_2 u \tag{4}$$

with the positive parameters.

3.1 Some Remarks on Ishlinskii's Operator

Here, we present some details on the Ishlinskii's operator technique (details of application of this technique can be found, e.g., in [1, 7, 10] and related references). First, let us introduce the necessary definitions.

The stop is an operator $W[t_0, x_0, E, h]$ which connects every continuous input $u(t)$ $(t \geqslant 0)$ with output $x(t)$ by the following rule (for the monotonic inputs):

$$x(t) = \begin{cases} \min\{h, E[u(t) - u(t_0)] + x(t)\}, & \text{if } u(t) \text{ increase}, \\ \max\{-h, E[u(t) - u(t_0)] + x(t)\}, & \text{if } u(t) \text{ decrease}. \end{cases}$$

Here, E is the elastic modulus of the material (we understand $u(t)$ and $x(t)$ as stress and deformation, respectively).

For the piecewise monotonic inputs, the output can be determined using the semi-group identity

$$W[t_0, x_0, E, h]u(t) = W[t_1, W[t_0, x_0, E, h]u(t_1), E, h]u(t),$$

and then, using the special limit construction, such an operator can be redefined for all continuous inputs. A detailed description of this operator as well as its properties is presented in the book Krasnosel'skii and Pokrovskii [5].

Let $U(h) = W[t_0, x_0, 1, h]$ be a single-parameter kind of stop with the elastic modulus equal to 1 and the yield stress $\pm h$. Let us define the nondecreasing continuous function $\Omega = \Omega(h)$ $(h \geqslant 0)$ which satisfies the following condition:

$$\int_0^\infty |\Omega(h)|\mathrm{d}h < \infty. \tag{5}$$

In the following consideration, we will use the condition (5) in the form

$$\int_0^\infty h\mathrm{d}\Omega(h) < \infty. \tag{6}$$

Let us denote as Z the set of continuous functions $z(h)$ $(h \geqslant 0)$ that satisfy the inequality $|z(h)| \leqslant h$ $(0 \leqslant h < \infty)$. Then, the pairs $\{u_0; z(h)\}$ form the set which determines the state space of the operator U. The dynamics of input–output relations is determined as

$$x(t) = W[t_0, z_0(h), 1, \Omega(h)]u(t) = \int_0^\infty U[t_0, z_0(h), h]u(t)\mathrm{d}\Omega(h), \quad (t \geqslant t_0). \tag{7}$$

Here, the integral is understood in the sense of Riemann–Stieltjes. However, it should be noted that this relation is uncomfortable for calculations of the output of Ishlinskii's operator.

As it follows from the definition, the operator $U(h)$ describes the ideal plastic fiber with the elastic modulus $E = \xi$ and the plastic limits $\pm\xi h$. Let us consider also the so-called charge function

$$\chi_+(u, \xi, h) = \begin{cases} -\xi h, \text{ at } u \leqslant -h, \\ \xi u, \text{ at } \; -h < u(t) < h, \\ \xi h, \text{ at } u(t) \geqslant h. \end{cases}$$

The analog of this function for the Ishlinskii's operator is the function

$$\chi_+(u, \Omega) = \int_0^u |\Omega(|h|)| \mathrm{d}h, \ \ (-\infty < u < \infty), \tag{8}$$

and the discharge function

$$\chi_-(u, \Omega) = 2\chi_+(\frac{u}{2}, \Omega), \ \ (-\infty < u < \infty).$$

In this way, the Ishlinskii's operator is the model of plastic body which is composed of the continual number of ideal plastic fibers. As it follows from the definition, for a monotonic input and uncharged state the alternating stress can be expressed in terms of the charge function, namely $x(t) = \chi_+ (u(t) - u(t_0), \Omega)$.

On the piecewise monotonic inputs, the Ishlinskii's operator can be determined (in analogous manner) using the semigroup identity. Unfortunately, relation (8) allows to determine the output using the charge function only at zero initial state. However, in the considered case, this condition is not restrictive because at the initial moment the state of a nanomaterial is naturally supposed to be uncharged.

Finally, the model which describes the dynamics of the system under consideration is based on Eqs. (1)–(3) together with the relations

$$x_1(t) = W[t_0, z_{01}(h), 1, \Omega_1(h)]u(t), \tag{9}$$

$$x_2(t) = W[t_0, z_{02}(h), 1, \Omega_2(h)]u(t), \tag{10}$$

where $u(t)$ is an external force applied to the fullerene film; $z_{01}(h)$ and $z_{02}(h)$ correspond to initial uncharged states of polymerized and depolymerized fractions, respectively.

In the experiments described in [6], the external charge is determined as a piecewise linear function, namely,

$$u(t) = \begin{cases} at, \ t \in \left[0, \frac{T}{2}\right], \\ -a(t - T), \ t \in \left(\frac{T}{2}, T\right]. \end{cases}$$

4 Conclusions

The results of numerical simulations show that the qualitative behavior of the hysteretic curves (in the frame of the proposed model) significantly depends on a choice of the parameters c_1 and c_2 which determine the intensity of transitions from depolymerized state and back. Optimization of the model by these (and other) parameters allows to obtain the results that differ from the experimental results approximately within 3%.

References

1. M. Arnold, N. Begun, P. Gurevich et al., Dynamics of discrete time systems with a hysteresis stop operator. SIAM J. Appl. Dyn. Syst. **16**, 91–119 (2017)
2. E.G. Atovmyan, A.A. Grishchuk, T.N. Fedotova, Russ. Chem. Bull. **60**, 1505–1507 (2011)
3. B.M. Darinsky, M.E. Semenov, A.M. Semenov, P.A. Meleshenko, in *Progress in Electromagnetics Research Symposium Proceedings* (Prague, 2015), pp. 1716–1719
4. F. Ikhouane, J. Rodellar, *Systems with Hysteresis: Analysis, Identification and Control Using the Bouc–Wen Model* (Wiley, 2007)
5. M.A. Krasnosel'skii, A.V. Pokrovskii, *Systems with Hysteresis* (Springer, Berlin, 1989)
6. O.V. Penkov, V.E. Pukha et al., Self-healing phenomenon and dynamic hardness of C_{60}-based nanocomposite coatings. Nano Lett. **14**, 2536–2540 (2014)
7. M. Ruderman, D. Rachinskii, Use of Prandtl-Ishlinskii hysteresis operators for Coulomb friction modeling with presliding. J. Phys. Conf. Ser. **811**(1), 012013 (2017)
8. M.E. Semenov, P.A. Meleshenko, A.M. Solovyov, A.M. Semenov, Hysteretic nonlinearity in inverted pendulum problem. Springer Proc. Phys. **168**, 463–506 (2015)
9. M.E. Semenov, A.M. Solovyov et al., Hysteretic damper based on the Ishlinsky-Prandtl model. MATEC Web Conf. **83**, 01008 (2016)
10. M.E. Semenov, A.M. Solovyov, P.A. Meleshenko, J.M. Balthazar, Nonlinear damping: from viscous to hysteretic dampers. Springer Proc. Phys. **199**, 259–275 (2017)
11. Y. Wang, H. Zettergren, P. Rousseau et al., Formation dynamics of fullerene dimers C_{118}^{+}, C_{119}^{+}, and C_{120}^{+}. Phys. Rev. A **89**, 062708 (2014)

Depinning of Traveling Waves in Ergodic Media

Sergey Tikhomirov

Abstract We study speed of moving fronts in bistable spatially inhomogeneous media at parameter regimes where the speed tends to zero. We provide a set of conceptual assumptions under which we can prove power law asymptotics for the speed, with exponent depending a local dimension of the ergodic measure near extremal values. We also show that our conceptual assumptions are satisfied in a context of weak inhomogeneity of the medium and almost balanced kinetics, and compare asymptotics with numerical simulations. The presentation is based on a joint work with Arnd Sheel.

1 Pinning in Traveling Wave Equations

Reaction–diffusion equations describe natural phenomenon in chemistry, biology, physics, and economics and are intensively studied in the last decades. In the simplest one-dimensional form, it can be written as follows:

$$\begin{aligned} u_t &= u_{xx} + f(u), \\ u &\to U_{\pm 1}, \quad \text{as } x \to \pm\infty, \end{aligned} \tag{1}$$

where $u\colon \mathbb{R}_x \times \mathbb{R}_t \to \mathbb{R}$, $f\colon \mathbb{R} \to \mathbb{R}$, $U_{\pm 1} \in \mathbb{R}$, $f(U_{\pm 1}) = 0$. One of the most studied cases is when f is a derivative of a double-well potential $f(u) = F'(u)$ with two wells in values $u = U_{\pm 1}$. In that case, term u_{xx} in Eq. (1) pushes function u to become constant in space, whereas term $f(u)$ pushes $u(x,t) \to U_{\pm 1}$. Usually, a solution $u(x,t)$ converges, as $t \to \infty$, to a traveling wave solution $u(x,t) = v(x - ct)$. In that case, a lot of information can be picked up from a single parameter c, which describes the speed of the traveling wave. The special case when $c = 0$ corresponds to a stationary front and appears in a case of a symmetric potential F.

The work of the author was supported by RFBR grants 18-01-00230a and 17-01-00678-a.

S. Tikhomirov (✉)
Saint-Petersburg State University, 7/9 Universitetskaya nab., St. Petersburg 199034, Russia
e-mail: s.tikhomirov@spbu.ru

A. Korobeinikov et al. (eds.), *Extended Abstracts Spring 2018*,
Trends in Mathematics 11, https://doi.org/10.1007/978-3-030-25261-8_42

The situation changes in a discrete environment. Let us consider a spatial discretization of a reaction–diffusion equation with a traveling wave solution, for instance, the discrete Nagumo/Schlogel equation

$$\dot{u}_n = \frac{1}{h^2}(u_{n+1} - 2u_n + u_{n-1}) + f(u_n), \tag{2}$$
$$f(u) = (u-a)(u^2-1),$$

where $a \in (-1, 1), a \neq 0$, which corresponds to a nonsymmetric double-well potential. If the step of discretisation h is too large, there appears a stationary front. This phenomenon is called pinning. In this research, we are interested in a bifurcation of a stationary front to a traveling wave with a change of the step of discretization h. For a review of the topic, see [6].

The pinning phenomenon is quite universal and appears in various contexts, such as periodic environments and forces [3], nonlocal interactions [1], etc. While plenty of heuristics are known near the bifurcation point (see, for instance, [2, 4]), there are only few rigorous results. In particular, the speed of a traveling wave after de-pinning is not known.

One of the known rigorous results on depinning is proved for the case of spatially periodic forces:

$$u_t = u_{xx} + (1-u)u(1+u) + \delta(l(x) + F), \tag{3}$$

where l is a 1-periodic function and $F \in \mathbb{R}$ is a parameter. In this case, there exists $F_c > 0$ such that if $F > F_c$ there exists a traveling wave, and for $F \leq F_c$ there exists a stationary solution of (3). Dirr and Yip proved asymptotics for the speed of the traveling wave for small δ and $F - F_c \sim \delta$, but still separated from 0; we refer for exact condition to [3]. In that case

$$\text{speed} \sim (F - F_c)^{1/2}. \tag{4}$$

See [9] for an overview.

2 Depinning Transition in Ergodic Media

We provide an abstract framework to study speed of a traveling wave in continuous inhomogeneous environment under depinning transition.

Consider an system in an abstract space

$$U_t = F(U, \theta; \mu), \tag{5}$$

where $U \in X$, a Banach space corresponding to $u(\cdot, t)$, $\Theta \in M$ a variable describing the environment, M is a smooth compact manifold, and μ is a depinning parameter. We assume that the system satisfies the following conditions (see [8]):

(C1) There exists a smooth flow S_ζ on M such that Eq. (5) have the symmetry $(T_\zeta U)_t = F(T_\zeta U, S_\zeta(\theta); \mu)$, where $T_\zeta : X \to X$ corresponds to a translation of U. Note that S_ζ could be interpreted as translation of the environment.

(C2) There exists a family of smooth one-dimensional manifolds $\mathcal{N}_\mu \subset X$ invariant under translation T_ζ and the flow restricted to it is generated by a C^2-vector field

$$\xi' = s(S_\xi(\theta); \mu), \tag{6}$$

where ξ is a coordinated on one-dimensional manifolds $\mathcal{N}_\mu$. Note that the form of (6) follows from condition (C1). Typically, existence of $\mathcal{N}_\mu$ can be obtained by establishing normal hyperbolicity in (5).

(C3) The function s is nondegenerate in the following sense: there is a unique $\theta_* \in \mathcal{M}$ such that $s(\theta; 0) > 0$ for $\theta \neq \theta_*$, $s(\theta_*; 0) = 0$, $\partial_\mu s(\theta_*; 0) > 0$, and $D^2_\theta s(\theta_*; 0) > 0$. This condition is the most difficult to be verified.

(C4) The flow S_ζ is ergodic with respect to an invariant measure ν on $\mathcal{M}$ with local dimension κ at point θ_*. In case when ν is the Lesbeugue measure, κ coincides with the dimension of M.

The term *ergodic media* refers to condition (C4).

Theorem 1 (Scheel–Tikhomirov, [8]) *If conditions (C1)–(C4) are satisfied, then for ν-almost $\theta \in M$ and small enough $|\mu|$ solution is pinned for $\mu < 0$ (i.e., $\xi(t)$ is bounded), solution is depinned for $\mu > 0$ (i.e., $\xi(t) \to \infty$ as $t \to \infty$), the speed $c(\mu) = \lim_{t\to\infty} \xi(t)/t$ have the following asymptotics:*

$$c(\mu) \sim \begin{cases} \mu^{1-\kappa/2}, & \kappa < 2, \\ (|\log(\mu)|^{-1}, & \kappa = 2, \\ 1, & \kappa > 2. \end{cases} \tag{7}$$

The proof is based on a skew product structure and notion of relative equilibria [5, 7].

The easiest example of an ergodic media, satisfying assumptions of Theorem 1, is a quasiperiodic media. Consider a modified Nagumo/Schlogel equation

$$u_t = u_{xx} + (u + \mu)(1 - u^2) + \varepsilon\alpha(x; \theta)g(u) \tag{8}$$

with a quasiperiodic inhomogeneity

$$\alpha(y; \theta) = \sum_{j=1}^{\kappa} \alpha_j \cos(\omega_j y + 2\pi\theta_j)$$

with rationally independent $(\omega_j)_{j=1,\dots,\kappa}$ and $\alpha_j \neq 0$. The function $g(u)$ satisfies technical assumptions $g(\pm 1) = g'(\pm 1) = 0$, $g \in C^2$; see [8].

3 Depinning Conjecture in Discrete Quasiperiodic Media

Note that results of a previous section do not provide a rigorous proof of depinning speed asymptotic in discrete media. The author was not able to find an asymptotic behavior $c(h)$ of a traveling wave solution $u_n(t) = v(n - ct)$ of (2) near the depinning transition either.

Analogy between pinning in discrete media and continuous periodic media and asymptotics (7) suggests to study depinning in nonhomogeneous discrete media. Consider nonhomogeneous discrete Nagumo–Schlogel equation

$$\begin{aligned} \dot{u}_n &= d(u_{n+1} - 2u_n + u_{n-1}) + (u_n - a_n)(1 - u_n^2), \\ u_n &\to \pm 1, \quad \text{as } n \to \pm 1, \end{aligned} \tag{9}$$

where $a_n = a + \varepsilon \sum_{j=1}^{k} b_n \cos(2\pi \omega_j n + \theta_j)$ is a quasiperiodic sequence, with ω_j rationally independent with $1, b_j, \theta_j \in \mathbb{R}$ and k is a number of additional frequencies. Results of numerical simulations (se Fig. 1 and [8]) show that the behavior of speed of wave propagation strongly depends on the value of k, where speed is defined as

$$\text{speed} = \frac{1}{2} \lim_{t \to \infty} \frac{1}{t} \sum_{n=-\infty}^{+\infty} u_n(t) - u_n(0).$$

While it is hard to make a good conjecture based only on numerical simulations, Theorem 1 suggests the following.

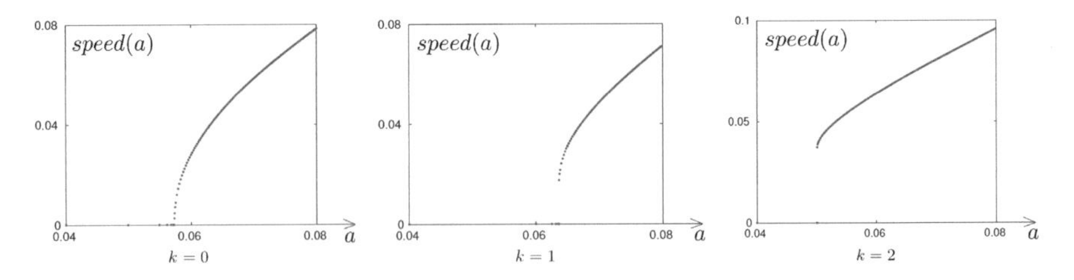

Fig. 1 Speed of wave propagation as a function of a for $k = 0, 1, 2$

Conjecture 2 *There exists $a_c > 0$ such that if $a \in (0, a_c)$ then Eq. (9) admits a stationarily solution. If $a > a_c$, then there exists a moving solution with an average speed behavior as $a \to a_c$*

$$speed(a) \sim \begin{cases} (a - a_c)^{1/2}, & k = 0, \\ (|\log(a - a_c)|^{-1}, & k = 1, \\ 1, & k \geq 2. \end{cases} \tag{10}$$

References

1. T. Anderson, G. Faye, A. Scheel, D. Stauffer, Pinning and unpinning in nonlocal systems. J. Dyn. Diff. Equat. **28**, 897–923 (2016)
2. A. Carpio, L. Bonilla, Depinning transitions in discrete reaction-diffusion equations. SIAM J. Appl. Math. **63**, 1056–1082 (2003)
3. N. Dirr, N. Yip, Pinning and de-pinning phenomena in front propagation in heterogeneous media. Interfaces Free Bound. **8**, 79–109 (2006)
4. T. Erneux, G. Nicolis, Propagating waves in discrete bistable reaction-diffusion systems. Phys. D **67**(1–3), 237–244 (1993)
5. B. Fiedler, B. Sandstede, A. Scheel, C. Wulff, Bifurcation from relative equilibria of noncompact group actions: skew products, meanders, and drifts. Doc. Math. J. DMV **1**, 479–505 (1996)
6. J. Mallet-Paret, Traveling waves in spatially discrete dynamical systems of diffusive type. Lecture Notes in Mathematics, vol. 1822 (Springer, Berlin, 2003), pp. 231–298
7. B. Sandstede, A. Scheel, C. Wulff, Dynamics of spiral waves on unbounded domains using center-manifold reduction. J. Differ. Equ. **141**, 122–149 (1997)
8. A. Scheel, S. Tikhomirov, Depinning asymptotics in ergodic media, in *Springer Proceedings in Mathematics & Statistics*, vol. 205 (Springer, Berlin, 2017), pp. 88–108
9. J. Xin, Front propagation in heterogeneous media. SIAM Rev. **42**, 161–230 (2000)

Cancer Evolution: The Appearance and Fixation of Cancer Cells

Stefano Pedarra and Andrei Korobeinikov

Abstract Cancer appears as a result of mutation of normal tissue cells. In this paper, we consider the initial stage of the cancer appearance and development. In particular, we study the conditions that are necessary for an initial fixation of the mutant cells in a patient tissue and their further successful development. In order to do this, we are using a reasonably simple mutation-selection model composed of two interacting populations, namely, the normal cells and the mutant cells. Conditions for persistence of the mutant cells are found.

1 Introduction

Cancer is characterized by uncontrolled growth of abnormal cells that appear as a result of a series of mutations of normal cells. To develop into cancer, the mutant cells should be able to successfully compete with the normal cells. It is highly surprising that this issue attracted significant attention; see, e.g., [3, 4], and literature therein.

In this paper, we focus at an initial stage of cancer and explore the conditions that are necessary for the initial fixation of the malignant mutant cells in a patient. Accordingly, we consider the dynamics of a simple mutation-selection model that comprises two interaction populations, namely, normal cells and mutant cancer cells.

A. Korobeinikov is supported by Ministerio de Economía y Competitividad of Spain via grant MTM2015-71509-C2-1-R.

S. Pedarra (✉)
Dipartimento di Matematica "Tullio Levi–Civita", Università degli Studi di Padova, Padova, Italy
e-mail: stefano.pedarra@studenti.unipd.it

A. Korobeinikov
Centre de Recerca Matemàtica, Campus de Bellaterra, Edifici C, 08193 Barcelona, Spain
e-mail: akorobeinikov@crm.cat

Departament de Matemàtiques, Universitat Autònoma de Barcelona, Barcelona, Spain

A. Korobeinikov et al. (eds.), *Extended Abstracts Spring 2018*,
Trends in Mathematics 11, https://doi.org/10.1007/978-3-030-25261-8_43

To address this issue, in this paper, we consider a version of the cancer evolution model suggested in [1, 2], where we consider only two classes, namely, the normal cells and the malignant cells.

2 Model

Let us consider a model described in [1]. The model postulates the existence of the normal cells and n cancer cell genotypes in the system. Let us denote the population size of the i-th genotype cell at time t by $C_i(t)$ and the population size of the normal cells at the same time by $C_0(t)$. The model is based upon the Lotka–Volterra model of competing populations and postulates that (i) all cells reproduce and die, (ii) there is a resource that limit populations growth, (iii) cells of different genotypes have to compete for this limited resource, (iv) in the process of mitosis, with some probability p_{ij}, a cell of the i-th genotype can produce a mutant daughter cell of the j-th genotype, which subsequently goes to the j-th population, and (v) as a result of somatic mutation, with probability $q_{i,j}$ a cell of the i-th genotype can move to the j-th genotype. This situation can be described by the following system of ordinary differential equations:

$$\dot{C}_i = \sum_{j=0}^{n}\left(p_{ji}a_jC_j\left(1-\frac{h_j}{K}\sum_{k=0}^{n}b_{jk}C_k\right)\right) - d_iC_i\left(1+\frac{g_i}{K}\sum_{k=0}^{n}b_{ik}C_k\right) + \sum_{j=0}^{n}q_{ji}C_j - \sum_{j=0}^{n}q_{ij}C_i\,. \tag{1}$$

Here, $i = 0, 1, \ldots, n$, a_i are the replication rates, d_i the death rates, K the carrying capacity, b_{ij} the competition factors, and h_i and g_i reflect the competition effects on the birth and death, respectively.

Our goal is to study the initial appearance and fixation of the mutant cells. Therefore, we consider only one type of mutant cells, or, what is the same, assume that all mutant cells are the same, and consider interaction of these with the normal tissue cells. Thus, we assume that $n = 1$ in model (1), and the model reduces to a two-dimensional system. Moreover, for simplicity, we assume that $h_0 = h_1 = 0$. Mathematically, this preserves the positive invariance of the first quadrant of the phase space; biologically, this means that the lack of resources increases the death rate, but does not inhibit the proliferation. For this case, denoting

$$x = b_{00}C_0/K, \qquad y = b_{11}C_1/K, \qquad \tau = d_0g_0t,$$

$$A = p_{00}a_0 - d_0, \quad B = (a_1p_{10} + q_{10})b_{00}/b_{11}, \quad C = (a_0p_{01} + q_{01})b_{11}/b_{00}, \quad D = p_{11}a_1 - d_1,$$

$$\alpha = p_{00}a_0h_0 + d_0g_0, \qquad \beta = d_0g_0b_{01}/b_{11}, \qquad \gamma = d_1g_1b_{10}/b_{00}, \qquad \delta = p_{11}a_1h_1 + d_1g_1,$$

the system (1) can be written as

$$\frac{dx}{d\tau} = \frac{1}{\alpha}(Ax + By - \alpha x^2 - \beta xy)\,, \tag{2}$$

$$\frac{dy}{d\tau} = \frac{1}{\alpha}(Cx + Dy - \gamma xy - \delta y^2)\,. \tag{3}$$

Further, we analyze this system.

3 Equilibrium States of the Model

Equilibrium states of model (2) satisfy the following system of algebraic equations:

$$Ax + By - \alpha x^2 - \beta xy = 0\,, \qquad Cx + Dy - \gamma xy - \delta y^2 = 0\,. \tag{4}$$

Accordingly, for this system, the equilibrium states correspond to the intersections of two conic curves defined by equalities (4). Of course, since the variables x and y represent sizes of populations, we are interested only in the intersections, which are located in the first quadrant of the phase space.

Let us start with some trivial observations expressed by the following lemmas:

Lemma 1 *The origin* $P_0 = (0, 0)$ *is always an equilibrium state of the system and is the only equilibrium state located on the coordinate axes, when* B *and* C *are strictly positive.*

By this lemma, the system always has at least one nonnegative equilibrium state.

Lemma 2 *The system (2) has from one to four equilibrium states.*

Through some geometrical observation, it is possible to obtain the following results about the nullclines (4):

Lemma 3 *Each of the nullclines (4) is either a hyperbola or a degenerate hyperbola (two intersecting straight lines).*

Lemma 4 *One of the two branches of each of the nullclines (4) has no points in the first quadrant.*

Consequently,

Lemma 5 *The system* (2) *has either one or two nonnegative equilibrium states.*

That is, the system either has no positive coexisting equilibrium states at all, or have only one such equilibrium state.

3.1 Stability of Equilibrium State P_0

The local analysis of the system (2) near equilibrium state P_0 immediately reveals that the eigenvalues of the linearized system are real numbers. Consequently, P_0 cannot be a focus. Moreover, no Hopf bifurcation is possible at the origin. There are the following three possibilities:

(i) if $AD - BC < 0$, then the origin is a saddle point;
(ii) if $AD - BC > 0$ and $(A + D) > 0$, then the origin is a repulsive (unstable) node;
(iii) if $AD - BC > 0$ and $(A + D) < 0$, then the origin is an attractive (stable) node.

In terms of the physical parameters, $(A + D) > 0$ is equivalent to condition $a_0 p_{00} + a_1 p_{11} > d_0 + d_1$. Please note that $A + D < 0$ is not impossible, as there can be a situation where one of the d_i is very large, whereas the corresponding $a_i p_{ii}$ is sufficiently small. This means that the corresponding genotype could appear, but cannot sustain even without competition.

Biologically, this condition means that the sum of rates of birth to the same genotype cells is larger than the sum of death rates. Condition $AD - BC < 0$ is equivalent to the inequality $(a_0 p_{01} + q_{01})(a_1 p_{10} + q_{01}) > (a_0 p_{00} - d_0)(a_1 p_{11} - d_1)$.

3.2 Existence and Properties of the Positive Fixed Point

Let us consider the existence and the location of the possible positive equilibrium state $P_* = (x_*, y_*)$.

Lemma 6 *If either $A\beta - B\alpha \geq 0$, or $D\gamma - C\delta \geq 0$, then the positive equilibrium state P_* exists.*

If $A\beta - B\alpha < 0$ and $D\gamma - C\delta < 0$, then the existence of P_* is uncertain. However, in either case, equilibrium state P_* can appear only as the result of a transcritical bifurcation that occurs at the origin. Therefore, P_* is generated when the two hyperbolas have a common tangent at the origin. This implies that the transcritical bifurcation occurs when $AD - BC = 0$. An intriguing fact is that the equilibrium state P_* can exist as when $AD - BC < 0$ as when $AD - BC > 0$.

Linearizing the system around P_*, it is easy to see that this point can be either a saddle point, or an attractive node, or an attractive focus. It is possible to state the following result:

Lemma 7 *No Hopf bifurcation is possible at P_*.*

Therefore, recalling that no Hopf bifurcation is possible at (0, 0), one can deduce the following.

Corollary 8 *The model* (2) *admits no Hopf bifurcation.*

Moreover, a sufficient (but not necessary) condition for the stability of P_* obtained via linearization is that $\alpha\delta - \beta\gamma > 0$. Thus, there are two sufficient conditions for the stability of P_*:

(i) $\alpha\delta - \beta\gamma > 0$, or
(ii) $AD - BC < 0$ and $A + D < 0$.

We already analyzed the second condition discussing the stability of equilibrium state P_0. In terms of the physical parameters, the first condition is equivalent to $b_{00}b_{11} - b_{01}b_{10} > 0$, that is, $\det B > 0$. Biologically, it means that the system achieves the stable equilibrium when the competition effects among cells with the same genotype are stronger than the ones between the two different genotypes. It is very unlikely to occur in the case of cancer.

3.3 Global Properties of the Model

Denoting $P(x, y) = \big((A - \alpha x)x + (B - \beta x)y\big)/\alpha$ and $Q(x, y) = \big((C - \gamma y)x + (D - \delta y)y\big)/\alpha$, we can rewrite model (2) as follows:

$$\frac{dx}{d\tau} = P(x, y), \qquad \frac{dy}{d\tau} = Q(x, y)\,. \tag{5}$$

Let us study the direction of the vector field (P, Q) at a point (x, y) in the first quadrant. It is easy to see that, if $x > \max\{A/\alpha, B/\beta\}$, then $P(x, y) < 0$; analogously, if $y > \max\{C/\gamma, D/\delta\}$, then $Q(x, y) < 0$. Moreover, on the positive semi-axes, the vector flow is directed inside of the first quadrant. This means that the compact square

$$S = \{[0, \max(A/\alpha, B/\beta)] \times [0, \max(C/\gamma, D/\delta)]\}$$

is a positive-invariant set and an attractive region of the system. Therefore, by the Poincaré–Bendixson Theorem, it contains at least one stable limit cycle or at least one stable fixed point. For this system, it is possible to exclude the existence of a limit cycle. Moreover, by choosing $\varphi(x, y) = \alpha$, one immediately obtains for the system (2):

$$\operatorname{div}(\varphi P, \varphi Q) = A + D - (2\alpha + \gamma)x - (\beta + 2\delta)y\,,$$

which is always negative if $A + D < 0$. Therefore, since S is a simply connected region, by the Bendixson–Dulac theorem, no limit cycle is possible within S when $A + D < 0$. It implies that all orbits tend to a stable fixed point. Please recall that for $A + D < 0$ the origin P_0 is stable if and only if $AD - BC > 0$. Hence, for $AD - BC < 0$, positive equilibrium state P_* must exist and be stable. This means that for the case when $A + D < 0$ the necessary and sufficient condition for the existence of P_* is $AD - BC < 0$.

References

1. A. Korobeinikov, S. Pedarra, A discrete variant space cancer evolution model, in *Proceedings of MURPHYS-HSFS-2018 Workshop*. Trends in Mathematics: Research Perspectives CRM Barcelona, Summer 2018 (Springer-Birkhäuser, Basel, 2019)
2. A. Korobeinikov, K.E. Starkov, P.A. Valle, Modeling cancer evolution: evolutionary escape under immune system control. J. Phys. IOP Conf. Ser. **811** (2017)
3. J. Sardanyés, R. Martínez, C. Simó, R.V. Solé, Abrupt transitions to tumor extinction: a phenotypic quasispecies model. J. Math. Biol. **74**, 1589–1609 (2017)
4. R.V. Solé, S. Valverde, C. Rodríguez-Caso, J. Sardanyés, Can a minimal replicating construct be identified as the embodiment of cancer? Bioessays **36**, 503–512 (2014)

Cheap Control Problem in a Critical Case

Vladimir Sobolev

Abstract A specific class of cheap control problems is considered. The micro-drone quadrocopter model is used as an illustration.

1 Introduction

Consider the control system

$$\varepsilon \dot{x} = A(t)x + \bar{B}(t)u, \quad x \in R^n, \ u \in R^m \ x(0) = x_0 \tag{1}$$

with the cost functional

$$J = \frac{1}{2}x^T(t_f)Fx(t_f) + \frac{1}{2}\int_0^{t_f}(x^T(t)Q(t)x(t) + \mu^2 u^T(t)R(t)u(t))dt, \tag{2}$$

where $A, \ F, \ Q$ are $(n \times n)$-matrices, $\bar{B}$ is $(n \times m)$-matrix, R is $(m \times m)$-matrix, and μ is a small positive parameter. This problem was considered in [4], where it was shown that the solution of this problem can be obtained by use of asymptotic expansions in fractional powers of the small parameter $\varepsilon = \mu^{1/L}$, where L can be found from

$$B_j^T Q B_j = 0, \qquad B_{L-1}^T Q B_{L-1} > 0,$$

for $j = 0, \dots, L-2$, with $B_0 = \bar{B}$ and $B_j = AB_{j-1} - \dot{B}_{j-1}$, $j \geq 1$. In this case, one has to consider the problem in a much larger dimension, namely, $n + Lr$ instead of n. The method of integral manifolds [3, 6] for the analysis of such problems was

This work was funded by RFBR and Samara Region (project 16-41-630524-p) and the Ministry of Education and Science of the Russian Federation under the Competitiveness Enhancement Program of Samara University (2013–2020).

V. Sobolev (✉)
Department of Differential Equations and Control Theory, Samara National Research University, Moskovskoye shosse, 34, Samara 443086, Russian Federation
e-mail: v.sobolev@ssau.ru

A. Korobeinikov et al. (eds.), *Extended Abstracts Spring 2018*,
Trends in Mathematics 11, https://doi.org/10.1007/978-3-030-25261-8_44

applied in [5]. In this paper, it is shown that for a natural class of problems one can analyze such a problem without increasing the dimensionality and, moreover, without solving differential equations.

2 Construction of Control Law

We consider the problem of constructing a control law for a second-order vector differential equation

$$\ddot{x} + G(t)\dot{x} + N(t)x = B(t)u$$

with the quadratic performance index

$$J = \frac{1}{2}x^T(t_f)F_1x(t_f) + \frac{1}{2}\dot{x}^T(t_f)F_2\dot{x}(t_f) +$$

$$\frac{1}{2}\int_0^{t_f}(x^T(t)Q_1(t)x(t) + \dot{x}^T(t)Q_2(t)\dot{x}(t) + \mu^2u^T(t)R(t)u(t))dt.$$

Introducing the small parameter $\mu = \varepsilon^2$ we can rewrite this problem in the form (1) and (2), where the corresponding matrices are

$$A = \begin{pmatrix} 0 & I \\ -N & -G \end{pmatrix}, \quad \bar{B} = \begin{pmatrix} 0 \\ B \end{pmatrix}, \quad R = (1), \quad Q = \begin{pmatrix} Q_1 & 0 \\ 0 & Q_2 \end{pmatrix},$$

$$F = \begin{pmatrix} F_1 & 0 \\ 0 & F_2 \end{pmatrix}, \quad \bar{S} = \begin{pmatrix} 0 & 0 \\ 0 & S \end{pmatrix}, \quad S = BR^{-1}B^T.$$

The optimal control is

$$u = -\varepsilon^{-4}R^{-1}(0 \;\; B^T)P\begin{pmatrix} x \\ \dot{x} \end{pmatrix},$$

and the matrix

$$P = \begin{pmatrix} \varepsilon P_1 & \varepsilon^2 P_2 \\ \varepsilon^2 P_2^T & \varepsilon^3 P_3 \end{pmatrix}$$

satisfies the matrix differential Riccati equation

$$\dot{P} + A^T P + PA + Q - \varepsilon^{-4}PSP = 0$$

with boundary condition

$$P(t_f) = \begin{pmatrix} \varepsilon F_1 & 0 \\ 0 & \varepsilon^2 F_2 \end{pmatrix}.$$

The corresponding equations are

$$\varepsilon\dot{P}_1 - \varepsilon^2(P_2N + N^TP_2^T) - P_2SP_2^T + Q_1 = 0,$$
$$\varepsilon\dot{P}_2 + P_1 - \varepsilon P_2G - \varepsilon^2N^TP_3 - P_2SP_3 = 0,$$
$$\varepsilon\dot{P}_3 + P_2 + P_2^T - \varepsilon(P_3G + G^TP_3) + Q_2 - P_3SP_3 = 0,$$

with boundary condition

$$\varepsilon P_1(t_f) = F_1,\ P_2(t_f) = 0,\ \varepsilon P_3(t_f) = F_2.$$

The Riccati equation in the equivalent form

$$\varepsilon^4(\dot{P} + A^TP + PA + Q) - PSP = 0$$

is singularly perturbed, since when the small parameter is equal to zero the ability to specify an arbitrary initial or boundary conditions is lost. Such systems play an important role as mathematical models of numerous nonlinear phenomena in different fields (see, e.g., [3, 6]. Moreover, we have so-called critical case, since the limiting equation $PSP = 0$ has multiple zero solution [3, 6]. Setting the small parameter equal to zero, we obtain the equations

$$-P_2SP_2^T + Q_1 = 0,\quad P_1 - P_2SP_3 = 0,\quad P_2 + P_2^T + Q_2 - P_3SP_3 = 0.$$

Suppose that these equations have the solution $P_1 = M_1(t), P_2 = M_2(t), P_3 = M_3(t)$, such that all eigenvalues $\lambda(\varepsilon)$ of the matrix

$$D = \begin{pmatrix} 0 & I \\ -\varepsilon^{-2}SM_2^T - N & -\varepsilon^{-1}SM_3 - G \end{pmatrix}$$

have the negative real parts $-\nu(t, \varepsilon)/\varepsilon, \nu(t, 0) > \nu_0 > 0$. Then we can neglect the boundary conditions, and a solution of the corresponding system of matrix equations takes the regular part of the solution of this system, which can be regarded as a zero-dimensional integral manifold of slow motions. In the stationary case, the role of this solution is played by a positive definite solution of the corresponding matrix algebraic Riccati equation. In the nonautonomous case matrices, $P_1,\ P_2,\ P_3$ can be found as asymptotic expansions [3, 5, 6]

$$P_1 = M_1 + \varepsilon P_{11} + \varepsilon^2 P_{12} + \cdots,\quad P_2 = M_2 + \varepsilon P_{21} + \varepsilon^2 P_{22} + \cdots,\quad P_3 = M_3 + \varepsilon P_{31} + \varepsilon^2 P_{32} + \cdots.$$

Note that in the case of time-invariant problem coefficients of these expansions can be found from corresponding algebraic matrix equations.

3 Control of Micro-drone Quadcopter

Consider the optimal control problem in the case $n = 3$, $m = 4$ with the following matrices [1, 2]:

$$G = N = 0, \quad B = \begin{pmatrix} -\alpha & -\alpha & \alpha & \alpha \\ \beta & -\beta & 0 & 0 \\ 0 & 0 & \beta & -\beta \end{pmatrix},$$

$$Q_1 = diag(q_1, q_2, q_3), \quad Q_2 = diag(0, q_5, q_6), \quad R = I.$$

The easy algebra shows that in the case under consideration we obtain $P_2 = diag(b_1, b_2, b_3)$, where

$$b_1 = \sqrt{q_1}/2\alpha, b_2 = \sqrt{q_2}/(\beta\sqrt{2}), b_3 = \sqrt{q_3}/(\beta\sqrt{2}).$$

Thus, we obtain

$$M_2 + M_2^T + Q_2 = diag(2b_1, q_5 + 2b_2, q_6 + 2b_3),$$

and $P_3 = M_3 = diag(c_1, c_2, c_3)$, where

$$c_1 = \sqrt{2b_1}/2\alpha, \quad c_2 = \sqrt{2b_2 + q_5}/(\beta\sqrt{2}), \quad c_3 = \sqrt{2b_3 + q_6}/(\beta\sqrt{2}).$$

These relationships give the expression for P_1:

$$P_1 = M_1 = diag(a_1, a_2, a_3),$$

where $a_1 = 4\alpha^2 b_1 c_1, a_2 = 2\beta^2 b_2 c_2, a_3 = 2\beta^2 b_3 c_3$.

It should be noted that the matrix M is positively definite if and only if $a_i c_i - b_i^2 > 0$, for $i = 1, 2, 3$.

It is easy to check that the matrix D has the following eigenvalues:

$$\lambda_{1,2} = -\varepsilon{-}14\alpha^2 c_1 \pm \varepsilon^{-1} 2i\alpha\sqrt{(b_1},$$

$$\lambda_3, 4 = -\varepsilon{-}14\beta^2 c_2 \pm \varepsilon^{-1} 2i\beta\sqrt{(b_2},$$

$$\lambda_5, 6 = -\varepsilon{-}14\beta^2 c_3 \pm \varepsilon^{-1} 2i\beta\sqrt{(b_3}.$$

The real parts of all eigenvalues are negative. This means that we can use the control law in the form

$$u_1 = -\varepsilon^{-2}(-\alpha b_1 x_1 + \beta b_2 x_2) - \varepsilon^{-1}(-\alpha c_1 \dot{x}_1 + \beta c_2 \dot{x}_2),$$

$$u_2 = -\varepsilon^{-2}(-\alpha b_1 x_1 - \beta b_2 x_2) - \varepsilon^{-1}(-\alpha c_1 \dot{x}_1 - \beta c_2 \dot{x}_2),$$

$$u_3 = -\varepsilon^{-2}(\alpha b_1 x_1 + \beta b_3 x_2) - \varepsilon^{-1}(-\alpha c_1 \dot{x}_1 + \beta c_3 \dot{x}_2),$$

$$u_4 = -\varepsilon^{-2}(\alpha b_1 x_1 - \beta b_3 x_2) - \varepsilon^{-1}(\alpha c_1 \dot{x}_1 - \beta c_3 \dot{x}_2).$$

A somewhat unexpectedly the controlled system is splitted into three independent equations

$$\ddot{x}_1 + \varepsilon^{-1} 4\alpha^2 c_1 \dot{x}_1 + \varepsilon^{-2} 4\alpha^2 b_1 x_1 = 0,$$

$$\ddot{x}_2 + \varepsilon^{-1} 4\alpha^2 c_2 \dot{x}_2 + \varepsilon^{-2} 4\alpha^2 b_2 x_2 = 0,$$

$$\ddot{x}_3 + \varepsilon^{-1} 4\alpha^2 c_3 \dot{x}_3 + \varepsilon^{-2} 4\alpha^2 b_3 x_3 = 0.$$

It is clear that the solutions of these equations decay sufficiently rapidly and it is not necessary to give any numerical calculations.

References

1. P. Castillo, R. Lozano, A. Dzul, Stabilisation of a mini rotorcraft with four rotors. IEEE Control Syst. Mag. **25**(6), 45–55 (2005)
2. P. Martin, E. Salaun, The true role of accelerometer feedback in quadrotor control, in *IEEE International Conference on Robotics and Automation* (2010), pp. 1623–1629
3. M.P. Mortell, R.E. O'Malley, A. Pokrovskii, V.A. Sobolev (eds.), *Singular Perturbation and Hysteresis* (SIAM, 2005)
4. R.E. O'Malley, A. Jameson, Singular perturbations and singular arcs I, II. Trans. Autom. Control **AC-20**, 218–226 (1975) and **AC-22**, 328–337 (1977)
5. E. Smetannikova, V. Sobolev, Regularization of cheap periodic control problems. Autom. Remote Control **66**(6), 903–916 (2005)
6. E. Shchepakina, V. Sobolev, M.P. Mortell, *Singular Perturbations: Introduction to System Order Reduction Methods with Applications*. Lecture Notes in Mathematics, vol. 2114 (Springer, Berlin, 2014)

Printed in the United States
By Bookmasters